Jenkins Administrator's Guide

Jenkins Administrator's Guide

Install, manage, and scale a CI/CD build and release system
to accelerate your product life cycle

Calvin Sangbin Park
Lalit Adithya
Samuel Gleske

BIRMINGHAM—MUMBAI

Jenkins Administrator's Guide
Calvin Sangbin Park, Lalit Adithya, & Samuel Gleske

Production reference: 1111121

Group Product Manager: Vijin Boricha
Publishing Product Manager: Vijin Boricha
Senior Editor: Hayden Edwards
Content Development Editor: Nihar Kapadia
Technical Editor: Nithik Cheruvakodan
Copy Editor: Safis Editing
Project Coordinator: Neil Dmello
Proofreader: Safis Editing
Indexer: Pratik Shirodkar
Production Designer: Aparna Bhagat
Senior Designer: Joseph Runnacles

First published: December 2021

Published by Packt Publishing Ltd.
Livery Place
35 Livery Street
Birmingham
B3 2PB, UK.

ISBN 978-1-83882-432-7

www.packt.com

To my parents, who will have the hardest time bragging about their son's book on Jenkins.
– Calvin Sangbin Park

I dedicate this to my wife, Kristie, whose support for me made this possible.
Also, my mother, Audrey, and Kristie's mother, Tina, who helped us tremendously
during this period of our pregnancy and the birth of our son, Cory.
– Samuel Gleske

Foreword

In 2019, I had the pleasure of working closely with Kohsuke Kawaguchi, creator of Jenkins. That year, Jenkins was celebrating its 15th anniversary, and Kohsuke reached out to the community to solicit Jenkins success stories. Each response was more impressive than the last, with user testimonies of how Jenkins had transformed industries, as well as success stories of how Jenkins had helped propel the careers of individual practitioners. My favorite was learning how Jenkins was used to build the World Anti-Malarial Network.

I came to see firsthand how Jenkins is one of the most used and loved technologies in the software industry. For over 15 years, Jenkins has had a huge impact on developer productivity in a wide range of industries, everything from aerospace to retail, through to education and finance. Jenkins is an open source project that is backed by one of the most dedicated open source communities, enabling it to continue a very long trend of pioneering work in the CI/CD space. As a result, Jenkins is continuously evolving and improving to serve its users' needs.

Software is increasingly playing a key role in various organizations and industries. We are delivering more software than ever before, and software delivery is a key differentiator for every organization. Jenkins remains at the heart of this transformation. As Jenkins continuously innovates, it is important to be able to keep up with the latest changes and understand the best ways to use this powerful technology. That is why I am delighted that Calvin Park, Lalit Adithya, and Sam Gleske have come together to write this book, with support from Vijin Boricha and all at Packt. I had the pleasure of meeting Sam Gleske at one of the Jenkins World contributor summits a few years ago and I appreciate his many and continuous contributions to the project.

In *Jenkins Administrator's Guide*, many key topics are covered. From the ever-important script security to tackling shared libraries, as well as using the ever-so-powerful Jenkins Configuration as Code. The book also covers emerging trends such as GitOps and very practical information for optimizing your Jenkins setup (complete with warnings not to overengineer your setup if you don't need to!).

This book is an invaluable resource for those who want to make the most of Jenkins and keep up with recent improvements and make the most of the amazing productivity you can unlock with an optimal Jenkins setup.

Tracy Miranda
Executive Director, Continuous Delivery Foundation

Contributors

About the authors

Calvin Sangbin Park is a CI/CD DevOps engineer at NVIDIA. He's been using Jenkins throughout his career to automate builds for Arduino maker boards, Android tablets, enterprise software packages, and even firmware for an industrial laser for etching CPUs. Lately, he's been focusing on Kubernetes, monitoring, and process visualizations. He plans to contribute to the open source community by developing a plugin that optimizes Kubernetes cluster management.

Behind every great book is a great family. Thank you, Eunyoung and Younghwan. This book is as much yours as it is mine. I love you.

Also, many thanks to my brother and editor, Sangyoon; my brilliant coauthors, Lalit and Sam; the insightful technical reviewers, Huo, Dominic, and Ray; and the wise mentors, Madhusudan N and Sebass van Boxel.

Lalit Adithya is a software engineer with the DevOps team at NVIDIA. He has built code-commit-to-production pipelines using Jenkins and GitHub Actions. He has built and scaled business-critical applications that serve several thousand requests every minute. He has also built frameworks that have boosted developer productivity by abstracting away the complexities of networking, request/response routing, and more. He knows the ins and outs of several public cloud platforms and can architect cost-effective and scalable cloud-native solutions.

I thank my parents for all their love, support, and encouragement. I also thank all my mentors and friends who supported me, encouraged me to be the best version of myself, and helped me strive for perfection.

Samuel Gleske has been a Jenkins user since 2011, actively contributing to documentation and plugins, and discovering security issues in the system. Some notable plugins that Sam has maintained include the Slack plugin, the GHPRB plugin, the GitHub Authentication plugin, and a half dozen others. Sam has presented on and shared scripts for the Script Console documentation and is the primary author of its Wiki page. Since 2014, Sam has been developing Jervis – Jenkins as a service – which enables Jenkins to scale to more than 4,000 users and 30,000 jobs in a single Jenkins controller. Jervis emphasizes full self-service within Jenkins for users while balancing security.

I thank my wife, Kristie, whose support for me made this possible.

About the reviewers

Huo Bunn has been working professionally in the tech field for over 15 years. He started as a computer specialist at a computer help desk center; he progressed to be a web developer and eventually moved into the DevOps space. At the time of reviewing this book, he is currently a senior DevOps engineer who works with and provides support for multiple teams of software developers. He provides accounts and security governance via automation for the three major cloud providers, Google Cloud Platform, Azure, and Amazon Web Services. Huo is also responsible for maintaining a Kubernetes cluster for a testing framework that relies heavily on Jenkins as the interface for other teams to run workload test suites against their code changes.

I would like to thank Calvin, the author of this book, for giving me the opportunity to be a part of this new experience. I have learned a great deal from reviewing the book and know that what I have learned will come in handy in my everyday operations as a DevOps engineer. I would also like to thank the Packt team for coordinating and providing an easy way to review the book.

Dominic Lam is a senior manager in cloud infrastructure at NVIDIA. He has worked in software development for more than 20 years in different disciplines ranging across kernel drivers, application security, medical imaging, and cloud infrastructure. He has been using Jenkins, fka Hudson, since 2007 and has an interest in making software development seamless through CI/CD. He served as a board member on the Industry Advisory Board of the computer science department at San Jose State University for 2 years. Dominic holds a master's degree in computer science from Stanford University and a bachelor's degree in electrical engineering and computer science from the University of California, Berkeley.

Raymond Douglass III holds a BSc in computer science and an MSc in information technology, specializing in software engineering. He has been in the software development industry for over 10 years, in positions ranging from software developer to DevOps engineer to manager. He has more than 5 years of experience as a Jenkins administrator as well as experience writing and maintaining Jenkins plugins.

I'd like to thank both my parents for all their love and support throughout my life. I'd also like to thank Calvin and Packt for giving me the opportunity to review this fantastic book. Finally, I'd like to thank all the co-workers I've had through the years for helping me to learn new things and grow as a developer and as a person.

Contents

GitOps-Driven CD Pipeline with Docker Hub and More Jenkinsfile Features

Headfirst AWS for Jenkins

⑥ Jenkins Configuration as Code (JCasC)

⑦ Backup and Restore and Disaster Recovery

8 Upgrading the Jenkins Controller, Agents, and Plugins

9 Reducing Bottlenecks

10 Shared Libraries

11 Script Security

Preface

Jenkins is a renowned name among build and release CI/CD DevOps engineers because of its usefulness in automating builds, releases, and even operations. Despite its capabilities and popularity, it's not easy to scale Jenkins in a production environment. *Jenkins Administrator's Guide* will not only teach you how to set up a production-grade Jenkins instance from scratch, but also cover management and scaling strategies.

This book will guide you through the steps for setting up a Jenkins instance on AWS and inside a corporate firewall, while discussing design choices and configuration options, such as TLS termination points and security policies. You'll create CI/CD pipelines that are triggered through GitHub pull request events, and also understand the various Jenkinsfile syntax types to help you develop a build and release process unique to your requirements. For readers who are new to Amazon Web Services, the book has a dedicated chapter on AWS with screenshots. You'll also get to grips with Jenkins Configuration as Code, disaster recovery, upgrading plans, removing bottlenecks, and more to help you manage and scale your Jenkins instance.

By the end of this book, you'll not only have a production-grade Jenkins instance with CI/CD pipelines in place, but also knowledge of best practices from industry experts.

Who this book is for

This book is for both new Jenkins administrators and advanced users who want to optimize and scale Jenkins. Jenkins beginners can follow the step-by-step directions, while advanced readers can join in-depth discussions on Script Security, removing bottlenecks, and other interesting topics. Build and release CI/CD DevOps engineers of all levels will also find new and useful information to help them run a production-grade Jenkins instance, following industry best practices.

What this book covers

Chapter 1, Jenkins Infrastructure with TLS/SSL and Reverse Proxy, introduces Jenkins and discusses its strengths, along with a little bit of history and important keywords. The chapter describes the architecture of the Jenkins infrastructure that we will be building in the coming chapters, one for Jenkins on AWS and another for Jenkins inside a corporate firewall. It discusses the architecture of the controllers, reverse proxy, agents, and the Docker cloud by listing the required virtual machines, operating system, and software packages we'll use, the ports that need to be opened, and other required components. The chapter continues to discuss frequently asked questions for the AWS infrastructure, such as EC2 instance types and sizes, Regions and Availability Zones, routing rules, and Elastic IPs. Then, the chapter discusses the TLS/SSL certificate choices and goes through the steps for using Let's Encrypt in detail to create a free certificate. Finally, the chapter discusses the importance of storage backend choices. It discusses the different options by benchmarking performance and going through the pros and cons of the popular storage backend solutions.

Chapter 2, Jenkins with Docker on HTTPS on AWS and inside a Corporate Firewall, goes through the entire journey of setting up the Jenkins controller, reverse proxy for HTTPS connections, agents, and the Docker cloud. It shows a way to create a directory on the host machine and mount it in a Docker container running Jenkins, so that the state is preserved across container restarts. It also shows three different ways of terminating the TLS to provide HTTPS connections. Once Jenkins is running on HTTPS, the chapter goes through the basic configuration options for login methods, pipeline default speed, user permissions, and other useful default settings. It continues to show the steps for attaching agents and creating and attaching a Docker cloud, so that we end up with production-grade Jenkins.

Chapter 3, GitOps-Driven CI Pipeline with GitHub, shows the steps for creating premerge CI pipelines that are triggered from a GitHub pull request activity. It first creates four example users, then assigns them various permissions for the two example projects, adder and subtractor, to demonstrate the Jenkins permission model. It then goes through the steps for creating the CI pipelines in detail, demonstrating and discussing each step as we progress. It shows the two different ways to configure a CI pipeline, one for AWS Jenkins using push hooks and another for firewalled Jenkins using the GitHub Pull Request Builder plugin. It finishes by showing the optional steps to allow the CI pipeline to build an arbitrary branch, along with the steps to require a successful build for merging a pull request.

Chapter 4, GitOps-Driven CD Pipeline with Docker Hub and More Jenkinsfile Features, shows the steps for creating postmerge CD pipelines that are triggered from a GitHub pull request merge activity. Along the way, it discusses various Jenkinsfile techniques such as running external scripts, passing variables across steps, several ways of using Docker-outside-of-Docker (DooD), using bare-metal agents, using credentials, and interacting with GitHub and Docker Hub. Similar to *Chapter 3, GitOps-Driven CI Pipeline with GitHub*, it goes through the detailed steps for creating CD pipelines for both AWS Jenkins using push hooks and firewalled Jenkins using polling.

Chapter 5, Headfirst AWS for Jenkins, shows detailed instructions on using AWS. In the earlier chapters, we have skipped the details on most of the AWS operations in order to keep the focus on Jenkins, and in this chapter, we discuss them in full detail so that new users can follow the steps click by click while referring to the numerous screenshots. It starts by discussing the basics of logging into AWS, then continues to the steps for creating an SSH key pair, managing security groups, creating EC2 instances with Elastic IPs, using Let's Encrypt to generate TLS/SSL certificates, creating and configuring Elastic Load Balancers (ELBs), using AWS Certificate Manager to generate TLS/SSL certificates, setting up routing rules, and finally configuring Route 53 to point the Jenkins URL to the controller.

Chapter 6, Jenkins Configuration as Code (JCasC), discusses JCasC in detail by creating a whole new Jenkins instance using a configuration file we generate throughout the chapter. It starts by installing the JCasC plugin and discussing the limitations and the boundaries of what JCasC can manage. It then continues to read the configuration details of the Jenkins we set up in *Chapter 1, Jenkins Infrastructure with TLS/SSL and Reverse Proxy*, through *Chapter 4, GitOps-Driven CD Pipeline with Docker Hub and More Jenkinsfile Features*. It discusses each section of the configuration, and builds a new JCasC configuration file based on the entries from the existing Jenkins. Once the configuration file is built for the controller, agent, and Docker cloud, it creates a new Jenkins instance using the configuration file. It revisits each configuration item and discusses how well (or not) it was restored. Finally, it shows an optional way to use Groovy scripting to work around some of the issues found during the restoration.

Chapter 7, Backup and Restore and Disaster Recovery, discusses the backup strategies for different scenarios, and goes over the exact steps to set up an automated backup system. It first discusses the pros and cons of a disk snapshot backup and a file-level backup. Then it looks at the content of $JENKINS_HOME and identifies the files and folders that need to be backed up at a high frequency, as opposed to the ones that need to be backed up only once a day. Once we have determined which files to back up, the chapter goes through ThinBackup plugin configurations. It first provides an off-site backup solution using NFS and Docker volume mount, then goes into the specifics of configuring the ThinBackup plugin to generate the backup files effectively. Once backup files are generated, the chapter goes through a disaster scenario where we restore a pipeline that a user mistakenly deleted. It shows various ways to identify the correct backup snapshot to restore, then goes into deep discussions on how to restore a backup effectively. In addition to restoring the mistakenly deleted pipeline, it teaches the fundamental mechanism for backup and restore by demonstrating a way to restore a pipeline that didn't exist. Finally, it goes through an infrastructure failure disaster scenario and provides a recovery playbook that you can follow.

Chapter 8, Upgrading the Jenkins Controller, Agents, and Plugins, discusses the upgrade strategies for both small and large Jenkins instances. It first discusses the pitfalls of upgrading plugins, and provides various ways of upgrading the plugins and controller effectively. In addition to the upgrade process, the chapter goes through an SRE runbook for an upgrade scenario where you are taught when and how to communicate with users about the upgrade. The runbook covers not only the success path but also the failure scenario and discusses the restore and rollback strategies.

Chapter 9, Reducing Bottlenecks, teaches you various ways to optimize your Jenkins, such as picking the right EC2 instance size, reducing the Jenkins memory footprint, not using periodic triggers in favor of webhook triggers, tracking the AWS costs, optimizing GitHub Pull Request Builder options, and removing the weather icon from the home page. It continues on to discuss the various Groovy scripts that terminate long-running pipelines, release stale locks, and clean up logs. It also discusses the best practices for writing Jenkins pipeline code, such as reducing the use of the echo step and using NonCPS code for faster execution. It then talks about reducing the agent startup time by baking the plugin archives into the EC2 AMI, and then finally discusses the ways to manage various logs effectively.

Chapter 10, Shared Libraries, starts by discussing the directory structure and the content of a shared library. Afterward, it creates an example shared library that uses many common features, then explains the differences between providing the shared library as a global shared library versus a folder-level shared library. Once the shared library is available to be used, the chapter teaches you several different methods of loading it and discusses the use case for each method. Afterward, the chapter goes through a hands-on example of creating shared library functions that use the Slack messenger app to provide standardized messaging wrappers. Finally, the chapter dives deeper into a more advanced use case of creating custom domain-specific languages (DSLs) using shared libraries.

Chapter 11, Script Security, starts with an explanation of the role of an administrator versus a non-administrator in Jenkins. It continues by explaining the concept of the Groovy sandbox, and discusses running outside and inside the sandbox. It teaches the dangers of running pipelines outside of the sandbox and provides a use case of using a global shared library to wrap dangerous method calls. It then continues to explain the Jenkins permission model by discussing running inside the sandbox and teaches you how to use method signature approvals effectively. The chapter takes a deep dive into explaining the approve assuming permissions check button, and explains the SYSTEM user and the dangers of the default Jenkins permission model. Finally, it discusses an alternate Jenkins design that doesn't rely on the Script Security plugin's protection.

To get the most out of this book

We will be using Git, Docker, systemd, OpenSSL, NGINX, and other tools on Linux, so you need basic familiarity with the Linux command line. For AWS Jenkins, you need an AWS account where you can create and manage EC2 instances, ELBs, AWS Certificate Manager certificates, and Route 53 entries. For the Jenkins inside a corporate firewall, you need three virtual machines and optionally access to a company Public Key Infrastructure (PKI) where you can generate TLS/SSL certificates. You also need a GitHub and a Docker Hub account.

Software/hardware covered in the book
Windows, macOS, Linux, or any other operating system that you can use to SSH into a Linux machine
Ubuntu 20.04 for the virtual machines and EC2 instances
Docker 18 or higher, used in the Ubuntu 20.04 hosts
Git and OpenSSL, which are preinstalled in Ubuntu 20.04
Jenkins 2.263.1-LTS or higher

Jenkinsfiles and Shared Libraries code are written in the Groovy programming language. You can follow along without prior experience with Groovy, but familiarity with Groovy will help you understand the shared libraries chapter more easily. It's very similar to Java.

If you are using the digital version of this book, we advise you to type the code yourself or access the code via the GitHub repository (link available in the next section). Doing so will help you avoid any potential errors related to the copying and pasting of code.

Download the example code files

You can download the example code files for this book from GitHub at https://github.com/PacktPublishing/Jenkins-Administrators-Guide. In case there's an update to the code, it will be updated on the existing GitHub repository.

We also have other code bundles from our rich catalog of books and videos available at https://github.com/PacktPublishing/. Check them out!

Download the color images

We also provide a PDF file that has color images of the screenshots/diagrams used in this book. You can download it here: https://static.packt-cdn.com/downloads/9781838824327_ColorImages.pdf

Conventions used

There are a number of text conventions used throughout this book.

Code in text: Indicates code words in text, folder names, filenames, dummy URLs, and user input. Here is an example: "For example, builds for a pipeline that specifies agent { label 'ubuntu2004-agent' } would run only on ubuntu2004-agent, even if you didn't label the agent with its own name."

A block of code is set as follows:

```
$ ssh ubuntu@52.53.150.203
$ docker ps
CONTAINER ID   IMAGE   COMMAND   CREATED   STATUS   PORTS   NAMES
```

When we wish to draw your attention to a particular part of a code block, the relevant lines or items are set in bold:

```
$ sudo usermod -aG docker $USER
$ exit
logout
```

Screen text: Indicates words that you see onscreen. Here is an example: "Click Save and Finish to continue (we will change this soon), then click Start using Jenkins."

Italics: Indicates an important word a phrase. Here is an example: "Most importantly, *you are responsible* for the restoration in the event of a disaster."

> **Tips or important notes**
> Appear like this.

Get in touch

Feedback from our readers is always welcome.

General feedback: If you have questions about any aspect of this book, mention the book title in the subject of your message and email us at customercare@ packtpub.com.

Errata: Although we have taken every care to ensure the accuracy of our content, mistakes do happen. If you have found a mistake in this book, we would be grateful if you would report this to us. Please visit www.packtpub.com/support/errata, selecting your book, clicking on the Errata Submission Form link, and entering the details.

Piracy: If you come across any illegal copies of our works in any form on the Internet, we would be grateful if you would provide us with the location address or website name. Please contact us at copyright@packt.com with a link to the material.

If you are interested in becoming an author: If there is a topic that you have expertise in and you are interested in either writing or contributing to a book, please visit authors.packtpub.com.

Reviews

Please leave a review. Once you have read and used this book, why not leave a review on the site that you purchased it from? Potential readers can then see and use your unbiased opinion to make purchase decisions, we at Packt can understand what you think about our products, and our authors can see your feedback on their book. Thank you!

For more information about Packt, please visit packt.com.

Part 1

Securing Jenkins on AWS and inside a Corporate Firewall for GitOps-Driven CI/CD with GitHub and Docker Hub

In this section, you will learn about the Jenkins infrastructure architecture, the application architecture, and the project architecture. You will learn how to set up Jenkins securely from scratch with static agents, as well as dynamic agents from the Docker cloud. You will create pipelines and connect them to GitHub for CI builds, and to Docker Hub for CD builds.

This part of the book comprises the following chapters:

- *Chapter 1, Jenkins Infrastructure with TLS/SSL and Reverse Proxy*

- *Chapter 2, Jenkins with Docker on HTTPS on AWS and inside a Corporate Firewall*

- *Chapter 3, GitOps-Driven CI Pipeline with GitHub*

- *Chapter 4, GitOps-Driven CD Pipeline with Docker Hub and More Jenkinsfile Features*

- *Chapter 5, Headfirst AWS for Jenkins*

1

Jenkins Infrastructure with TLS/SSL and Reverse Proxy

In this chapter, we will learn about the foundational components of Jenkins: the controller, agents, cloud, domain name, TLS/SSL certificates, and reverse proxy. First, we will learn where each component fits into the architecture, and then prepare the VMs and TLS/SSL certificates. Finally, we will learn the importance of choosing the right storage medium for the Jenkins controller and discuss the pros and cons of some of the popular storage options. By the end of this chapter, we will understand the Jenkins architecture and have the necessary components ready so that we can put them together in the next chapter.

In this chapter, we're going to cover the following main topics:

- Why Jenkins?
- Searching for answers online with Jenkins keywords
- Understanding the Jenkins architecture
- AWS: FAQs, routing rules, EC2 instances, and EIPs
- Installing Docker on our VMs
- Acquiring domain names and TLS/SSL certificates
- Storage concerns (very important!)

Technical requirements

You need a domain name (for example, jenkins.example.com) for your Jenkins instance and one or both of the following:

- An AWS account with permission to create three EC2 instances, create an Application ELB, create certificates via AWS Certificate Manager, create AWS access keys via IAM, and modify domain records in Route 53.

- Three VMs running Ubuntu 20.04, access to domain records for your domain, and optionally a company public key infrastructure (PKI).

The files for this chapter are available in this book's GitHub repository at https://github.com/PacktPublishing/Practical-Jenkins-2.x/blob/main/ch1.

Why Jenkins?

A Continuous Integration (CI) build runs when a pull request (PR) is created or updated so that a developer can build and test the proposed change before merging the change. In the 2020s, a CI system is a normal fact of life for any software development, and it is difficult to imagine developing a software product without it.

But finding a good CI solution remains difficult, particularly when the validation process is more complex than a simple use case of building a Go binary and creating a Docker image with it. For example, if we need to build a phone firmware and need to flash a physical phone in one of the three test phones as a part of the validation, we would need a CI system that can handle physical connections to the hardware, as well as manage resource locks to make sure that the validation does not disrupt other validations that are already in progress.

The use cases are even more complex during the Continuous Delivery or Continuous Deployment (CD) process, where the end product of a build is either stored in an archive (such as Docker registry or Artifactory) or deployed to a production system. The CD builds' more frequent interactions with external systems require credential management and environment preparation for a deployment. For example, it's not uncommon for a production system to be in an isolated network inaccessible from the corporate network, and a jumpbox session must be prepared before an update can be deployed. It's also common that the deployments are handled in a different set of pipelines that are not tied to the code changes in the Git repository, which means that we would need a CD solution made up of free-standing pipelines that are detached from any PR activities.

Jenkins, known for its supreme flexibility, can handle such complex use cases easily. Managing a physical hardware connection, handling resource locks, managing credentials, and handling session data across multiple stages is all built into Jenkins or can be made available easily through one of 1,500+ plugins[1].

In addition to the rich feature set, Jenkins supports real programming languages such as Java and Groovy to specify the build steps. This allows us to write real code to express our build logic using variables, conditionals, loops, exceptions, and sometimes even classes and methods, rather than being bound by a domain-specific language (DSL) that is often limiting.

For those of us who prefer a more structured solution, Jenkins also supports a DSL to provide a uniform development experience. The most typical use cases are covered by the plugins that provide a wrapper to the common code, created by the dedicated user base who continues to contribute to the open source platform. CloudBees, which does an excellent job maintaining the project, also contributes to the open source plugin ecosystem. The vast number of available plugins indicates that developing a plugin is easy if we need to create a specific solution for our business use case.

Finally, a big advantage of Jenkins is that it's free. The Jenkins source code uses the MIT License[2], which is one of the most permissive open source licenses. We can scale Jenkins vertically by having one very powerful shared instance, or we can develop a "Jenkins vending machine" infrastructure, which creates a new Jenkins instance for each team, all without paying any licensing fees. We can even embed Jenkins in a commercial product and sell the product.

Let's learn how to use Jenkins, the most flexible and powerful CI system.

Searching for answers online with Jenkins keywords

Let's start with the most important aspect of software engineering: searching for answers online. Jenkins has a development history of over a decade and many ideas and keywords have come and gone. Let's go over the keywords so that we can search for them online effectively.

The Jenkins project was initially released in 2005 under the name Hudson. After an ownership dispute about the name with Oracle Corporation, in 2011, it was officially renamed and released as Jenkins. *Hudson* rarely comes up during normal use, so casual Jenkins users will not come across the term. However, a large part of the code is under the hudson Java package (https://github.com/jenkinsci/jenkins/tree/master/

---

1 https://plugins.jenkins.io/

2 https://www.jenkins.io/license/

core/src/main/java/hudson), so an admin or an author of a plugin should understand that Hudson is a precursor of Jenkins.

The Jenkins architecture has a controller that acts as a central server and an agent that runs build steps. The controller and agent were originally called master and slave until they were renamed for Jenkins 2.0[3]. This is a recent change, so you'll still see references to master throughout Jenkins. For example, the first agent is named master and it can't be changed (https://jenkins.example.com/computer/(master)/). In addition, as of 2021, there are more search results for *jenkins master* than *jenkins controller*, so debugging a controller issue may require we search with the term *master*. There is also a node, which means a computer. Both the controller and agent are nodes, but sometimes, an agent is mistakenly called a node.

The two most popular agent types are the SSH agent and inbound agent. SSH agents are typically used when the controller can reach out to the agent, whereas the inbound agents are typically used when the controller cannot reach the agent and the agent must initiate the connection back to the controller. The inbound agent was originally called JNLP slave (https://hub.docker.com/r/jenkins/jnlp-slave), and we may find many references to it as we search online for help.

In Jenkins, a configuration is created to perform a task (for example, build software), and this configuration is called a project. In the past, a project was called a job. Even though the term *job* had been deprecated, it is still widely used, which leads us to the next term...

Job DSL is one of the first plugins that was created for codifying a Jenkins project. DSL is a fancy term for *custom programming language*, so *Job DSL* means *a custom programming language for a Jenkins project*. It is used to define and create a Jenkins project through code rather than through a GUI. Job DSL is not Pipeline DSL. They are entirely different solutions and an answer for a question for one of them will not be applicable to the other.

Pipeline DSL is the custom programming language for a Jenkins Pipeline (or simply *Pipeline* with a capital *P* (https://www.jenkins.io/doc/book/pipeline/#overview)). Pipelines are enabled through the Pipeline plugin (https://plugins.jenkins.io/workflow-aggregator/), which used to be called Workflow, a term you may come across if you're creating a plugin.

There are two flavors of Pipeline DSL syntax: Scripted Pipeline and Declarative Pipeline (https://www.jenkins.io/doc/book/pipeline/syntax/#compare). Both are Pipeline DSLs that are written into a text file named Jenkinsfile, and Jenkins can process either flavor of the syntax. Scripted Pipeline syntax was created first, so older posts on online forums will likely be in Scripted Pipeline syntax, while

---

3   https://www.jenkins.io/blog/2020/06/18/terminology-update

Declarative Pipeline syntax will typically be found on newer posts. The two syntaxes are not directly interchangeable, but an answer to a problem for one syntax flavor can usually be converted into the other. The underlying language of both is Groovy, a fully featured programming language independent of Jenkins.

Finally, the verb for executing a project is *build*. For example, we *build* a pipeline to execute its steps. Some pipelines are generic tasks that don't really build software, so many people say *run* a pipeline or even *run a build* for a pipeline instead. However, regardless of what the pipeline does, the button for executing a pipeline is captioned Build Now. Most people use the terms project, job, and pipeline interchangeably, so building a project, running a pipeline, and running a build for a job all mean the same thing.

With that, you should have all the keywords for the web searches in case you have questions that this book doesn't answer. Next, let's understand the Jenkins architecture.

Understanding the Jenkins architecture

Before we set up Jenkins, let's go over the blueprint of what we will build.

Controller

As we mentioned previously, the controller is the central server for Jenkins. We'll set up the controller as follows:

- There are two Jenkins instances – one on AWS and the other in a corporate firewalled network. AWS Jenkins is built with an open source project in mind – builds need to run on the internet, where all contributors can see them. Firewalled Jenkins is for a typical corporation setting where all development and testing happens inside the corporate firewall. You can create one that fits your use case or create both to see which fits your needs better, even though, ultimately, you'll only need only one of the two. The settings for the two are not always interchangeable, so be sure to pick the correct sections for each type as you read ahead.

- Both Jenkins controllers use a VM running the latest Ubuntu LTS, which is 20.04 at the time of writing.

- The Jenkins controller runs as a Docker container listening on port 8080 for HTTP and port 50000 for inbound agents.

- The Jenkins controller container bind mounts a directory on the host to store jenkins_home.

- The AWS controller's hostname is `aws-controller` and it runs as the `ubuntu` user. Therefore, the commands that are run on it will start with `ubuntu@aws-controller:<path>$`.

- The firewalled controller's hostname is `firewalled-controller` and it runs as the `robot_acct` user. Therefore, the commands that are run on it will start with `robot_acct@firewalled-controller:<path>$`.

- The commands starting with `controller:<path>$` indicate the commands that can run on either controller.

- Finally, the AWS controller has an Elastic IP (EIP) attached to it. The EIP provides a static IP that doesn't change, and the inbound agent will connect to the EIP's address.

This is what it looks like:

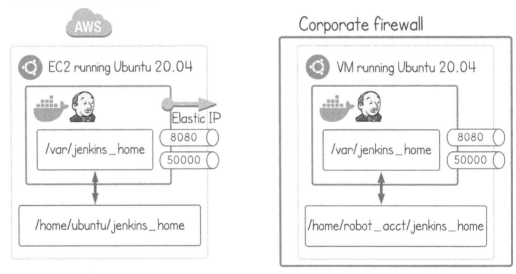

Figure 1.1 – Architecture of the VMs and containers for Jenkins controllers

With that, we have learned about the architectures of the AWS controller and firewalled controller. Now, let's move on and cover some more components.

Domain name, TLS/HTTPS, load balancer, and reverse proxy

The endpoint for a production-grade web service should be on HTTPS with a domain name rather than on HTTP with an IP. This requires that we have domain names, TLS certificates, a load balancer, and a reverse proxy. We'll set them up as follows:

- jenkins-aws.lvin.ca points to the AWS Jenkins controller.
- jenkins-firewalled.lvin.ca points to the firewalled Jenkins controller.
- HTTPS is provided through a load balancer, reverse proxy, or directly on Jenkins.
- The AWS Jenkins controller uses an Elastic Load Balancer (ELB) for TLS termination.
- The firewalled Jenkins controller uses an NGINX reverse proxy or the Jenkins controller itself for TLS termination.
- The load balancer and reverse proxy receive traffic from HTTP port 80 and redirects it to HTTPS port 443 for secure communication.
- The load balancer and reverse proxy receive traffic from HTTPS port 443, terminate the TLS, and proxy it to HTTP port 8080, which is where the Jenkins controller listens.

Here's how it's arranged:

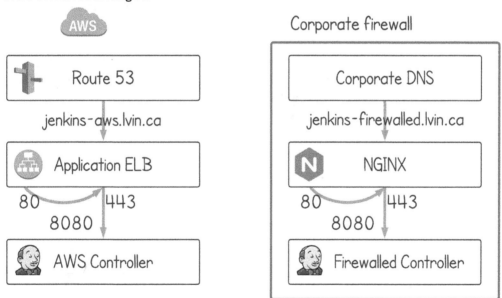

Figure 1.2 – Architecture of the DNS and reverse proxy for Jenkins controllers

With that, we have learned about the network architecture of DNS, TLS, the load balancer, and the reverse proxy. Let's continue and look at agents.

Agents

If a controller is where the pipelines are managed, then agents are where the actual builds run. We'll set up our agents as follows:

- There are two nodes for agents – one on AWS and another inside a corporate firewall.

- The agent nodes also run on VMs running Ubuntu 20.04.

- Each agent node connects to both Jenkins instances (yes, it's possible to connect a node to multiple Jenkins controllers).

- The AWS agent has an EIP attached to it. The EIP provides a static IP that doesn't change, and the firewalled controller will connect to the EIP's address.

- The firewalled agent connects to the AWS Jenkins controller on port 50000 as an inbound agent. For all other agents, the controller initiates the SSH connection on port 22 to configure the agents as SSH agents.

- The AWS agent's hostname is aws-agent, and it runs as the ubuntu user. Therefore, the commands that are run on it will start with ubuntu@aws-agent:<path>$.

- The firewalled agent's hostname is firewalled-agent, and it runs as the robot_acct user. Therefore, the commands that are run on it will start with robot_acct@firewalled-agent:<path>$.

- The commands starting with agent:<path>$ indicate the commands that can run on either agent.

Here's what it looks like:

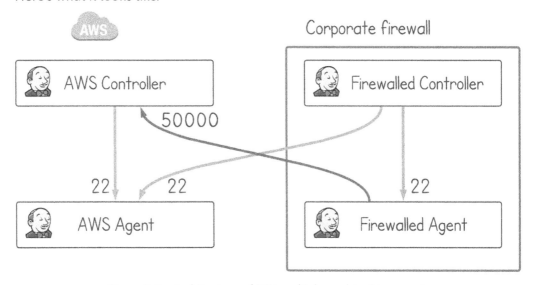

Figure 1.3 – Architecture of SSH and inbound Jenkins agents

With that, we have learned about the architectures of two possible agent connection types that can cross a corporate firewall. We have just one component remaining: Docker cloud.

Docker cloud

Docker cloud is used to dynamically generate an agent using Docker containers. There needs to be a Docker host where the containers will run, and this is how we will set it up:

- There are two Docker hosts – one on AWS and another inside a corporate firewall.

- The Docker hosts also run on VMs running Ubuntu 20.04.

- *The Docker hosts are not Jenkins agents*. Each provides a TCP endpoint for ephemeral Docker agents to be dynamically generated.

- A controller communicates with a Docker host on TCP port 2376, which is secured with an X.509 certificate. We will follow the steps in the official document[4].

- The AWS Docker host's hostname is aws-docker-host, and it runs as the ubuntu user. Therefore, the commands that are run on it will start with ubuntu@aws-docker-host:<path>$.

- The firewalled Docker host's hostname is firewalled-docker-host, and it runs as the robot_acct user. Therefore, the commands that are run on it will start with robot_acct@firewalled-docker-host:<path>$.

- The commands starting with docker-host:<path>$ indicate the commands that can run on either Docker host.

---

4   https://docs.docker.com/engine/security/protect-access/

Here's what it looks like:

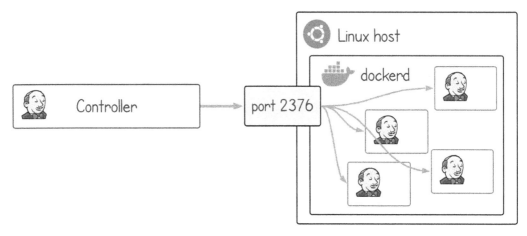

Figure 1.4 – Architecture of the Docker cloud host

With that, we've learned how a Docker host is configured for a Docker cloud and how a controller connects to the Docker host. Now, let's take a step back and look at it in its entirety.

Bringing it all together

There are six machines all running Ubuntu 20.04. The NGINX reverse proxy runs on the same machine running the firewalled Jenkins controller:

- AWS Jenkins controller (ubuntu@aws-controller)
- AWS Jenkins agent (ubuntu@aws-agent)
- AWS Docker host (ubuntu@aws-docker-host)
- Firewalled Jenkins controller (robot_acct@firewalled-controller)
- Firewalled Jenkins agent (robot_acct@firewalled-agent)
- Firewalled Docker host (robot_acct@firewalled-docker-host)

Here's how it all stacks up:

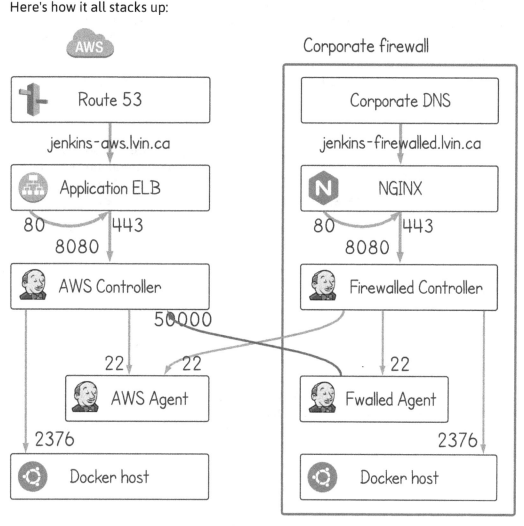

Figure 1.5 – Overview of the complete Jenkins architecture

That is the complete picture of what we will build. Let's continue and get our hands dirty by preparing the VMs and the TLS certificates.

AWS: FAQs, routing rules, EC2 instances, and EIPs

AWS is used heavily in this book, but it is such a vast ecosystem that we can't sufficiently discuss all the details without taking the focus away from Jenkins. Rather than trying to guess what level of detail we should provide, instead we have dedicated a separate chapter to discussing AWS in depth. *Chapter 5, Headfirst*

AWS for Jenkins, features step-by-step instructions with plenty of screenshots, best practices you should follow, and more. The rest of this book will still cover the AWS topics at a high level, and you can turn to *Chapter 5, Headfirst AWS for Jenkins*, for a deeper dive.

Now, let's cover some common pitfalls that everyone should watch out for.

EC2 instance types and sizes

You can start with an EC2 instance as small as t2.micro – I used t2.micro for the AWS Jenkins build for this book and it worked just fine. For a production controller, you can start with a larger end of the T2 type and then switch to a more powerful C5 type if you need to. The agent can be a general-purpose T2 type since it serves various pipelines.

Next, let's look at where to put them.

Regions and Availability Zones

Putting all the VMs in the same Availability Zone yields the best performance, but it's not strictly necessary. It would be a good idea to put them in the same region, but it's not an invalid setup to use an agent in a different region if the agent needs to be geographically closer to the resources that are used during a build. For example, if your HQ is in the US but the lab equipment that's used for testing is in India, the agent should be in India to minimize any latency-related issues during the build. Beware that in some cases, AWS charges extra for transferring data across regions. For example, copying data from an S3 bucket to an EC2 instance within the same region is free of charge, whereas accessing the S3 bucket from a different region incurs costs[5]. If there is a large amount of data transfer, it may make sense to plan the agent's locations based on the data transfer cost.

Next, let's check out the routing rules.

Routing rules

VPC routing is a very complex topic that will be discussed more fully in *Chapter 5, Headfirst AWS for Jenkins*. For now, the most important rules are as follows:

* The VMs and ELB can talk to one another.
* Anyone can reach the ELB on port 80 for HTTP and port 443 for HTTPS.
* Only the inbound agent can reach the controller on port 50000.

---

5   https://aws.amazon.com/blogs/mt/using-aws-cost-explorer-to-analyze-data-transfer-costs/

- Only we can reach the VMs on port 22 for SSH.

- Only we can reach the VMs on port 8080 to test Jenkins without going through ELB.

The easiest way is to have three security groups – one for internal connections, another for the VMs, and the last for the ELB.

On EC2 Dashboard, click Security Groups under Network & Security. Find the security group named default – this is your default security group. All resources (such as EC2 or ELB) with this security group attached to them can talk to one other. For the internal connections, attach this security group to all your VMs and ELBs. Let's create the other two security groups.

First, create a new security group named jenkins-vm that accept traffic to port 22, 8080, and 50000. All three ports should accept traffic from just My IP. This way, only we can SSH to the hosts or connect to port 8080 to debug. This configuration assumes that the inbound agent's IP is the same as the IP of your laptop. If the inbound agent's IP is not from the same network as your laptop, then change the source IP for the port 50000 to match the inbound agent's IP. All VMs we create should have this security group attached in addition to the default security group as discussed.

Inbound rules should look like this (the source CIDR will be different for yours):

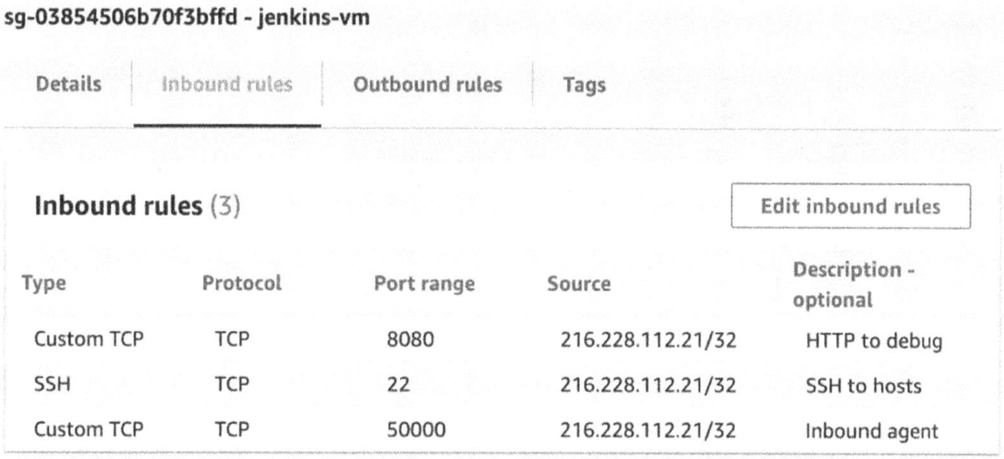

sg-03854506b70f3bffd - jenkins-vm

Details	Inbound rules	Outbound rules	Tags

Inbound rules (3)   [Edit inbound rules]

Type	Protocol	Port range	Source	Description - optional
Custom TCP	TCP	8080	216.228.112.21/32	HTTP to debug
SSH	TCP	22	216.228.112.21/32	SSH to hosts
Custom TCP	TCP	50000	216.228.112.21/32	Inbound agent

Figure 1.6 – Security group inbound rules for jenkins-vm

Next, create a new security group named `jenkins-elb` that accepts traffic to port 80 and 443. Both ports should accept traffic from Anywhere to allow anyone on the internet to access Jenkins through HTTP and HTTPS. HTTP will be redirected to HTTPS as we'll see in the next chapter. The ELB we create should have this security group attached in addition to the default security group. Inbound rules should look like this:

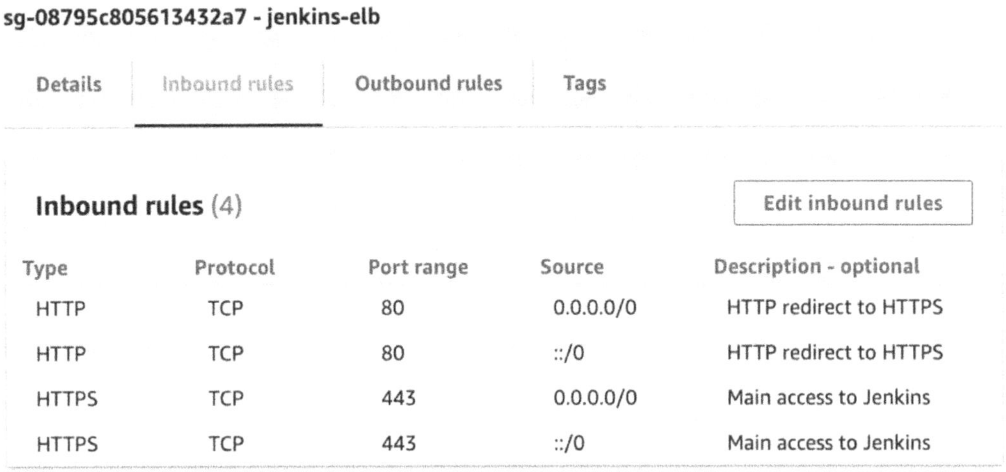

sg-08795c805613432a7 - jenkins-elb

Details	Inbound rules	Outbound rules	Tags

Inbound rules (4)   Edit inbound rules

Type	Protocol	Port range	Source	Description - optional
HTTP	TCP	80	0.0.0.0/0	HTTP redirect to HTTPS
HTTP	TCP	80	::/0	HTTP redirect to HTTPS
HTTPS	TCP	443	0.0.0.0/0	Main access to Jenkins
HTTPS	TCP	443	::/0	Main access to Jenkins

Figure 1.7 – Security group inbound rules for jenkins-elb

Finally, let's create the EC2 instances and the EIPs.

EC2 instances and EIPs

Create three EC2 instances to the following specifications. If you're unsure of the steps, check out *Chapter 5, Headfirst AWS for Jenkins*, for a detailed guide:

- **Amazon Machine Image (AMI)**: Ubuntu Server 20.04 LTS (HVM), SSD Volume Type with the 64-bit (x86) architecture.

- Instance Type: t2.micro. Once you get the hang of running Jenkins, you will be able to create a larger one for your production server.

- Instance Details / Auto-assign Public IP: Disable. We will use an EIP instead.

- Storage: For a test instance, 8 GiB is fine. For a production instance, increase it to 100 GiB. Keep the type as the gp2 type because io1 and io2 are very expensive. If the performance becomes a problem, check out the tips and tricks from *Chapter 9, Reducing Bottlenecks*, for the solutions.

- Tags: Set Name as `Jenkins Controller`, `Jenkins Agent`, or `Jenkins Docker cloud host` for each of the three hosts:

Key	Value	Instances	Volumes	Network Interfaces
Name	Jenkins Controller	☑	☑	☑

Figure 1.8 – Name tag for Jenkins Controller

- Security Group: *This part is important.* Click Select an existing security group, then check both the default and jenkins-vm security groups as shown here:

Step 6: Configure Security Group

Assign a security group: ○ Create a **new** security group

◉ Select an **existing** security group

Security Group ID	Name	Description
■ sg-8cf9b1b8	default	default VPC security group
sg-004231f17d99949a1	jenkins-elb	Allow inbound for HTTP and HTTPS
■ sg-0494ae671a6534028	jenkins-vm	Allow access for debugging and inbound agent

Figure 1.9 – Selecting both the default and jenkins-vm security groups

- Finally, create an EIP and attach it to the instance. The instance details page should show its public IP matching its EIP:

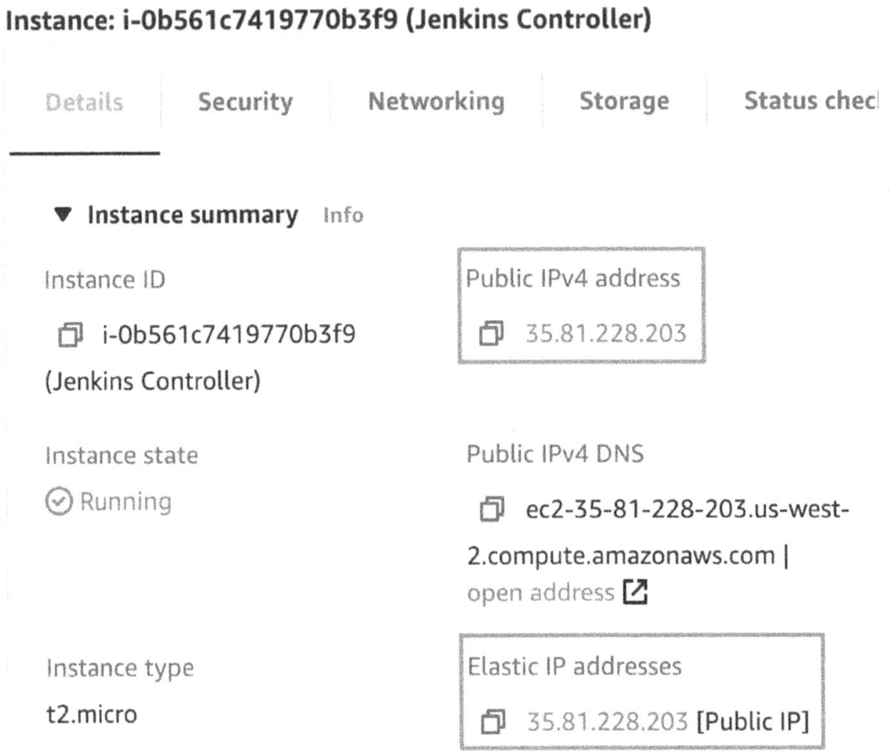

Instance: i-0b561c7419770b3f9 (Jenkins Controller)

Details	Security	Networking	Storage	Status chec

▼ **Instance summary**   Info

Instance ID

🗇   i-0b561c7419770b3f9

(Jenkins Controller)

Public IPv4 address

🗇   35.81.228.203

Instance state

⊘ Running

Public IPv4 DNS

🗇   ec2-35-81-228-203.us-west-

2.compute.amazonaws.com |

open address ↗

Instance type

t2.micro

Elastic IP addresses

🗇   35.81.228.203 [Public IP]

Figure 1.10 – EIP is attached to Jenkins Controller EC2 instance

Once the three EC2 instances are ready, let's continue to install Docker.

Installing Docker on our VMs

Docker is a fundamental tool in modern software engineering. It provides a convenient way of establishing a preconfigured isolated environment that is defined as text in a Dockerfile.

Docker is used for everything in our Jenkins setup, so we need to install Docker on all our VMs. Follow the installation steps in Docker's official documentation (https://docs.docker.com/engine/install/ubuntu/#install-using-the-repository).

Once the installation is complete, be sure to add the user to the docker group, *log out, and log back in*. 52.53.150.203 is the IP of one of my VMs. You should use your VM's IP instead:

```
$ sudo usermod -aG docker $USER
$ exit
logout
Connection to 52.53.150.203 closed.
$ ssh ubuntu@52.53.150.203
$ docker ps
CONTAINER ID   IMAGE   COMMAND   CREATED   STATUS   PORTS   NAMES
```

With Docker installed and the user added to the docker group, the VMs are ready to be used for Jenkins. The first task in setting up Jenkins is acquiring the TLS certificates. Let's go through that together.

Acquiring domain names and TLS/SSL certificates

A production-grade web service should use a domain name and HTTPS, even if it's an internal tool. Let's examine their role in our architecture.

Domain names

Two domain names are needed for the two Jenkins instances. If you are using a subdomain of your company's domain (for example, jenkins.companyname.com), be sure that you can modify the A record, CNAME, and TXT record for the domain name. A new .com domain name can be purchased from AWS for around $12. For the AWS Jenkins instance, the DNS configuration is simpler if the domain is managed through Route 53. In our setup, we will be using jenkins-aws.lvin.ca and jenkins-firewalled.lvin.ca.

TLS/SSL certificates

TLS (also commonly referred to as SSL, which is TLS's predecessor technology) enables HTTPS, which allows secure communication. A TLS certificate can be obtained in several different ways:

- AWS Certificate Manager provides free public certificates to be used by AWS resources. This is useful if your Jenkins instance is on AWS, but the free certificates cannot be exported to be used in more advanced ways. For example, the certificate can be used on ELB, but cannot be exported for your own NGINX reverse proxy running on an EC2 instance.

- In a corporate setting, sometimes, there is a PKI at pki.companyname.com where you can generate TLS certificates for the domain names that the company owns. These are often internal certificates that are signed by the company's own certificate authority (CA), which are only accepted by the machines where the company's CA is pre-installed. This is useful if your Jenkins instance is behind a corporate firewall and will only be accessed by the company's equipment.

- Let's Encrypt provides free public certificates. This is useful when the Jenkins instance is not running on AWS and your company doesn't provide a PKI. The certificates, however, are valid only for 90 days and it requires additional configuration to auto-renew.

- Commercial vendors such as Comodo or RapidSSL sell public certificates. There are resellers who sell the same certificates for a fraction of the original cost, so search online for deals.

Try the various methods and get the TLS certificate for your Jenkins URLs.

Let's Encrypt

Let's Encrypt certificates can be generated and renewed using Certbot. Since the certificates are used in the controller, the certificates should be generated on the controller host.

The certificates expire in 90 days, and the same commands can be run again to regenerate the same certificates with a renewed expiry date. The generation is *rate limited*, so plan to minimize the number of requests to Let's Encrypt by waiting about 80 days before regenerating the updated certificates and not sooner. Creating new certificates doesn't invalidate the existing certificates.

Prepare the work directories, like so:

```
robot_acct@firewalled-controller:~$ mkdir -p ~/letsencrypt/
{certs,logs,work}
```

When we request Let's Encrypt to generate a certificate for a domain, Let's Encrypt asks us to verify that we own the domain. We can verify it either by manually modifying the TXT record on the domain or by letting Certbot automatically modify it on Amazon Route 53 for us.

Manual verification

A manual verification method requires that we modify the TXT record to prove that we own the domain. This approach is used when the domain is not managed through Route 53. The following is a sequence diagram for the manual verification process:

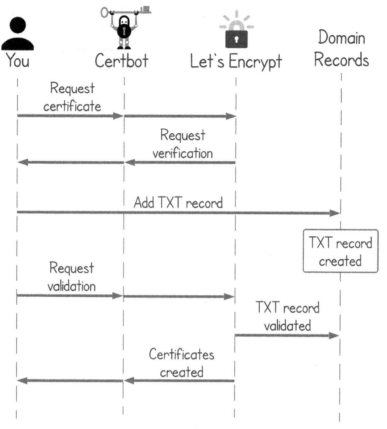

Figure 1.11 – Let's Encrypt workflow for manual verification

Run the following commands with your own values for EMAIL and DOMAIN to generate new certificates:

```
robot_acct@firewalled-controller:~$ export EMAIL=calvinspark@
gmail.com
robot_acct@firewalled-controller:~$ export DOMAIN=jenkins-
firewalled.lvin.ca
robot_acct@firewalled-controller:~$ docker run -it --rm \
    -u $(id -u):$(id -g) \
    -v ~/letsencrypt/work:/work \
    -v ~/letsencrypt/logs:/logs \
    -v ~/letsencrypt/certs:/certs \
    --name certbot \
    certbot/certbot:v1.10.1 \
    certonly \
    --work-dir /work \
    --logs-dir /logs \
    --config-dir /certs \
    --agree-tos \
    --manual \
    --preferred-challenge dns \
    -m $EMAIL \
    -d $DOMAIN
```

The process pauses for us to verify that we own the domain:

```
- - - - - - - - - - - - - - - - - - - - - - - - - - - -
Please deploy a DNS TXT record under the name

_acme-challenge.jenkins-firewalled.lvin.ca with the following
value:

kLmhtIfqI5PZFuk-lXna13Z4_oIYDmaJoPd6RaFgwqQ

Before continuing, verify the record is deployed.

- - - - - - - - - - - - - - - - - - - - - - - - - - - -
Press Enter to Continue
```

Once we have added the TXT record and verified our ownership, the certificates are issued. If the verification is failing, wait 10 minutes to let the DNS propagate and then try again:

```
IMPORTANT NOTES:
- Congratulations! Your certificate and chain have been saved at:
    /certs/live/jenkins-firewalled.lvin.ca/fullchain.pem

    Your key file has been saved at:

    /certs/live/jenkins-firewalled.lvin.ca/privkey.pem
```

The certificates are now ready for use. Let's see how the process differs when the verification step is automated.

Automated verification for Amazon Route 53

An automated verification method allows Certbot to modify the TXT record on Amazon Route 53 on our behalf to prove that we own the domain. This approach is used when the domain is managed through Route 53. The following is a sequence diagram for the automated verification process:

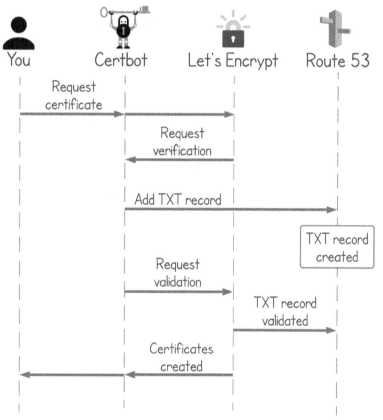

Figure 1.12 – Let's Encrypt workflow for automated verification for AWS Route 53

Follow these steps on an AWS account where the domain name is managed through Route 53:

1. Create an IAM user with this IAM policy, and then create AWS access keys. Be sure to replace YOURHOSTEDZONEID in the policy with your own hosted zone ID:

```
{
    "Version": "2012-10-17",
    "Statement": [
        {
            "Effect": "Allow",
            "Action": [
                "route53:GetChange",
                "route53:ListHostedZones"
            ],
            "Resource": "*"
        },
        {

            "Effect": "Allow",
            "Action": "route53:ChangeResourceRecordSets",
            "Resource": "arn:aws:route53:::hostedzone/YOURHOSTEDZONEID"
        }
    ]
}
```

2. Create the letsencrypt/aws_config file with the AWS access keys:

```
robot_acct@firewalled-controller:~$ cat letsencrypt/aws_config
[default]
aws_access_key_id=AKIAZB45TXBWE4AFMPM5
aws_secret_access_key=lrl6A3U39...xmoiytJFSBlE
```

3. Run the following commands with your own values for EMAIL and DOMAIN to generate the certificates. The verification process is handled automatically:

```
robot_acct@firewalled-controller:~$ export EMAIL=calvinspark@
gmail.com

robot_acct@firewalled-controller:~$ export DOMAIN=jenkins-
firewalled.lvin.ca

robot_acct@firewalled-controller:~$ docker run -it --rm \
    -u $(id -u):$(id -g) \
    -v ~/letsencrypt/work:/work \
    -v ~/letsencrypt/logs:/logs \
    -v ~/letsencrypt/certs:/certs \
    -v ~/letsencrypt/aws_config:/aws_config \
    -e AWS_CONFIG_FILE=/aws_config \
    --name certbot \
    certbot/dns-route53:v1.10.1 \
    certonly \
    --work-dir /work \
    --logs-dir /logs \
    --config-dir /certs \
    --agree-tos \
    --dns-route53 \
    -m $EMAIL \
    -d $DOMAIN
```

Certbot uses the AWS access keys to respond to the challenge and automatically verifies that we own the domain name. Here is the abridged output with the most important lines shown:

```
Credentials found in config file: /aws_config

Plugins selected: Authenticator dns-route53, Installer None

Requesting a certificate for jenkins-firewalled.lvin.ca

Performing the following challenges:

dns-01 challenge for jenkins-firewalled.lvin.ca

Waiting for verification...

Cleaning up challenges

- Congratulations! Your certificate and chain have been saved at:

    /certs/live/jenkins-firewalled.lvin.ca/fullchain.pem

    Your key file has been saved at:

    /certs/live/jenkins-firewalled.lvin.ca/privkey.pem
```

Since the domain validation process is automated, the certificate renewal process can also be automated by simply running the docker run command for Certbot again.

Applying the renewed certificate requires us to restart the controller container, so plan for updating the certificate during an upgrade period.

The domain and TLS certificates are ready. Let's turn our focus to our next major component – storage.

Storage concerns

The following is very important:

> *Everything* in Jenkins is a flat file on the controller.
> Everything in Jenkins is a *flat file* on the controller.
> Everything in Jenkins is a flat file on the *controller.*

The controller's storage is *extremely important* to the performance of Jenkins and the backup/restore process. Jenkins doesn't use a database as a backend. Instead, it uses directories and files to store and look up everything. On a mid-sized Jenkins installation with a hundred pipelines, the controller runs dozens of builds concurrently and keeps track of tens of thousands of build history. This results in *hundreds of open files* and *hundreds of thousands of small files* stored for history tracking. *The performance of Jenkins is dictated almost entirely by the storage's input/output operations per second (IOPS)*. The storage medium determines how far you can scale the Jenkins instance in terms of both its performance and maintenance costs.

Can we just put everything in a fast SSD and call it a day? Not always. Since all the files are stored on the controller, a backup process must either copy the files from the controller or take a snapshot of the storage. It is often simpler to take a snapshot of the storage than to copy the files, but in some cases, the snapshot requires the disk to be unmounted, which is not an option because it stops Jenkins.

Now, let's consider the popular storage options. We'll examine the characteristics and the performance of each option to determine the best solution for each use case.

IOPS benchmarks using fio

To understand each storage backend option's performance, I have run benchmark tests using fio[6] to measure the IOPS. The following is the exact command that was used for the test, and the results in the following tables are a median from five runs:

```
$ fio --name=jenkins --ioengine=posixaio --iodepth=64 \
    --size=1G --readwrite=randrw
```

---

6   https://github.com/axboe/fio

Like all benchmarks, it is important to understand that this isn't the perfect way to model the disk usage pattern of Jenkins. Use the benchmark result to understand the comparative performance across the different storage options, rather than relying on the numbers as an absolute measure that can be used in a different context.

Based on my experience, a small-scale Jenkins instance supporting one or two teams would work fine on 1,000 IOPS from this test. A medium-scale Jenkins with over 100 pipelines should aim to get over 2,000 IOPS, while a large-scale Jenkins with over 1,000 pipelines should aim to get over 3,000 IOPS.

EC2 and EBS

If you're running on AWS, EC2 with EBS is a good choice. There are EBS-optimized EC2 instance types that provide very high IOPS, and EBS allows snapshotting without unmounting the disk, which simplifies the backup process. This is the recommended approach for running on AWS.

Here is the result of a basic benchmark with four variants of EC2 and EBS:

	t2.2xlarge + gp2	t2.2xlarge + io2	c5.2xlarge + gp2	c5.2xlarge + io2
IOPS Read / Write	1177 / 1178	1614 / 1615	1819 / 1820	2501 / 2502

Going by the IOPS guidelines I listed in the previous section, a t2.2xlarge instance would be sufficient for a small-scale Jenkins instance. A c5.2xlarge with gp2 EBS may be okay for a medium-scale Jenkins instance. A large-scale Jenkins instance would need an even bigger configuration. A word of caution is that the io2 storage type is very expensive, so do some thorough research before adopting it. Also, take note that the volume size also affects the IOPS so run your own benchmarks to get the actual value for your setup.

Let's continue to see the available options for firewalled Jenkins.

The IT VM's disk

In a corporate setting, sometimes, IT provides a VM as a service with a disk attached. These are often backed up daily by IT, which relieves you of backup duties (only partially, though, as we'll see in *Chapter 7, Backup and Restore and Disaster Recovery*). IT VM is often provided in an HA environment, so it's a good choice for stability. This is the recommended approach for running inside the corporate firewall if the service is available.

The performance of an IT VM's disk is difficult to generalize because the technology and the configuration of the VMs vary widely. For example, the disk can be a slice of the disk that's available on the same physical hypervisor, SAN, NAS, or something entirely different – the performance will vary widely, depending on the technology and the configuration. Despite the opaqueness of the configuration, IT generally tries its best to provide a configuration that's performant, so by using the IT VM's disk, we're leaning on the expertise of IT.

This is the result of the same benchmark we used previously on a couple of my IT VMs:

	VM 1	VM2
IOPS Read / Write	5797 / 5878	2736 / 2732

In this case, VM 1 outperformed VM 2 by more than double the IOPS. Because of these differences, it's worth mentioning to IT that you're looking for VMs with a high disk performance, in case they choose the backend from multiple configuration options.

Let's continue to see the available options for network drives.

NFS/SAN

In a corporate setting, sometimes, IT provides NFS or SAN as a service. These are often backed up daily by IT, which relieves you of backup duties. If the VM's own disk is not providing sufficient IOPS, SAN can be a good alternative due to its high IOPS. NFS is generally slower, but it can still be a good option since NFS can be mounted on multiple hosts at the same time.

Similar to the IT VM's disk, the performance of NFS or a SAN is difficult to generalize because their technology and configuration vary widely. Even with the same storage service, a VM that's located in the same data center as the service will get a much higher IOPS than a VM that's located in a different geographical region. Because of this, you should work with the IT team to make sure that your VMs and the storage services are closely located.

This is the result of the same tests I ran previously on the NFS and SAN storage that I have.

	NFS	SAN
IOPS Read / Write	1991 / 1988	3259 / 3253

If IT happens to provide a SAN that's significantly faster than the VM's disk, mount the SAN on the VM and use that for the controller's storage. If the SAN is not that much faster, just use the VM's disk to reduce the number of dependent services – this will help you if there is a system or network failure.

NFS is usually slower than the VM's disk, but it is still useful in the case of a VM failure. For example, we can have a second VM as a hot-standby with the NFS already mounted – in the event of a disaster where the controller VM is dead, the hot-standby can start a Jenkins container and restore the service quickly, thereby providing a speedy disaster recovery path. If NFS is not significantly slower than the VM's disk, or if NFS's IOPS meets the requirements for the size of your Jenkins instance, consider using NFS as the controller storage.

If none of these options are provided as a service, then we should look at physical disks.

Physical disks

Physical disks are fast – faster than any network drive. The downside of physical disks is that they can physically die, so an appropriate hardware preparation such as RAID is required. Another significant downside is that they can't be snapshotted while they're in use if you're using the very common ext filesystem. Being unable to create a daily snapshot backup is a significant disadvantage, so physical disks should only be used if you're using a filesystem such as ZFS or BTFS that does allow a snapshot, and also have a good backup and restore procedure.

Here is the result of the same tests that I ran on a physically attached SSD and M.2:

	SSD	M.2
IOPS Read / Write	40.6k / 41.0k	68.6k / 68.8k

The performance advantage is staggering – in my case, M.2 is over three hundred times faster than NFS. However, it still comes with severe disadvantages that are hard to overlook. The performance of Jenkins is important but lowering its maintenance cost and making disaster recovery easier are also very important aspects to consider. This is especially true on a large-scale Jenkins instance where the business suffers greatly when Jenkins is down. In such a case, it may make sense to purchase a SAN as its cost is justified as a part of the business continuity plan.

Review

We have paid lots of attention to the performance of each medium, but the decision you make will likely depend on the platform you choose and the available services from IT. As we'll see in the upcoming chapters, switching the storage medium is not very complex – since everything in Jenkins is a flat file on the controller, the migration is a simple file copy. Start with the one that you're most comfortable with, then consider a switch if you need to.

> **Other Platforms and Backends**
>
> There are, of course, ECS, Kubernetes, and other platforms to choose from if you're familiar with the technology. While you are deciding on which platform you wish to use, keep IOPS and backup processes in mind.

Summary

In this chapter, we explored the different use cases of CI/CD builds and learned that the flexibility of Jenkins allows us to handle even the most complex scenarios. We learned that Jenkins supports a DSL if we prefer a more structured approach, and also learned that there are many plugins to handle most common tasks.

We went through a brief history of Jenkins to understand the keywords and the old names of the components so that we can search for answers online effectively.

Then we learned about the major building blocks of Jenkins and where each component belongs in the Jenkins architecture.

We learned that AWS is a prominently used technology in this book and went through the commonly asked questions. We created the routing rules and EC2 instances together, then installed Docker on the EC2 instances in preparation for using them to build Jenkins.

We learned several different ways of acquiring TLS certificates, and generated certificates using Let's Encrypt so that they can be used in the controllers.

Finally, we learned the importance of the controller's storage and went through the common storage backend options to understand their pros and cons.

In the next chapter, we will use these components to build a fully functional Jenkins instance that's ready to take on production workloads.

2

Jenkins with Docker on HTTPS on AWS and inside a Corporate Firewall

With an understanding of the architecture from *Chapter 1, Jenkins Infrastructure with TLS/SSL and Reverse Proxy*, we are ready to run Jenkins. We'll learn two ways of running Jenkins – one on AWS and another inside a corporate firewall. We'll see that they're generally similar, but the reverse proxy and the port opening for inbound agent configuration are different. Don't worry – we'll configure not only a Jenkins controller but also the reverse proxy, HTTPS, agents, and even Docker Cloud together. By the end of the chapter, we will have a fully functioning Jenkins instance that's ready to take on a production workload.

In this chapter, we're going to cover the following main topics:

- Running a Jenkins controller with Docker on HTTPS
- Reverse proxy and TLS/SSL termination options
- Installing plugins and configuring Jenkins
- Attaching SSH and inbound agents
- Creating a secure Docker Cloud

Technical requirements

From *Chapter 1, Jenkins Infrastructure with SSL and Reverse Proxy*, we carry over the same set of VMs running Ubuntu 20.04 with Docker installed. We need a domain for each Jenkins instance (for example, jenkins-aws.lvin.ca, jenkins-firewalled. lvin.ca). For AWS Jenkins, using Route 53 to manage the domain is recommended as it simplifies the TLS certificate acquisition steps from AWS Certificate Manager.

Files in the chapter are available on GitHub at https://github.com/PacktPublishing/ Jenkins-Administrators-Guide/blob/main/ch2.

Running a Jenkins controller with Docker on HTTPS

Let's dive right into deploying a Jenkins controller with Docker on HTTPS.

Custom image to match the UID/GID for a bind mount

We may need to customize the UID/GID of the user inside the container.

Let's take a look at the Dockerfile for the jenkins/jenkins image, lines 20 to 26 (https://github.com/jenkinsci/docker/blob/jenkins-docker-packaging-2.235.1/ Dockerfile):

```
# Jenkins is run with user `jenkins`, uid = 1000
# If you bind mount a volume from the host or a data container,
# ensure you use the same uid
RUN mkdir -p $JENKINS_HOME \
  && chown ${uid}:${gid} $JENKINS_HOME \
  && groupadd -g ${gid} ${group} \
  && useradd -d "$JENKINS_HOME" -u ${uid} -g ${gid} -m -s /bin/bash ${user}
```

Figure 2.1 – UID mismatch warning in a Jenkins Dockerfile

Our Jenkins controller Docker container *does* bind mount a volume from the host, and therefore the directory on the host must be owned by a UID and GID that matches the user inside the container. The default user inside the container is jenkins, which has UID 1000 and GID 1000.

On an AWS EC2 Ubuntu 20.04 instance, the default ubuntu user is already 1000:1000, so no changes are necessary:

```
ubuntu@aws-controller:~$ echo "$(whoami) $(id -u) $(id -g)"
ubuntu 1000 1000
```

On a VM in the corporate network, the user for running Jenkins may not be 1000:1000. This is especially true if you are using a robot account with its UID/GID predefined in the corporate Active Directory. In our example, robot_acct has UID 123 and GID 30, which we cannot change:

```
robot_acct@firewalled-controller:~$ echo "$(whoami) $(id -u) $(id -g)"
robot_acct 123 30
```

In such a case, we need to extend the Docker image to change the UID/GID of the jenkins user to match our robot account, so that the files created from inside the container by the jenkins user are accessible on the host by the robot_acct user, as illustrated in *Figure 2.2*:

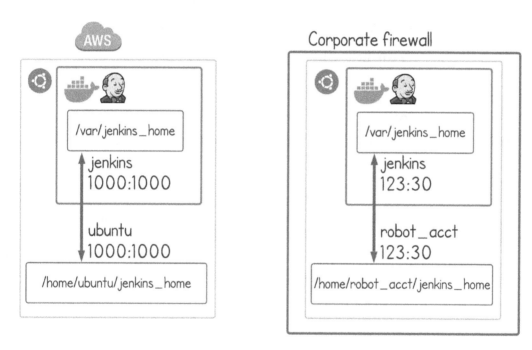

Figure 2.2 – Container user and host user with matching UID/GID

1. Create the following Dockerfile. Let's use the 2.263.1-lts tag for now – we will upgrade to the latest LTS in *Chapter 8, Upgrading the Jenkins Controller, Agents, and Plugins*:

jenkins.dockerfile

```
FROM jenkins/jenkins:2.263.1-lts
USER root
RUN  usermod -u 123 -g 30 jenkins
USER jenkins
```

2. Build the image and name it <Docker Hub ID>/jenkins:2.263.1-lts:

```
robot_acct@firewalled-controller:~$ docker build -t calvinpark/
jenkins:2.263.1-lts -f jenkins.dockerfile .
[...]
Successfully tagged calvinpark/jenkins:2.263.1-lts
```

All users now have a matching UID/GID and we avoid a headache with file permission issues.

Running Jenkins

Let's launch the Jenkins controller. This is what the architecture looks like:

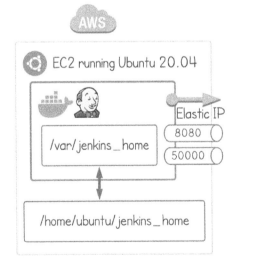
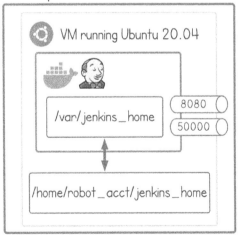

Figure 2.3 – Architecture of VMs and containers for Jenkins controllers

Let's begin:

1. Create `jenkins_home`. Skipping this will cause Docker to create the directory as root, causing Jenkins to fail with permission issues:

```
controller:~$ mkdir jenkins_home
```

2. Then, run the Jenkins container. We can use jenkins/jenkins on an AWS controller or our custom image on a firewalled controller:

```
controller:~$ docker run \
    --detach \
    --restart on-failure \
    -u $(id -u):$(id -g) \
    -v ~/jenkins_home:/var/jenkins_home \
    -p 8080:8080 -p 50000:50000 \
    --name jenkins_controller \
    calvinpark/jenkins:2.263.1-lts
```

There are a few flags in this command. Let's examine what they mean:

- `--detach`: Run in the background.

- `--restart on-failure`: Automatically restart if the container crashes or the machine reboots.

- `-u $(id -u):$(id -g)`: Run as the UID and GID of the host user.

- `-v ~/jenkins_home:/var/jenkins_home`: Bind mount the host ~/jenkins_home directory to /var/jenkins_home inside the container so that the Jenkins data is written on the host directory.

- `-p 8080:8080`: Bind TCP port 8080 on the host to port 8080 inside the container. Jenkins, by default, runs HTTP on 8080. Traffic to HTTP port 80 and HTTPS port 443 will be forwarded to port 8080 through a reverse proxy.

- `-p 50000:50000`: This port is for inbound agents. If your Jenkins will not have any inbound agents, this port doesn't need to be bound.

- `--name jenkins_controller`: Name the running container jenkins_controller for a memorable name and easier access.

3. Verify that the container is running and didn't crash:

```
controller:~$ docker ps
CONTAINER ID    IMAGE                       COMMAND
        CREATED         STATUS          PORTS
                                NAMES
597e509542eb    jenkins/jenkins:2.263.1-lts    "/sbin/tini -- /
usr/…"    5 seconds ago    Up 4 seconds    0.0.0.0:8080->8080/tcp,
0.0.0.0:50000->50000/tcp    jenkins_controller
```

4. In a minute, Jenkins should be up and running. Open a web browser to the IP of
   your VM on port 8080 to see a prompt to unlock Jenkins. If you're on AWS, the IP
   should be the public IP of your VM, such as http://54.70.250.76:8080. If you're
   using a VM inside the corporate firewall, the IP should be the IP of your VM, such
   as http://192.168.1.16:8080:

Figure 2.4 – First boot unlock prompt

5. We can find the initial admin password in the ~/jenkins_home/secrets/
   initialAdminPassword file:

```
controller:~$ cat ~/jenkins_home/secrets/initialAdminPassword
98b13e64bebf4003844baa863d1dc2fd
```

6. Copy and paste the value into the password box in the browser and click Continue.

7. At the Customize Jenkins screen, click Install suggested plugins to continue. We will install more packages soon.

In a few minutes, Jenkins finishes installing the plugins and asks you to create the admin user. *Do not skip this step!* The default password for admin is the long string that you had to look up before, and now is the best time to change it:

Figure 2.5 – First boot Create First Admin User screen

8. Finally, it asks us to configure the Jenkins URL. Click Save and Finish to continue (we will change this soon), and then click Start using Jenkins:

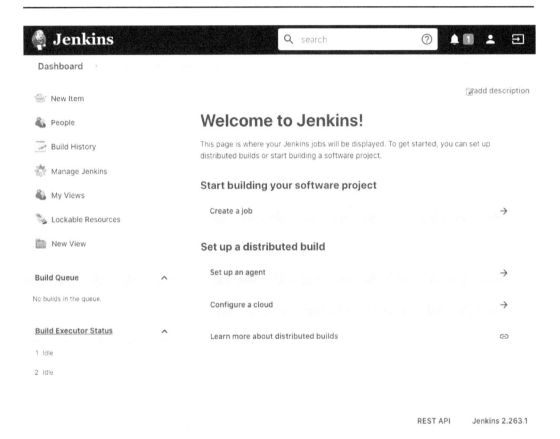

Figure 2.6 – Jenkins is ready

Jenkins is fully up and running. While we can start using Jenkins as is, let's continue to set up a reverse proxy and HTTPS so that the URL is easier to remember and the traffic is encrypted.

> **Locked out?**
>
> If you didn't change the admin password during the first boot, you'll need an admin password to get back into Jenkins. The password is the long string in ~/jenkins_home/secrets/jenkins_controller.

Reverse proxy and TLS/SSL termination options

TLS enables HTTPS to encrypt the message between two parties, but the message needs to be decrypted in the end so that each party can read it. In Jenkins, there are two places where the message can be decrypted, or in industry lingo, where the TLS is terminated.

TLS termination at the reverse proxy

A popular configuration is to set up a reverse proxy in front of Jenkins and terminate the TLS at the reverse proxy:

- DNS resolves the URL to the IP of the reverse proxy.
- If the traffic is on HTTP port 80, the reverse proxy redirects the traffic to HTTPS port 443.
- If the traffic is on HTTPS port 443, the reverse proxy decrypts the message and forwards it to the Jenkins controller on HTTP port 8080.
- The Jenkins controller processes the incoming message from HTTP port 8080.

This is what the configuration looks like:

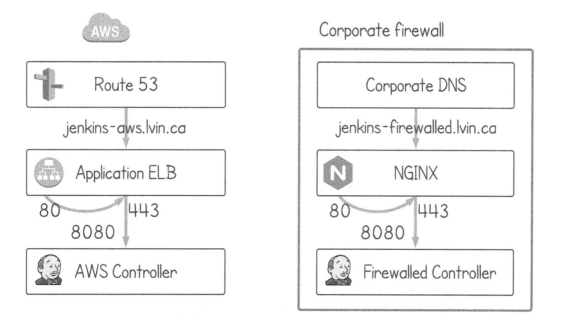

Figure 2.7 – Architecture of Jenkins with reverse proxy

There are several benefits to this configuration:

- The reverse proxy acts as a public face of the application. Therefore, the applications can stay in a private network where an attacker cannot directly access.

- Attacks are easier to mitigate since reverse proxies are built to handle attacks. In AWS, Elastic Load Balancer (ELB) can act as a reverse proxy and provide additional protections through AWS Shield.

- Traffic to HTTP is automatically redirected to HTTPS.

- Connections are pooled and compressed for higher performance.

- The processing cost of terminating the TLS is offloaded to the reverse proxy, however small it is in reality.

There are also drawbacks:

- It adds to the amount of infrastructure to manage.

- Communication between the reverse proxy and Jenkins can fail due to misconfigurations. Reverse proxy issues are, in fact, quite common.

- Communication between the reverse proxy and Jenkins is unencrypted, creating a window of opportunity for an attack, however small it is in reality.

Let's examine the steps for configuring a reverse proxy for each Jenkins controller.

Setting up an Application ELB for the AWS Jenkins controller

Setting up an ELB for Jenkins is largely a three-part exercise:

1. Create a TLS certificate through AWS Certificate Manager.

2. Create a target group that does the following:

   - Has the Jenkins controller EC2 instance

   - Listens to port 8080

   - Runs health checks on the /login path

3. Create a load balancer that does the following:

   - Listens to port 80, which is redirected to port 443

   - Listens to port 443, which is forwarded to the target group from *step 2*

   - Uses the TLS certificate from *step 1*

Here are the actual steps that we can follow:

1. From the AWS console, go to the EC2 service and click Load Balancers.

2. Create a new Application Load Balancer named jenkins. Add two listeners, HTTP and HTTPS. Choose the availability zones for the ELB and click Next.

3. Create a certificate in AWS Certificate Manager for our domain by selecting the first option, Choose a certificate from ACM, and then clicking the Request a new certificate from ACM link:

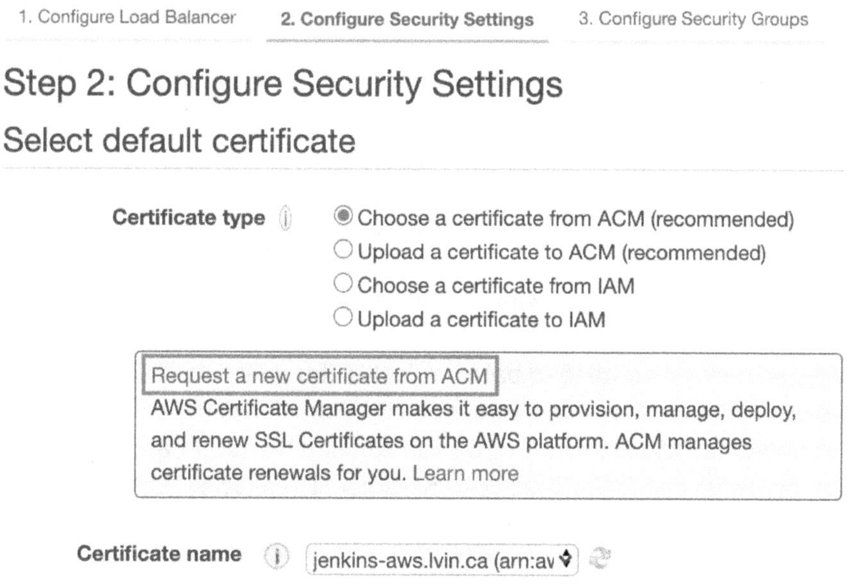

1. Configure Load Balancer	**2. Configure Security Settings**	3. Configure Security Groups

Step 2: Configure Security Settings

Select default certificate

Certificate type ⓘ   ● Choose a certificate from ACM (recommended)
○ Upload a certificate to ACM (recommended)
○ Choose a certificate from IAM
○ Upload a certificate to IAM

> Request a new certificate from ACM
> AWS Certificate Manager makes it easy to provision, manage, deploy, and renew SSL Certificates on the AWS platform. ACM manages certificate renewals for you. Learn more

Certificate name ⓘ   jenkins-aws.lvin.ca (arn:av ❖) ⬆

Figure 2.8 – Choose the first option and click the link below to create a new certificate from ACM

4. Once the certificate is created, choose the certificate from the drop-down menu for Certificate name and then click Next.

5. Choose the two security groups, default and jenkins-elb, which we made in *Chapter 1, Jenkins Infrastructure with TLS/SSL and Reverse Proxy*. Click Next.

6. Create a new target group named jenkins-controller-8080 and change the Port from 80 to 8080 so that it points to the Jenkins controller running on port 8080. Set the Health checks Path to /login. The default / fails the health check because unauthenticated access returns 403 authentication error, whereas /login always succeeds as long as Jenkins is running. Click Next.

7. Add the Jenkins controller instance to registered targets, and then click Next, Create, and Close.

8. Choose the new jenkins load balancer, click the Listeners tab, and then edit HTTP : 80. Delete the current rule of forwarding to jenkins-controller-8080 and replace it with a Redirect to HTTPS at port 443, so that the traffic to HTTP is redirected to HTTPS.

This is what we should see on the Application ELB Listeners configuration:

Listener ID	Security policy	SSL Certificate	Rules
HTTP : 80 arn...572bb7d1d24fec43 ▾	N/A	N/A	Default: redirecting to HTTPS://#{host}:443/#{path}?#{query} View/edit rules
HTTPS : 443 arn...53d0a43e0544df34 ▾	ELBSecurityPolicy -2016-08	Default: 6bf999e5-... -4fc7937ed8b7 (ACM) View/edit certificates	Default: forwarding to jenkins-controller-8080 View/edit rules

Figure 2.9 – Application ELB Listener configuration

9. We can now go to the load balancer's URL (for example, jenkins-1394494507. us-west-2.elb.amazonaws.com) on HTTP or HTTPS to see that Jenkins loads with a TLS certificate error. You can find the load balancer's URL at EC2 Dashboard | Load Balancing | Load Balancers | jenkins | Description | DNS name:

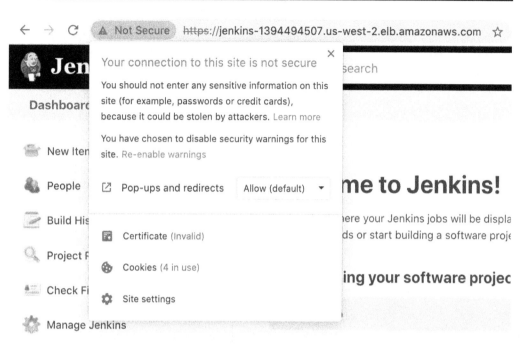

Figure 2.10 – AWS Jenkins on the load balancer URL showing a TLS certificate error

10. Create an A record alias on our domain for Jenkins (for example, `jenkins-aws.`
    `1vin.ca`) and point it to the ELB. Once the DNS cache refreshes, we can go to
    the Jenkins URL on HTTP or HTTPS and see it load without an error:

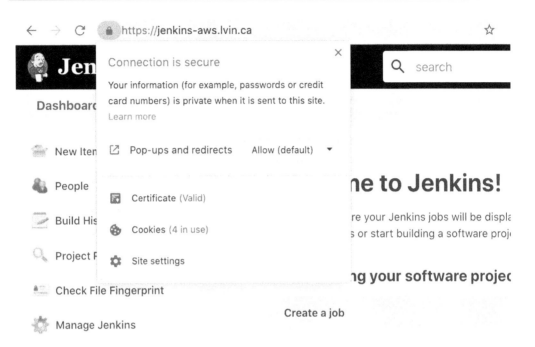

Figure 2.11 – AWS Jenkins on the Jenkins URL showing a valid TLS certificate

AWS Jenkins with ELB is up and running using HTTPS.

Setting up an NGINX reverse proxy for the firewalled Jenkins controller

Let's create an NGINX Docker container on the firewalled Jenkins controller host that acts as the reverse proxy for the Jenkins controller Docker container:

1. Copy the TLS certificate and the key that we made using Let's Encrypt (covered in *Chapter 1, Jenkins Infrastructure with TLS/SSL and Reverse Proxy*), and move them to ~/nginx/certs/ directory.

```
robot_acct@firewalled-controller:~$ mkdir -p nginx/certs/
```

```
robot_acct@firewalled-controller:~$ cp ~/letsencrypt/certs/live/
jenkins-firewalled.lvin.ca/{fullchain.pem,privkey.pem} ~/nginx/
certs/
```

2. Create the ~/nginx/nginx.conf NGINX configuration file as seen in the following snippet. Replace the domains in the two server/server_name sections with your own domains. This file can be downloaded from the book's GitHub repository (https://github.com/PacktPublishing/Jenkins-Administrators-Guide/blob/main/ch2/nginx.conf):

~/nginx/nginx.conf

```
server {
  listen        80;
  server_name   jenkins-firewalled.lvin.ca;
  return        301  https://$host$request_uri;
}

server {
  listen        443 ssl;
  server_name   jenkins-firewalled.lvin.ca;

  ssl_certificate       /certs/fullchain.pem;
  ssl_certificate_key   /certs/privkey.pem;

  location / {
    proxy_pass            http://localhost:8080;
    proxy_http_version    1.1;

    proxy_set_header  Host  $host:$server_port;
    proxy_set_header  X-Real-IP  $remote_addr;
    proxy_set_header  X-Forwarded-For  $proxy_add_x_forwarded_for;
    proxy_set_header  X-Forwarded-Proto  $scheme;

    # Max upload size. Useful for a custom plugin.
    client_max_body_size    10m;
    client_body_buffer_size  128k;

    proxy_buffering          off;
    proxy_request_buffering  off;
    proxy_set_header         Connection   "";
  }
}
```

3. With the three files in place, start the NGINX container:

```
robot_acct@firewalled-controller:~$ docker run \
    --detach \
    --restart on-failure \
    -v ~/nginx/certs:/certs:ro \
    -v ~/nginx/nginx.conf:/etc/nginx/conf.d/default.conf:ro \
    --network=host \
    --name nginx \
    nginx
```

There are a few flags in this command. Let's examine what they mean:

- --detach: Run in the background.

- `--restart on-failure`: Automatically restart if the container crashes or the machine reboots.

- `-v ~/nginx/certs:/certs:ro`: Bind mount the host `~/nginx/certs` directory to `/certs` inside the container so that the TLS certificates are available to the NGINX process. Mount as read-only because NGINX doesn't need to modify them.

- `-v ~/nginx/nginx.conf:/etc/nginx/conf.d/default.conf:ro`: Bind mount the host `~/nginx/nginx.conf` file to `/etc/nginx/conf.d/default.conf` inside the container so that the configuration file is available to the NGINX process. Mount as read-only because NGINX doesn't need to modify it.

- `--network=host`: Allow the container to have direct access to the host's network so that it can manage the traffic for the Jenkins controller container.

- `--name nginx`: Name the running container `nginx` for a memorable name and easier access.

4. We can now go to the NGINX container host's IP (for example, 192.168.1.16) on HTTP or HTTPS to see that Jenkins loads with a TLS certificate error:

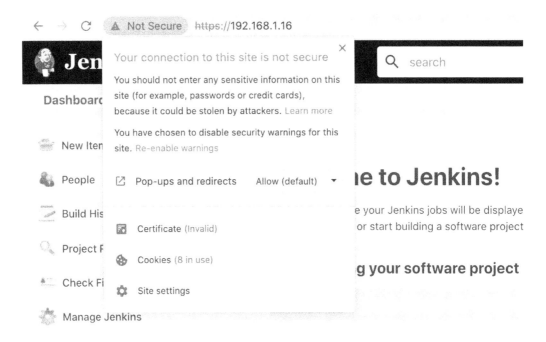

Figure 2.12 – Firewalled Jenkins on a reverse proxy IP showing a TLS certificate error

5. Create an A record on our domain for Jenkins (for example, `jenkins-firewalled.lvin.ca`) and point it to the reverse proxy. Once the DNS cache refreshes, we can go to the Jenkins URL on either HTTP or HTTPS and see it load without an error:

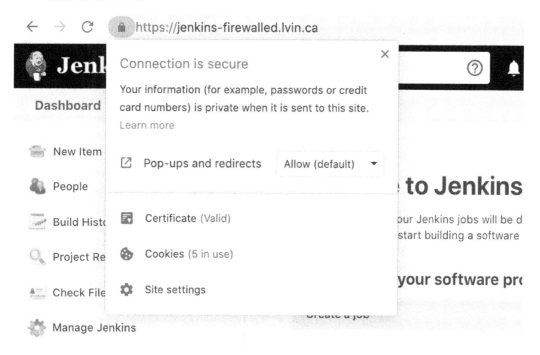

Figure 2.13 – Firewalled Jenkins on the Jenkins URL showing a valid TLS certificate

Firewalled Jenkins with an NGINX reverse proxy is up and running using HTTPS.

Terminating the TLS certificate directly on the Jenkins controller

It is also possible to terminate the TLS directly on the Jenkins controller instead of using a reverse proxy. This reduces the amount of infrastructure to manage; however, it is suitable only in a trusted environment for a small-scale Jenkins instance, due to the lack of protection and scaling that a reverse proxy provides. It also lacks the HTTP to HTTPS redirect, as illustrated in the following figure:

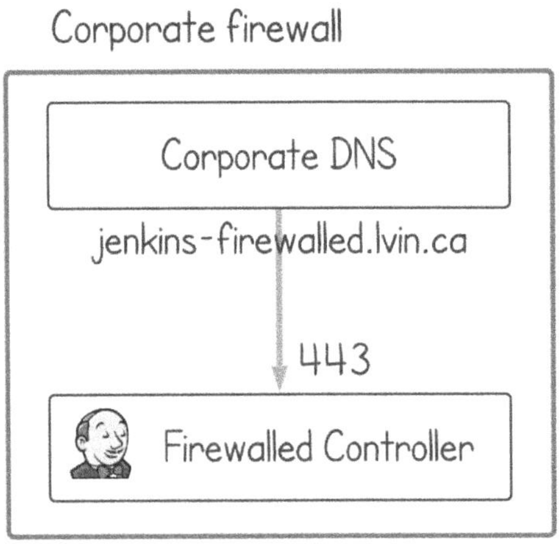

Corporate firewall

Figure 2.14 – Architecture of Jenkins terminating the TLS certificate directly on the controller

Let's begin:

1. Copy the TLS certificate and the key that we made in *Chapter 1, Jenkins Infrastructure with TLS/SSL and Reverse Proxy*, using Let's Encrypt in the ~/ certs/ directory:

```
robot_acct@firewalled-controller:~$ mkdir ~/certs/
```

```
robot_acct@firewalled-controller:~$ cp ~/letsencrypt/certs/live/
jenkins-firewalled.lvin.ca/{fullchain.pem,privkey.pem} ~/certs/
```

2. The private key must be converted to an old-styled PKCS #1 format due to an unimplemented feature[1]. Run the following command to convert the private key:

```
robot_acct@firewalled-controller:~$ openssl rsa \
    -in ~/certs/privkey.pem \
    -out ~/certs/privkey.pkcs1.pem
```

3. Then, restart Jenkins with these updated options:

---

1   https://issues.jenkins.io/browse/JENKINS-22448

```
robot_acct@firewalled-controller:~$ docker run \
    --detach \
    --restart on-failure \
    -u $(id -u):$(id -g) \
    -v ~/jenkins_home:/var/jenkins_home \
    -v ~/certs:/certs:ro \
    -e JENKINS_OPTS="
        --httpsPort=443
        --httpsCertificate=/certs/fullchain.pem
        --httpsPrivateKey=/certs/privkey.pkcs1.pem" \
    -p 443:443 -p 50000:50000 \
    --name jenkins_controller \
    calvinpark/jenkins:2.263.1-lts
```

There are three changes:

- -v ~/certs:/certs:ro
  Bind mount the ~/certs host directory to /certs inside the container so that the certificates are available to Jenkins. Mount as read-only because Jenkins doesn't need to modify it.

- -e JENKINS_OPTS="
      --httpsPort=443
      --httpsCertificate=/certs/fullchain.pem
      --httpsPrivateKey=/certs/privkey.pkcs1.pem"
  Configure Jenkins to run on HTTPS using the certificates for TLS termination. Take note that the lines don't end with backslashes – they are a multi-line string.

- -p 443:443
  Bind HTTPS port 443 on the host to HTTPS port 443 inside the container so that Jenkins is accessible on HTTPS.

4. We can now go to the controller host's IP (for example, 192.168.1.16) on HTTPS to see that Jenkins loads with a TLS certificate error. See *Figure 2.10*.

5. Create an A record alias on our domain for Jenkins (for example, jenkins-firewalled.lvin.ca) and point it to the controller host's IP. Once the DNS cache refreshes, we can go to the Jenkins URL on HTTPS and see it load without an error. See *Figure 2.11*.

6. Loading Jenkins on HTTP will not work since we don't have the reverse proxy that redirects HTTP to HTTPS. See Figure 2.15:

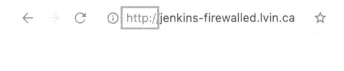

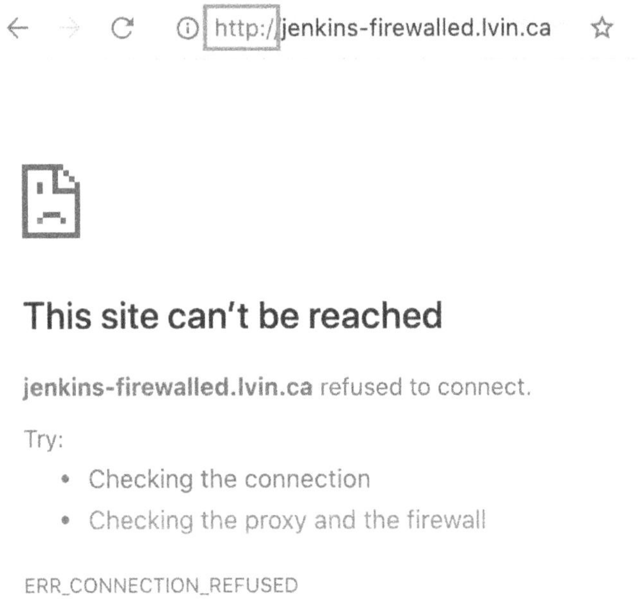

http://jenkins-firewalled.lvin.ca
This site can't be reached
jenkins-firewalled.lvin.ca refused to connect.
Try:
Checking the connection
Checking the proxy and the firewall
ERR_CONNECTION_REFUSED

Figure 2.15 – Accessing Jenkins using HTTP fails without a reverse proxy

Firewalled Jenkins without a reverse proxy is up and running using HTTPS.

Installing plugins and configuring Jenkins

Jenkins is now running – it's time to customize it to make it our own.

Installing more plugins

We will now learn how to install plugins. We can use the same technique to install additional plugins in other chapters. The following info box shows which plugins to install:

> **Required plugins**
> Active Directory

Click Manage Jenkins on the left | Manage Plugins | Available and then search for the plugin name. You can search for and check multiple plugins at once. There are many plugins with similar names – searching for docker shows over 10 different similarly named plugins. Be sure to pick the exact name from the info box:

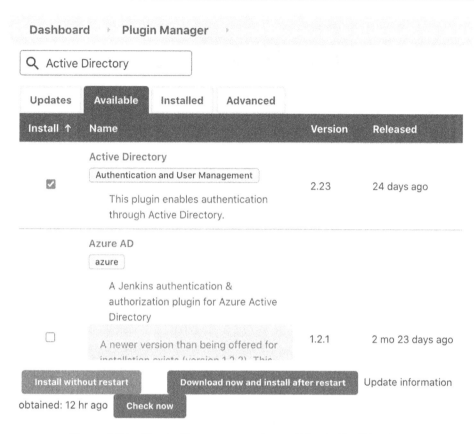

Figure 2.16 – Plugin Manager search result for Active Directory

Most plugins can be installed without a restart, but an upgrade always requires a restart. Click Install without restart to install the Active Directory plugin. If you are planning on using OAuth, search for the OAuth provider plugin and install that instead. If you are planning on not using any authentication provider, you can skip the install.

Configure System

With the basic plugins installed, let's continue to configure the system.

Click Manage Jenkins on the left and then click Configure System. This is the main configuration page for Jenkins, which unfortunately doesn't have a name. We will refer to this page as System Configuration throughout the book:

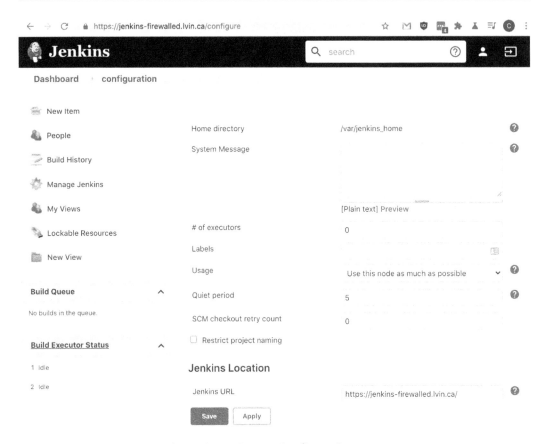

Figure 2.17 – System Configuration page

There are several configuration items here. In fact, this is where almost all the configurations are. When we install a new plugin, its global configurations will most likely be here. Some plugins require additional configurations on the pipelines, and those will be made on each pipeline's configuration page.

Read through and configure as needed, and be sure to configure these four items:

- # of executors: 0.
- Jenkins URL: https://<Jenkins URL>/ (for example, https://jenkins-firewalled. lvin.ca/).
- Pipeline Default Speed/Durability Level: Performance-optimized.
- Global Build Discarders: **Add a** Specific Build Discarder **that keeps a maximum of 100 builds.**

# of executors configures the number of executors on the controller to run the build steps. This is useful when there are no other agents. Since we are attaching two dedicated agents and a Docker Cloud, set it to 0 so that the controller does not run build steps.

Jenkins URL is updated from the IP address with temporary HTTP on port 8080 to the Jenkins URL with HTTPS.

Reducing Pipeline Default Speed/Durability Level significantly speeds up Jenkins at the cost of a possible data loss in case of an unexpected shutdown. The Jenkins documentation suggests Performance-optimized as the global default[2]. This can be overridden to a higher durability setting per pipeline for the pipelines that need a guaranteed record of executions. You can find more information about this in *Chapter 9, Reducing Bottlenecks*.

Everything in Jenkins is a flat file on the controller. It's important that Global Build Discarders is configured to prune the build history so that the number of files on the controller doesn't grow indefinitely. This can be overridden to a higher number per pipeline, and important builds can further be pinned so that they don't get pruned:

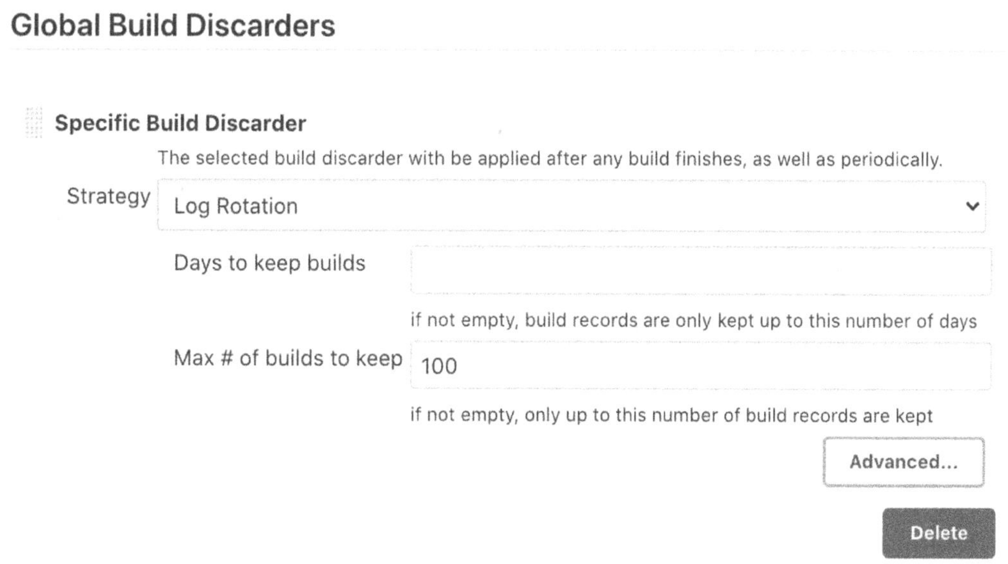

Global Build Discarders

Specific Build Discarder

The selected build discarder with be applied after any build finishes, as well as periodically.

| Strategy | Log Rotation ⌄ |

Days to keep builds

if not empty, build records are only kept up to this number of days

Max # of builds to keep   100

if not empty, only up to this number of build records are kept

Advanced...

Delete

Figure 2.18 – Global Build Discarders configured to keep a maximum of 100 builds by default

The basic Jenkins configuration is done. We can always come back to make additional changes. Let's now move on to the security configurations.

---

2   https://www.jenkins.io/doc/book/pipeline/scaling-pipeline/#suggested-best-practices-and-tips-for-durability-settings

Configure Global Security

Go to the Global Security page to configure the security-related items by clicking Manage Jenkins on the left and then click Configure Global Security:

- Choose the Authentication Security Realm that you plan to use. In a corporate setting, Active Directory or LDAP works well. For others, Jenkins' own user database is often sufficient. There are also many OAuth provider plugins such as GitHub and Google that could be useful.

- In Authorization Strategy, choose Project-based Matrix Authorization Strategy:

  - Give Job Discover permission to Anonymous Users. This redirects unauthenticated users to a login page rather than showing a 404 error page.

  - Give the following permissions to Authenticated Users:

    - Overall Read

    - Credentials View (this allows the authenticated users to see that a credential exists. It doesn't allow the users to see the actual secret. It helps non-administrator users in building their Jenkinsfiles by letting them see the name of the secrets they'd use.)

    - View Read

  - Also, add the admin user and give Administrator permission, as shown in the following screenshot:

  > **Lockout alert!**
  >
  > You will be locked out if you don't add the admin user and give Administrator permission.

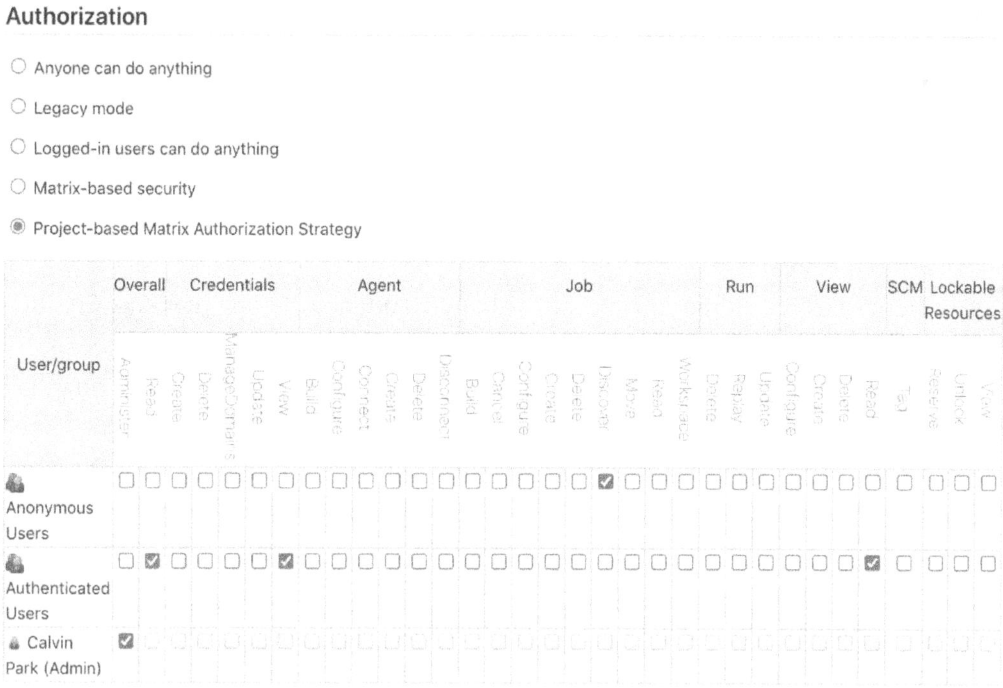

Authorization

○ Anyone can do anything

○ Legacy mode

○ Logged-in users can do anything

○ Matrix-based security

◉ Project-based Matrix Authorization Strategy

User/group	Overall	Credentials	Agent	Job	Run	View	SCM	Lockable Resources
Anonymous Users				Discover ✓				
Authenticated Users	Read ✓		View ✓	Read ✓		Read ✓		
Calvin Park (Admin)	Administer ✓							

Figure 2.19 – Project-based Matrix Authorization Strategy configuration

We are deliberately not giving the users permission to see the projects or run a build, as those will be configured for each project.

- **Change** Markup Formatter **to** Safe HTML **so that we can customize fonts or add links on the status messages.**

- **In** CSRF Protection, **check** Enable proxy compatibility **to prevent proxy compatibility issues.**

Click Save **to save and exit.**

With basic authentication and authorization configured, let's continue to manage secrets.

Configure Global Credentials

Secrets are an important part of automation. Click Manage Jenkins **on the left |** Manage Credentials | (global). **This is the** Global Credentials **page where we store secrets. Keep this page open on a tab because we will come back to** Global Credentials **very soon to create the secrets for authenticating with agents.**

Let's continue to wrap up the controller configuration.

Installing even more plugins

In addition to the required plugins, here are some optional plugins that are useful:

- AnsiColor: Colorize the build log. This is so fundamental I wonder why it's not installed by default.

- Blue Ocean: Enable a new UI for Jenkins. Great for visualizing pipelines with multiple stages.

- build-metrics: Helps you gather the build frequency of each pipeline. Great for gathering the metrics for a presentation.

- GitLab: Integrate with GitLab for webhooks and link-backs. There are plugins for other popular VCS as well.

- Jira: Link a Jira issue ID from Jira to Jenkins and back, updating the issue with the builds associated with the issue. There are plugins for other popular bug trackers as well.

- Line Numbers: Puts line numbers and links to the build log. This is useful for collaboration by allowing us to link to a specific line of the logs.

- Read-only configurations: Allows a non-administrator user to see the configurations of a pipeline. This helps users to debug a failing pipeline by inspecting the configurations without needing permission to modify it.

- Rebuilder: Re-run a build with the same parameters. Useful for pipelines with many parameters.

- Slack Notification: Great for keeping an eye on failed nightly builds. It can get chatty, so create a dedicated Slack channel for it. There are plugins for other popular messengers as well.

The complete list of available plugins can be found on the Plugins Index page at https://plugins.jenkins.io/.

Our controller configuration is now complete. Let's move on to agent configuration.

Attaching SSH and inbound agents

We have two Jenkins agents, one on AWS and another inside the corporate firewall. Each agent will connect to both Jenkins controllers, one on AWS and another inside the firewall.

The AWS agent is accessible by both the AWS controller and the firewalled controller. We will connect the AWS agent as an SSH agent, which is simple and effective:

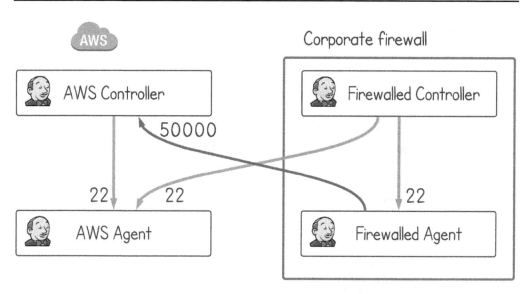

Figure 2.20 – Architecture of SSH and inbound agents

The firewalled agent, on the other hand, is not accessible by the AWS controller. The only way for them to communicate is by the agent initiating the connection, so we will connect the agent as an inbound agent. The firewalled controller can access the firewalled agent, therefore it will be connected as an SSH agent.

SSH agent

An SSH agent is the most widely used agent type for an agent running a Unix-based operating system. In our environment, there are three SSH agent connections, and the setup process is the same for all three:

- AWS controller → AWS agent
- Firewalled controller → AWS agent
- Firewalled controller → Firewalled agent

Let's begin:

1. An SSH agent host is a VM running Ubuntu 20.04, just like the controllers. We have already installed Docker on it in *Chapter 1, Jenkins Infrastructure with TLS/SSL and Reverse Proxy*. SSH into the agent and install JDK 11:

```
agent:~$ sudo apt update
agent:~$ sudo apt install -y openjdk-11-jdk
```

2. Still on the agent, use ssh-keygen to create an SSH public/private key pair, and have the current user (ubuntu on AWS agent and robot_acct on the firewalled agent) accept the key. Don't forget to set the correct file permission on the ~/.ssh/authorized_keys file:

```
agent:~$ ssh-keygen
agent:~$ cat ~/.ssh/id_rsa.pub >> ~/.ssh/authorized_keys
agent:~$ chmod 600 ~/.ssh/authorized_keys
agent:~$ ls -la ~/.ssh
total 20
drwx------ 2 ubuntu 4096 Dec 15 04:34 .
drwxr-xr-x 4 ubuntu 4096 Dec 15 04:31 ..
-rw------- 1 ubuntu  577 Dec 15 04:36 authorized_keys
-rw------- 1 ubuntu 2610 Dec 15 04:34 id_rsa
-rw-r--r-- 1 ubuntu  577 Dec 15 04:34 id_rsa.pub
```

3. We need to store the private key in the Jenkins credential store so that the controller can use it to connect to the agent. Output the content of the private key and keep it handy:

```
agent:~$ cat ~/.ssh/id_rsa

-----BEGIN OPENSSH PRIVATE KEY-----
b3BlbnNzaC1rZXktdjEAAAAABG5vbmUAAAAEbm9uZQAAAAAAAAABAAAB1
[...]
WM1+k8b+6GZJMAAAAXdWJ1bnR1QGlwLTE3Mi0zMS0xNS0xOTEBAgME
-----END OPENSSH PRIVATE KEY-----
```

4. **Go to** Global Credentials **and then click** Add Credentials:

- Kind: SSH Username with private key
- Scope: System (Jenkins and nodes only)
- ID: `aws-agent-ubuntu-priv` or `firewalled-agent-robot_acct-priv`
- Description: **Same as** ID
- Username: `ubuntu or robot_acct`
- Private Key: **Check the radio button for** Enter directly **and click** Add. **The gray box turns into a textbox – copy and paste the private key into this box.**
- Passphrase: **Enter if you've created the SSH keys with a passphrase:**

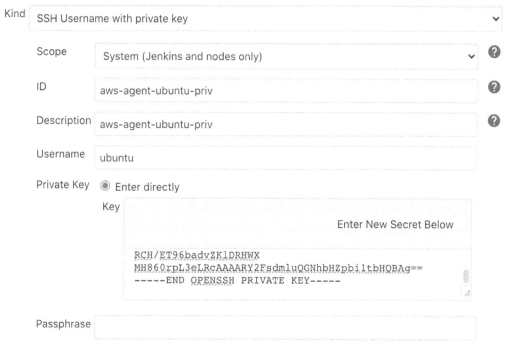

Figure 2.21 – Adding an SSH private key to the Jenkins credential store

5. Click OK to save. The SSH user and the private key are now stored as a secret in the Jenkins credential store:

Global credentials (unrestricted)

Credentials that should be available irrespective of domain specification to requirements matching.

	ID	Name	Kind	Description	
	aws-agent-ubuntu-priv	ubuntu (aws-agent-ubuntu-priv)	SSH Username with private key	aws-agent-ubuntu-priv	

Figure 2.22 – SSH private key for the Ubuntu user in the AWS agent saved as a credential

6. We also need to get the agent's SSH host key because SSHing to a host for the first time requires that you accept the host key. Print the host key from the agent and keep it handy as we will use it soon:

`agent:~$ cat /etc/ssh/ssh_host_rsa_key.pub`

`ssh-rsa AAAAB3N[...]iQwaAqsEDN1e+c= root@ip-172-31-15-191`

7. Now, let's add the agent to the controller using the SSH keys. Go back to the Jenkins home page by clicking the Jenkins icon in the upper-left corner, and then click Manage Jenkins | Manage Nodes and Clouds | New Node.

8. Enter Node name, choose the Permanent Agent radio button, and click OK. Populate the following fields:

   - Name: `aws-aws-agent`, `firewalled-aws-agent`, or `firewalled-firewalled-agent`

   - # of executors: `10` (this allows one agent to handle 10 concurrent builds. Since each build is isolated in a Docker container, there isn't a concern of concurrently running builds stepping on each other)

   - Remote root directory:

     - AWS controller → AWS agent: `/home/ubuntu/aws-aws-agent`

     - Firewalled controller → AWS agent: `/home/ubuntu/firewalled-aws-agent`

     - Firewalled controller → Firewalled agent: `/home/robot_acct/firewalled-firewalled-agent`

   - Labels: `docker`

   - Launch method: Launch agents via SSH

   - Host: IP of the agent

   - Credentials: `<user>` (`<env>-agent-<user>-priv`)

   - Host Key Verification Strategy: Manually provided key Verification Strategy

   - SSH Key: Copy and paste the host key into the box

This is what it should look like:

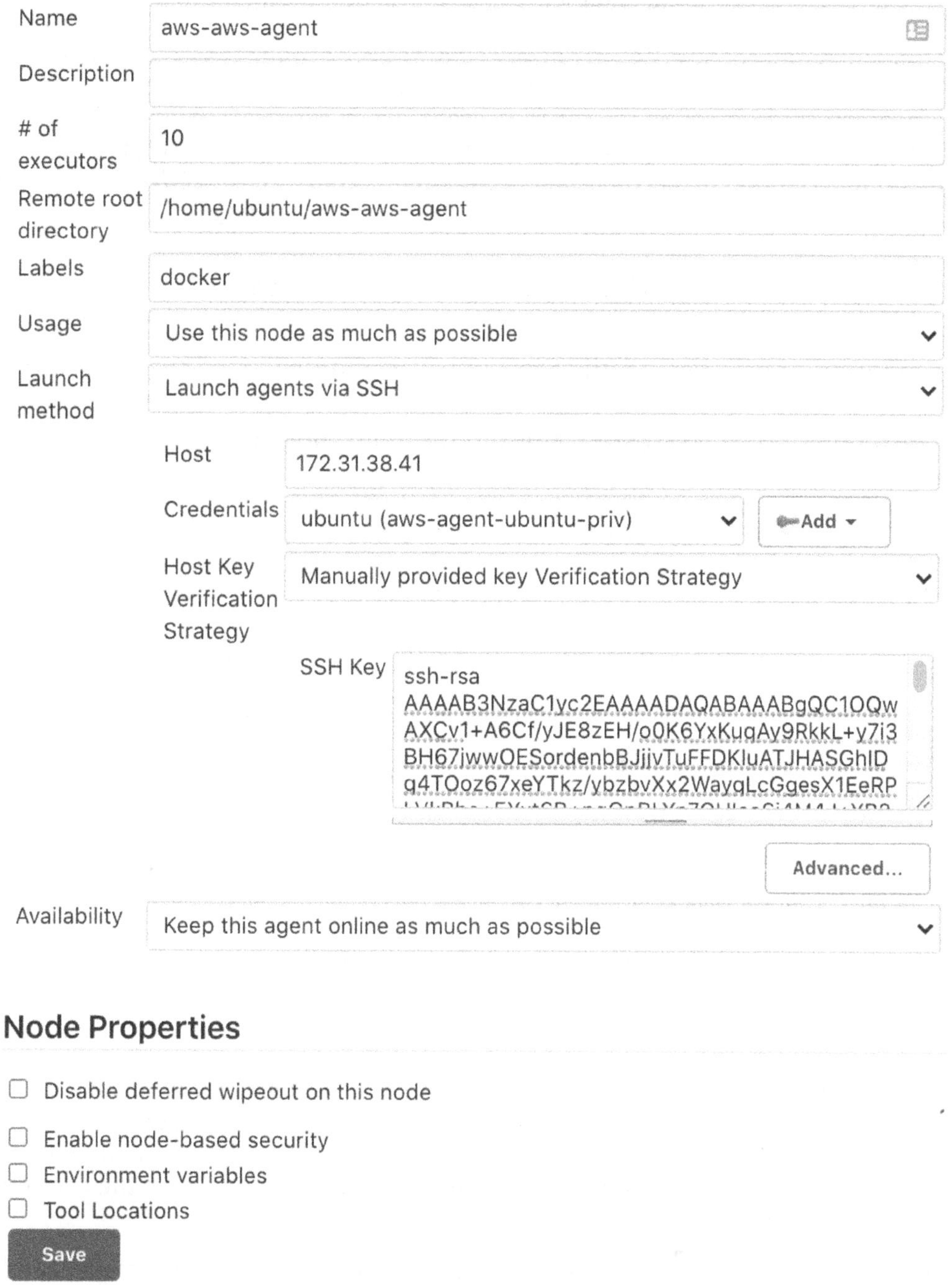

Name	aws-aws-agent
Description	
# of executors	10
Remote root directory	/home/ubuntu/aws-aws-agent
Labels	docker
Usage	Use this node as much as possible ⌄
Launch method	Launch agents via SSH ⌄

	Host	172.31.38.41
	Credentials	ubuntu (aws-agent-ubuntu-priv) ⌄ 🔑 Add ⌄
	Host Key Verification Strategy	Manually provided key Verification Strategy ⌄
	SSH Key	ssh-rsa AAAAB3NzaC1yc2EAAAADAQABAAABgQC1OQw AXCv1+A6Cf/yJE8zEH/oOK6YxKugAy9RkkL+y7i3 BH67jwwOESordenbBJjivTuFFDKluATJHASGhID g4TOoz67xeYTkz/ybzbvXx2WaygLcGgesX1EeRP

Advanced...

Availability	Keep this agent online as much as possible ⌄

Node Properties

☐ Disable deferred wipeout on this node

☐ Enable node-based security

☐ Environment variables

☐ Tool Locations

Save

Figure 2.23 – SSH agent creation page

9. Save to create the agent. In a few seconds, the SSH agent will be connected. Click the agent name and then Log to see an output similar to this:

```
SSHLauncher{host='172.31.38.41', port=22, credentialsId='aws-
agent-ubuntu-priv', jvmOptions='', javaPath='',
prefixStartSlaveCmd='', suffixStartSlaveCmd='',
launchTimeoutSeconds=60, maxNumRetries=10, retryWaitTime=15,
sshHostKeyVerificationStrategy=hudson.plugins.sshslaves.
verifiers.ManuallyProvidedKeyVerificationStrategy,
tcpNoDelay=true, trackCredentials=true}
```

```
[01/18/21 04:12:03] [SSH] Opening SSH connection to
172.31.38.41:22.
```

```
[01/18/21 04:12:03] [SSH] SSH host key matched the key required
for this connection. Connection will be allowed.
```

```
[01/18/21 04:12:03] [SSH] Authentication successful.
```

```
[...]
```

```
[01/18/21 04:12:04] [SSH] Starting agent process: cd "/home/
ubuntu/aws-aws-agent" && java -jar remoting.jar -workDir /home/
ubuntu/aws-aws-agent -jar-cache /home/ubuntu/aws-aws-agent/
remoting/jarCache
```

```
[...]
```

```
<===[JENKINS REMOTING CAPACITY]===>channel started
```

```
Remoting version: 4.5
```

```
This is a Unix agent
```

```
Evacuated stdout
```

Agent successfully connected and online

Congratulations! We have an agent and we can run builds now!

Inbound agent

An inbound agent allows us to add an agent even when the controller can't reach the agent due to a network restriction such as a firewall, but the agent can reach the controller. As the name suggests, an inbound agent initiates the connection from the agent to the controller, which is the opposite of the SSH agent, where the controller initiates the connection.

An inbound agent host is a VM running Ubuntu 20.04, just like the controllers and the SSH agents. Let's configure it:

1. We have already installed Docker on it in *Chapter 1, Jenkins Infrastructure with TLS/SSL and Reverse Proxy.* SSH into the agent and install JDK 11:

```
robot_acct@firewalled-agent:~$ sudo apt update
robot_acct@firewalled-agent:~$ sudo apt install -y openjdk-11-jdk
```

2. Next, create the work directory and download agent.jar from our Jenkins. The address for agent.jar is https://<Jenkins URL>/jnlpJars/agent.jar:

```
robot_acct@firewalled-agent:~$ mkdir inbound-agent
robot_acct@firewalled-agent:~$ cd inbound-agent/
robot_acct@firewalled-agent:~/inbound-agent$ wget https://
jenkins-aws.lvin.ca/jnlpJars/agent.jar
```

3. Go to Jenkins to create a placeholder for the inbound agent. Click Manage Jenkins | Manage Node and Clouds | New Node. Enter inbound-agent in the Node name field, choose the Permanent Agent radio button, click OK, and then populate the following fields:

   - Name: inbound-agent

   - Remote root directory: /home/robot_acct/inbound-agent

   - # of executors: 10 (this allows one agent to handle 10 concurrent builds. Since each build is isolated into a Docker container, there isn't the worry of concurrently running builds stepping on each other.)

   - Labels: docker

   - Launch method: Launch agent by connecting it to the master

   - Advanced | Tunnel connection through: <public IP of the controller>:50000 (In my case, it's 52.88.1.104:50000. Recall that in the *AWS: FAQs, routing rules, EC2 instances, and EIPs* section of *Chapter 1*, we opened the port 50000 on the EC2 instances directly, instead of on the ELB. *Figure 1.6* and *Figure 1.7* illustrate these configurations. In order for the inbound agent to connect to Jenkins on port 50000, we need to route the traffic through the controller.)

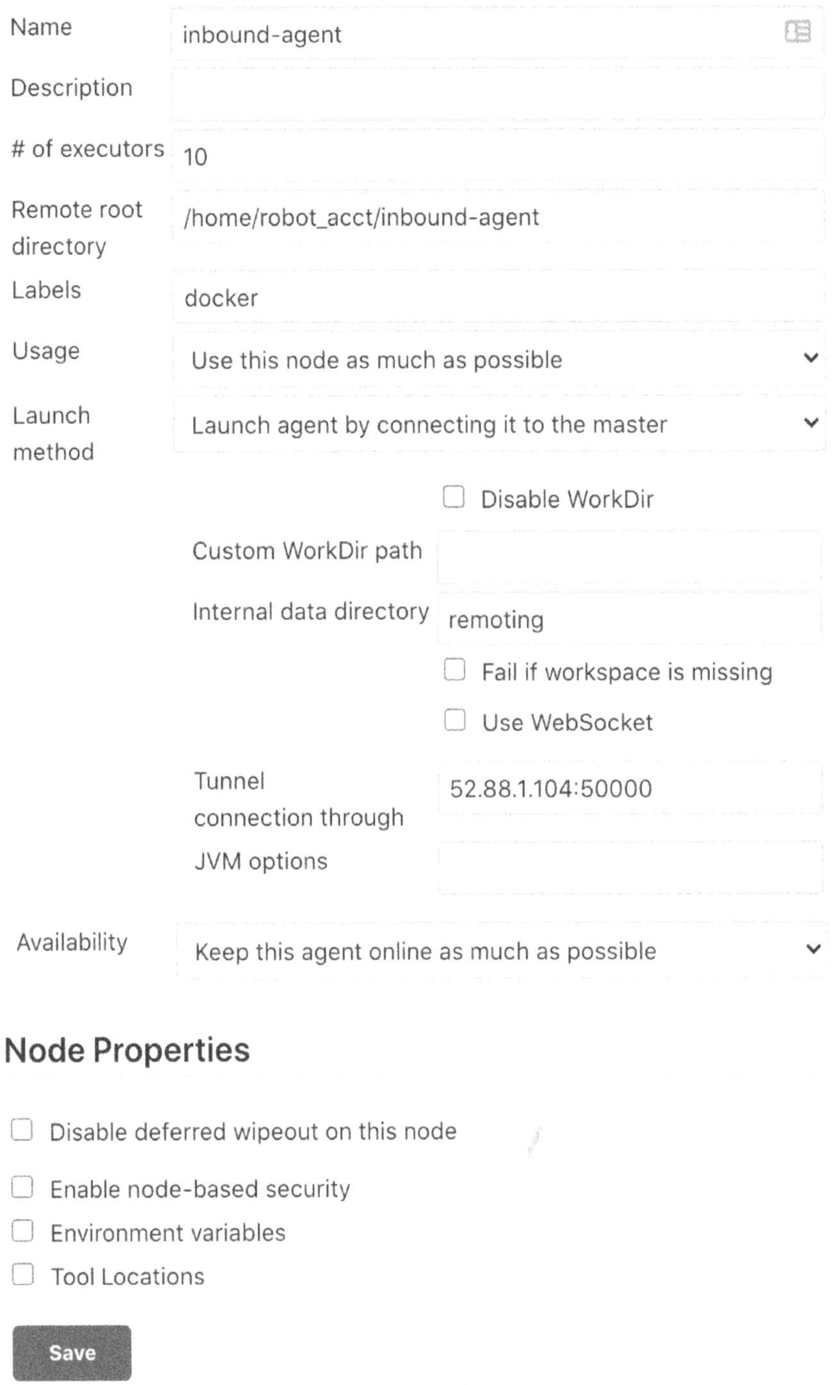

Name	inbound-agent
Description	
# of executors	10
Remote root directory	/home/robot_acct/inbound-agent
Labels	docker
Usage	Use this node as much as possible ⌄
Launch method	Launch agent by connecting it to the master ⌄

☐ Disable WorkDir

Custom WorkDir path

Internal data directory   remoting

☐ Fail if workspace is missing

☐ Use WebSocket

Tunnel connection through   52.88.1.104:50000

JVM options

Availability	Keep this agent online as much as possible ⌄

Node Properties

☐ Disable deferred wipeout on this node

☐ Enable node-based security

☐ Environment variables

☐ Tool Locations

Save

Figure 2.24 – Inbound agent creation page

4. Click Save and then click inbound-agent with the red x on the icon. The page shows us three different ways to connect:

 Agent inbound-agent   Mark this node temporarily offline

Connect agent to Jenkins one of these ways:

- Launch Launch agent from browser

- Run from agent command line:

```
java -jar agent.jar -jnlpUrl https://jenkins-aws.lvin.ca/computer/inbound-
agent/jenkins-agent.jnlp -secret
2677db7766582989f1f2333bceb056d474a56e88c793f03a795b4192cf782db6 -workDir
"/home/robot_acct/inbound-agent"
```

Run from agent command line, with the secret stored in a file:

```
echo 2677db7766582989f1f2333bceb056d474a56e88c793f03a795b4192cf782db6 >
secret-file
java -jar agent.jar -jnlpUrl https://jenkins-aws.lvin.ca/computer/inbound-
agent/jenkins-agent.jnlp -secret @secret-file -workDir
"/home/robot_acct/inbound-agent"
```

Figure 2.25 – Inbound agent connection options

We will use the third method. Simply copy and paste the two commands into the CLI on the inbound agent. Here is the abridged output:

```
robot_acct@firewalled-agent:~/inbound-agent$ echo
2677db7766582989f1f2333bceb056d474a56e88c793f03a795b4192cf782db6
> secret-file
```

```
robot_acct@firewalled-agent:~/inbound-agent$ java -jar agent.
jar -jnlpUrl https://jenkins-aws.lvin.ca/computer/inbound-agent/
slave-agent.jnlp -secret @secret-file -workDir "/home/robot_acct/
inbound-agent"
```

```
INFO: Using /home/robot_acct/inbound-agent/remoting as a
remoting work directory
```

```
[...]
```

```
INFO: Remoting TCP connection tunneling is enabled. Skipping
the TCP Agent Listener Port availability check
```

```
INFO: Agent discovery successful
  Agent address: 52.88.1.104
  Agent port:    50000
  Identity:      61:2a:ef:73:90:8d:40:ed:01:0d:c9:13:ee:76:f4
```

```
INFO: Handshaking
```

```
INFO: Connecting to 52.88.1.104:50000
```

```
INFO: Trying protocol: JNLP4-connect
```

```
INFO: Remote identity confirmed: 61:2a:ef:73:90:8d:40:ed:01:0d:c9:1
3:ee:76:f4:26
```

```
INFO: Connected
```

And that's all! Now you have a Jenkins instance that crosses the corporate firewall. You can refresh the Jenkins agent page to see that the agent is connected.

The agent connection is running in the foreground of the CLI. Stop the connection with *Ctrl + C*, and then run the command with the trailing ampersand (&) to send it to the background:

```
^Crobot_acct@firewalled-agent:~/inbound-agent$ java -jar agent.
jar -jnlpUrl https://jenkins-aws.lvin.ca/computer/inbound-agent/
slave-agent.jnlp -secret @secret-file -workDir "/home/robot_acct/
inbound-agent" &
```

```
[1] 8110
```

```
robot_acct@firewalled-agent:~/inbound-agent$ Dec 28, 2020
7:48:53 PM org.jenkinsci.remoting.engine.WorkDirManager
initializeWorkDir
```

```
INFO: Using /home/robot_acct/inbound-agent/remoting as a
remoting work directory
```

```
[...]
```

```
INFO: Connected
```

```
robot_acct@firewalled-agent:~/inbound-agent$
```

You can now close the SSH connection to the agent and the agent will stay connected to the controller.

Let's now see how we can use the systemd service to make this process restart when the VM reboots.

Creating a systemd service to auto-connect the agent

When the inbound agent connection process dies, perhaps due to a network or a memory issue, it needs to be reconnected. We can have the agent reconnect automatically by creating a service.

Create a new file in /etc/systemd/system/jenkins-inbound-agent.service with the following content. You can download this file from the book's GitHub repository

/etc/systemd/system/jenkins-inbound-agent.service

```
[Unit]
Description=Jenkins Inbound Agent
Wants=network.target
After=network.target

[Service]
ExecStart=java -jar /home/robot_acct/inbound-agent/agent.jar
-jnlpUrl https://jenkins-aws.lvin.ca/computer/inbound-agent/
slave-agent.jnlp -secret @/home/robot_acct/inbound-agent/secret-
file -workDir /home/robot_acct/inbound-agent

User=robot_acct
Restart=on-failure
RestartSec=10

[Install]
WantedBy=multi-user.target
```

Once the file is created, start the service to kick off the process, and then enable the service so that the service restarts upon a VM reboot. You can see the same console output in the logs using the `journalctl` command:

```
robot_acct@firewalled-agent:~$ sudo systemctl start jenkins-
inbound-agent
```

```
robot_acct@firewalled-agent:~$ sudo systemctl enable jenkins-
inbound-agent
```

```
robot_acct@firewalled-agent:~$ sudo journalctl -u jenkins-
inbound-agent
```

The inbound agent is now configured to auto-reconnect upon a failure. Try rebooting the VM to see that the inbound agent reconnects automatically.

There are additional ways to connect an inbound agent. See the online docs (https://github.com/jenkinsci/remoting/blob/master/docs/inbound-agent.md) for the latest information.

Labels and Usage

Here is some more information about the Labels and Usage fields on the agent configuration page. Let's talk about Labels first.

Agent labels

In the Labels textbox, we can put multiple keywords delimited by a space to describe an agent. Suppose these four agents have the following labels:

- `centos8-agent:` `docker linux centos centos8`
- `ubuntu2004-agent:` `docker linux ubuntu ubuntu2004`
- `windows-95-agent:` `windows windows95`
- `windows-10-agent:` `docker windows windows10`

A Jenkinsfile can specify one or more labels in the agent `label` directive to specify the agent that it requires. Builds for a pipeline that specifies agent `{ label 'docker' }` would run on `centos8-agent`, `ubuntu2004-agent`, or `windows-10-agent` because the three agents have the docker label. Builds for a pipeline that specifies agent `{ label 'windows' }` would run on any of the two Windows agents. In a more advanced use case, builds for a pipeline that specifies agent `{ label 'windows && docker' }` would run only on `windows-10-agent` because that is the only agent with both windows and docker labels. Similarly, builds for a pipeline that specifies agent `{ label 'centos || ubuntu' }` would run on any of the two Linux agents.

The agent labels were more useful before Docker was invented – in those dark days, each agent would be a VM with a specific set of tools preinstalled, and labels were used to identify its operating system, tools, and configurations. With Docker available, nearly all agents can have just one docker label as we have configured, and each pipeline can specify its requirements in the Dockerfile. Labels are, of course, still useful if you are using bare-metal agents (rather than Docker or Dockerfile agent) with specific hardware characteristics such as GPU availability or a non-x86 CPU architecture – a pipeline can request an agent with specific hardware using a label.

One last thing about agent labels is that each agent's own name acts as a label. For example, builds for a pipeline that specifies agent `{ label 'ubuntu2004-agent' }` would run only on `ubuntu2004-agent`, even if you didn't label the agent with its own name. If you need to pin a pipeline to a specific agent, simply use the agent's name as the label.

Next, let's look at Usage.

Agent usage

Usage configuration determines agent availability. In the previous section, we saw that a Jenkinsfile can request an agent with a specific set of labels. A Jenkinsfile can also request *any* available agent without any label specifications with agent any. In a Jenkins environment with a homogenous set of agents (such as all of them being Ubuntu 20.04), a simple operation such as a file copy doesn't really need a special capability that's noted by agent labels. Usage configuration of an agent determines whether the agent is eligible to be used by a pipeline that specifies agent any.

Choosing Use this node as much as possible makes the agent available to the builds for a pipeline with agent any.

Choosing Only build jobs with label expressions matching this node makes the agent unavailable to the builds for a pipeline with agent any. As the option message says, only build jobs with label expressions matching this node (as opposed to a build job requesting any agent) will run on this node. It's important to understand that this doesn't stop a build from running on this agent if a pipeline pins to this agent. Use the Job Restrictions plugin if you need to limit which pipelines can use an agent.

Also, take note that at least one agent must be configured with the Use this node as much as possible option for a pipeline with agent any to work. If you are setting up a Docker Cloud, a Docker Agent Template is usually a good choice for this configuration.

Let's continue to create a Docker Cloud.

Creating a secure Docker Cloud

Required plugins
Docker

In addition to the static agents, we will add a Docker Cloud in order to dynamically generate agents using Docker containers. We need to set up a Docker host where the containers will run. It's possible to reuse an existing agent to act as a Docker host; however, this is not recommended because the Docker engine is modified to require a certificate. Here is what the connection flow looks like:

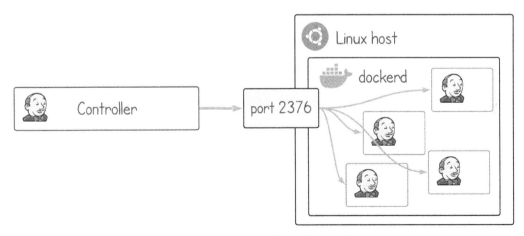

Figure 2.26 – Architecture of the Docker Cloud host

There are a few things to understand about the Docker Cloud. First, the Docker Cloud doesn't support an inbound setup. This means that a Docker host for the AWS controller must also be on AWS (or similar network locations where the controller can access). Second, setting up a secure connection is complex. It involves creating multiple self-signed SSL certificates and placing them in just the right places for both the host and the clients. Third, an insecure connection is *very* insecure. Not only is it unencrypted, but it also doesn't require authentication. In other words, it creates a server for anyone in the world to connect to freely. Finally, it can only run a specific set of images that we pre-populate. It can't run a random image that a pipeline needs, which means it's really only good for providing generic catch-all images.

> **Never create an insecure Docker host on AWS!**
> Anyone can connect without authentication and start mining bitcoin. Don't ask me how I know =(

Setting up a secure Docker Cloud is a four-step process that applies to both AWS and firewalled Jenkins:

1. Create a certificate authority (CA). Create server and client certificates signed by the CA. Only the client who presents a certificate from this CA will be accepted by the server.

2. Configure the Docker engine to use the host certificate and listen to TCP port 2376.

3. Configure Jenkins to use the client certificate.

4. Connect Jenkins to the Docker host using the client certificate.

Here is what the certificate architecture looks like:

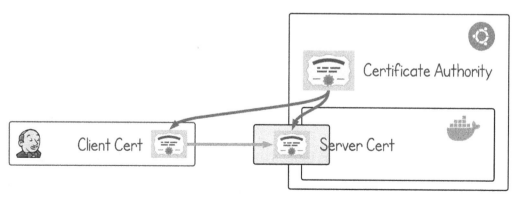

Figure 2.27 – Architecture of Docker certificates

Let's get started. SSH into the Docker host. It's a VM running Ubuntu 20.04 just like all other VMs. Docker was installed in *Chapter 1, Jenkins Infrastructure with TLS/SSL and Reverse Proxy*.

Generating a CA, server certificates, and client certificates

We'll start by generating the necessary certificates. First, we'll generate a CA, then, using the CA, we'll generate the certificates for the server and the client. Let's begin:

1. Generate a CA. This is done by first creating a private key, and then creating a public certificate signed by the private key:

```
docker-host:~$ openssl genrsa -aes256 -out ca.key 4096
```

```
docker-host:~$ openssl req -new -x509 -days 3650 -key ca.key
-sha256 -out ca.crt
```

2. Next, create server certificates for the Docker host. First, create a private key and create a certificate signing request (CSR):

```
docker-host:~$ openssl genrsa -out server.key 4096
```

```
docker-host:~$ openssl req -sha256 -new -key server.key -out
server.csr
```

3. Enter the domain and/or the IP for the Docker host. It's possible to add multiple as follows:

```
docker-host:~$ echo 'subjectAltName = DNS:firewalled-docker-
host.lvin.ca,IP:192.168.1.18,IP:127.0.0.1' > extfile.cnf
```

For the firewalled Docker host, enter just one IP. For the AWS Docker host, enter both the public and the private IPs, so that the controller can connect to the Docker host using either IP:

```
docker-host:~$ echo 'subjectAltName = IP:192.168.1.18' > extfile.cnf
```

4. Set extendedKeyUsage to serverAuth so that the certificate can only be used for a server. Notice the >> characters for appending to the file rather than overwriting:

```
docker-host:~$ echo 'extendedKeyUsage = serverAuth' >> extfile.cnf
```

5. Finally, sign the CSR to create a server certificate for the Docker host using the newly created CA. The resulting certificate is valid for 1 year. We can run

through the same steps again in about 300 days to generate a new certificate with an updated expiry date. Generating a new certificate doesn't invalidate the existing certificates:

```
docker-host:~$ openssl x509 -req -days 365 -sha256 -extfile
extfile.cnf -CA ca.crt -CAkey ca.key -CAcreateserial -in server.
csr -out server.crt
```

6. Next, create client certificates for the Jenkins Docker client. Create a private key and a CSR:

```
docker-host:~$ openssl genrsa -out client.key 4096
```

```
docker-host:~$ openssl req -subj '/CN=client' -new -key client.
key -out client.csr
```

7. Set extendedKeyUsage to clientAuth so that the certificate can only be used for a client. Notice the > character for overwriting the file:

```
docker-host:~$ echo 'extendedKeyUsage = clientAuth' > extfile.cnf
```

8. Finally, sign the CSR to create a client certificate for the Jenkins Docker client using the newly created CA. This certificate is also valid for 1 year:

```
docker-host:~$ openssl x509 -req -days 365 -sha256 -extfile
extfile.cnf -CA ca.crt -CAkey ca.key -CAcreateserial -in client.
csr -out client.crt
```

> **Client certificates are passwords**
> Anyone with client certificates can connect and launch a container. Treat the certificates like a password and store them securely.

All the required keys are created. Delete the intermediary files to clean up. Verify that the permissions for the certificates and the keys are 644 and 600, respectively:

```
docker-host:~$ rm -v ca.srl client.csr server.csr extfile.cnf
removed 'ca.srl'
removed 'client.csr'
removed 'server.csr'
removed 'extfile.cnf'
robot_acct@firewalled-docker-host:~$ ls -l
total 24
-rw-r--r-- 1 robot_acct dip 2199 Dec 29 04:41 ca.crt
-rw------- 1 robot_acct dip 3326 Dec 29 04:35 ca.key
-rw-r--r-- 1 robot_acct dip 1919 Dec 29 05:08 client.crt
-rw------- 1 robot_acct dip 3243 Dec 29 05:02 client.key
-rw-r--r-- 1 robot_acct dip 2114 Dec 29 05:01 server.crt
-rw------- 1 robot_acct dip 3247 Dec 29 04:57 server.key
```

The certificates are ready. Let's distribute them to the right places.

Storing the certificates

Save the CA and server certificates in /etc/ssl/docker-host/:

```
docker-host:~$ sudo mkdir /etc/ssl/docker-host
```

```
docker-host:~$ sudo mv ca.crt ca.key server.crt server.key /etc/
ssl/docker-host/
```

Save the client certificates in Jenkins. Go to the Global Credentials page, click Add Credentials, and then choose Kind as X.509 Client Certificate. Copy and paste the content of client.key, client.crt, and ca.crt into the three boxes. Enter docker-host-client for ID and Description, and then click OK to save:

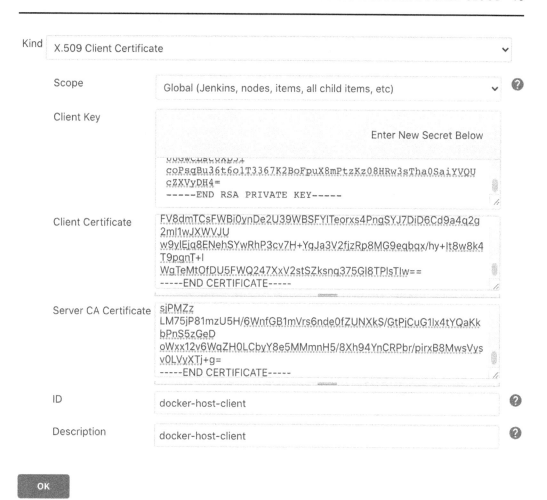

Figure 2.28 – Docker client keys stored in Jenkins

Both the server and client certificates are ready. Let's continue to configure the Docker service.

Configuring the Docker service

Configure docker.service on the Docker host to listen to TCP port 2376 while using the server certificates:

1. Open an override file for docker.service and paste in the following content. The syntax is a bit odd here. The line with ExecStart= seems useless but is actually required. Also, the last line with /usr/bin/dockerd is one long line without a line break. It may be better if you just download this file from the book's GitHub repository because writing this out can be very error-prone:

```
docker-host:~$ sudo systemctl edit docker.service

[Service]

ExecStart=

ExecStart=/usr/bin/dockerd -H tcp://0.0.0.0:2376 -H unix:///var/
run/docker.sock --tlsverify --tlscacert=/etc/ssl/docker-host/
ca.crt --tlscert=/etc/ssl/docker-host/server.crt --tlskey=/etc/
ssl/docker-host/server.key
```

2. Apply the changes: save, exit, `systemctl daemon-reload`, and restart Docker. Check that it's listening on port 2376:

```
docker-host:~$ sudo systemctl daemon-reload

docker-host:~$ sudo systemctl restart docker.service

docker-host:~$ sudo netstat -lntp | grep dockerd

tcp6   0   0   :::2376      :::   LISTEN       5351/dockerd
```

The Docker service is now ready to accept traffic on port 2376 using the server certificate. In order to update the certificates with a new expiry date, replace the certificate files and restart the Docker service.

Directions on docker-plugin documents are insecure!

Do not follow the directions on docker-plugin documents as they configure the Docker host to be open to everyone without authentication. This is equivalent to allowing SSH as root without requiring a password.

Configuring Jenkins

Let's put it all together by configuring the Docker Cloud on Jenkins:

1. Click Manage Jenkins | Manage Nodes and Clouds | Configure Clouds | Add a new cloud, and then choose Docker. A Docker configuration panel is created.

2. Click Docker Cloud details.

3. Enter the IP of the Docker host into the Docker Host URI field in the format of `tcp://<IP>:2376` and choose docker-host-client for Server credentials.

4. Clicking Test Connection should show the version and the API version of the Docker host.

5. Check Enabled and then click Apply to save the progress:

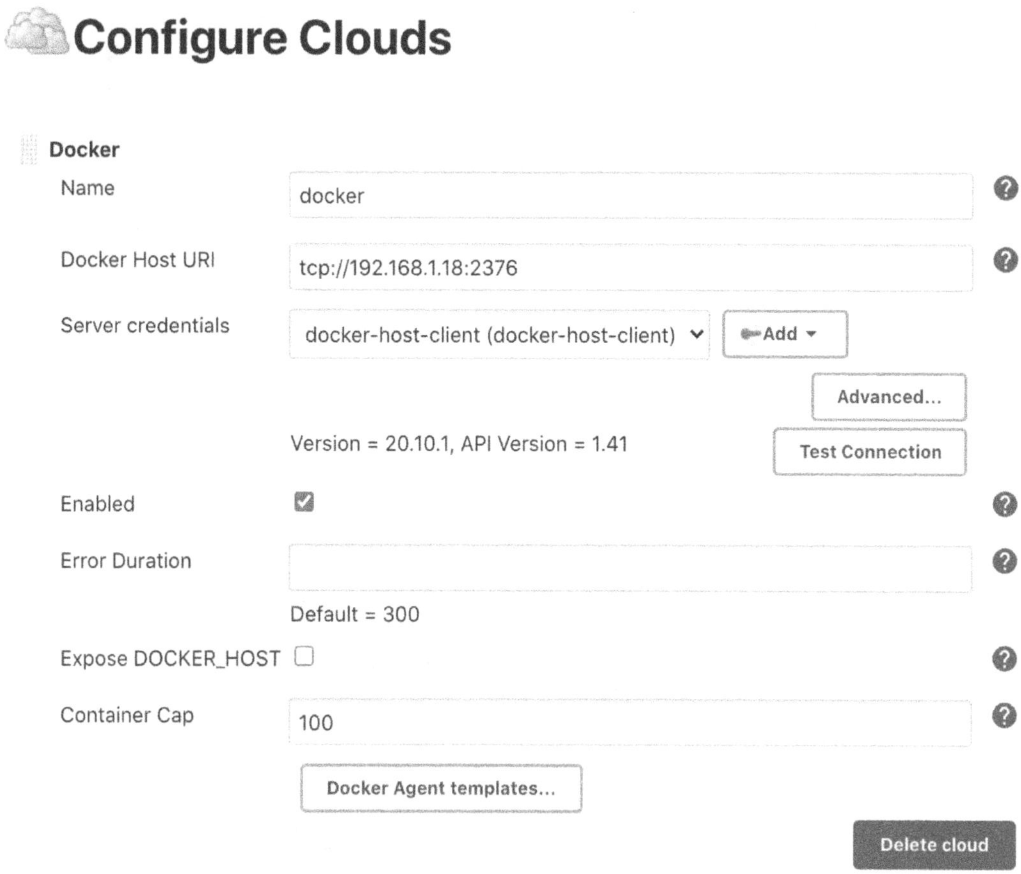

☁ Configure Clouds

Docker

Name	docker ❓
Docker Host URI	tcp://192.168.1.18:2376 ❓
Server credentials	docker-host-client (docker-host-client) ▾ ⟵ Add ▾
	[Advanced…]
	Version = 20.10.1, API Version = 1.41 [Test Connection]
Enabled	☑ ❓
Error Duration	❓
	Default = 300
Expose DOCKER_HOST	☐ ❓
Container Cap	100 ❓
	[Docker Agent templates…]
	[Delete cloud]

Figure 2.29 – Docker Cloud connection configured and tested

6. **Finally, add some agent templates that the builds can use. Click** Docker Agent templates **and** Add Docker Template:

   - Labels: `linux`
   - Enabled: **Check**
   - Name: `docker`
   - Docker Image: `jenkins/agent`
   - Remote File System Root: `/home/jenkins`

   We can leave everything else as is, as shown in the following screenshot, and then click Save:

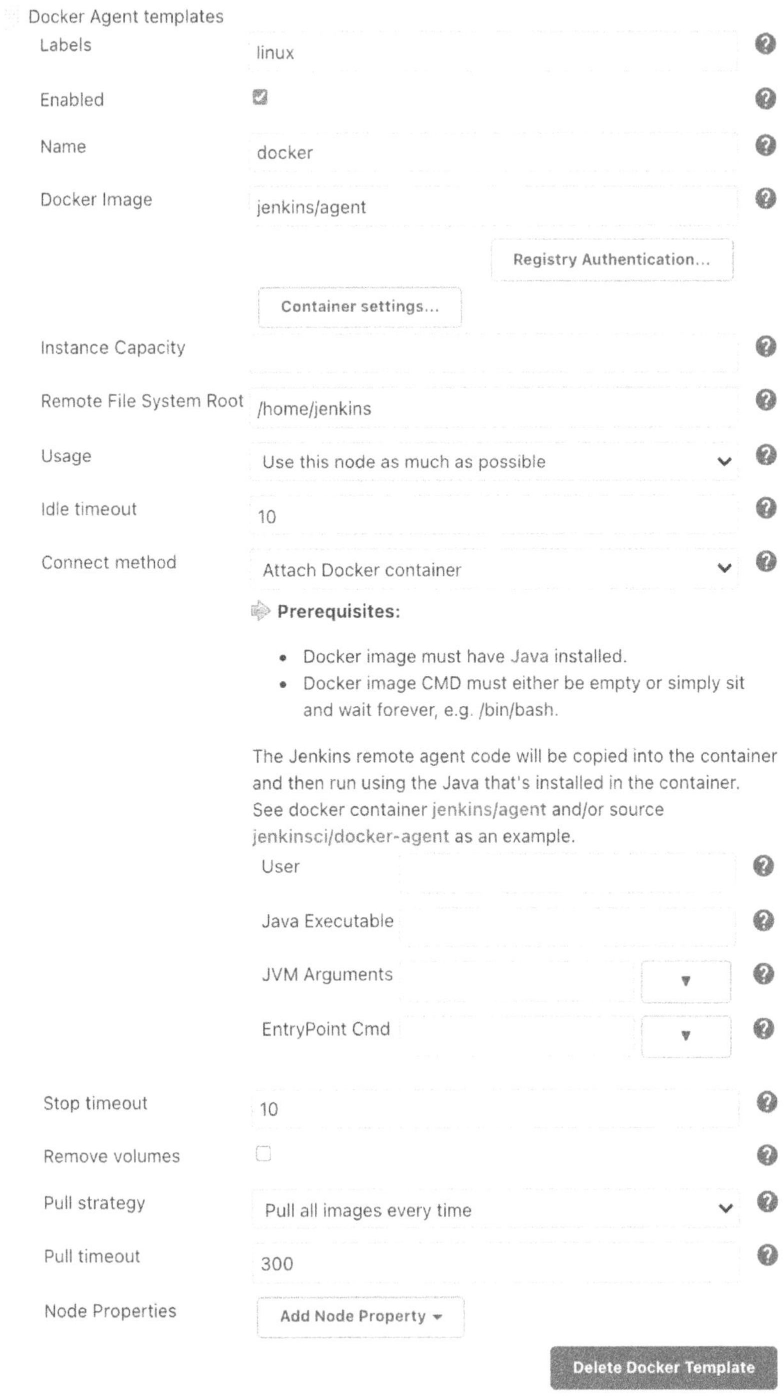

Docker Agent templates
Labels   linux   ❓

Enabled   ☑   ❓

Name   docker   ❓

Docker Image   jenkins/agent   ❓

                                          Registry Authentication...

                     Container settings...

Instance Capacity   ❓

Remote File System Root /home/jenkins   ❓

Usage   Use this node as much as possible   ⌄   ❓

Idle timeout   10   ❓

Connect method   Attach Docker container   ⌄   ❓

                     ⇨ **Prerequisites:**

                     • Docker image must have Java installed.
                     • Docker image CMD must either be empty or simply sit
                       and wait forever, e.g. /bin/bash.

                     The Jenkins remote agent code will be copied into the container
                     and then run using the Java that's installed in the container.
                     See docker container jenkins/agent and/or source
                     jenkinsci/docker-agent as an example.

                     User   ❓

                     Java Executable   ❓

                     JVM Arguments   ▾   ❓

                     EntryPoint Cmd   ▾   ❓

Stop timeout   10   ❓

Remove volumes   ☐   ❓

Pull strategy   Pull all images every time   ⌄   ❓

Pull timeout   300   ❓

Node Properties   Add Node Property ▾

                                          Delete Docker Template

Figure 2.30 – Docker Agent template configuration

The Docker Cloud is now ready. When we build a pipeline that uses the `linux` agent label, a new agent will be created from the Docker Cloud on the fly. In addition, since we've set Usage to Use this node as much as possible, pipeline builds using agent any will also use an agent created from this template.

Summary

In this chapter, we have set up the complete Jenkins instance both on AWS and inside the firewall. For each Jenkins controller, we've set up a reverse proxy and configured TLS certificates for HTTPS support. Then we've added two agents, one as an SSH agent and another as an inbound agent, to handle various network requirements. Finally, we've added the Docker Cloud so that agents can be dynamically generated from a Docker container. The Jenkins instances are ready to take on production workloads.

In the coming chapters, we'll use the Jenkins instances to set up GitOps-driven CI/CD pipelines.

3
GitOps-Driven CI Pipeline with GitHub

With Jenkins up and running, it's time to create pipelines.

We will create two projects, an adder and a subtractor, then create users with varying permissions to understand the roles and permissions model. We will also create a static pipeline that compiles the code, runs unit tests, and generates a code coverage report.

Afterward, we will convert the static pipeline into a premerge CI pipeline that is triggered by GitHub pull requests (PRs). Building a trigger for AWS Jenkins is different from firewalled Jenkins due to network restrictions, so we'll cover both options in depth.

Once the build is complete, the original GitHub PR that triggered the build is updated with the result. We'll also configure the GitHub repository to require a successful premerge build for a merge.

In this chapter, we're going to cover the following main topics:

- Project overview
- Creating two sets of projects and users in Jenkins
- Creating a static pipeline for build and unit tests
- Displaying test results and a code coverage report
- Creating a premerge CI pipeline with GitHub PR hooks
- Requiring a successful build for a merge

Technical requirements

You need a GitHub account with two Git repositories. The account is used to create a personal access token that is used by Jenkins to communicate with GitHub. You also need Git installed on your laptop to create and push commits to GitHub. Finally, you need Python 3 installed on your laptop to run the test scripts in the chapter.

Files in the chapter are available in the GitHub repository at https://github.com/PacktPublishing/Jenkins-Administrators-Guide/blob/main/ch3.

Project overview

In order to understand how the roles and permissions work, we will create two Python scripts that can add or subtract two numbers. Two GitHub repositories are created to contain each script:

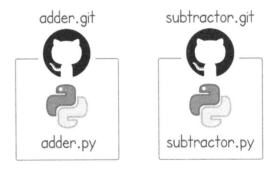

Figure 3.1 – Two GitHub repositories each containing a Python script and support files

Creating or updating a PR triggers a premerge build in Jenkins to compile the project, run unit tests, and generate a code coverage report. A premerge build is a CI build that validates a PR. Since a PR must be validated *before a merge*, the validation build is called a *premerge* build. At the end of the build, Jenkins updates the commit status in GitHub to indicate whether or not the premerge build was successful. The GitHub repo is configured to require a successful premerge build as a prerequisite to a merge:

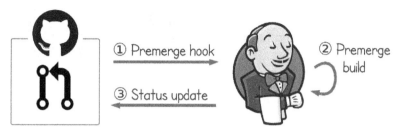

Figure 3.2 – Premerge build workflow

A successful premerge build followed by a merge triggers a postmerge build in Jenkins. A postmerge build is a CD build that packages and deploys software. Packaging and deployment builds run *after a merge*, therefore the build is called a *postmerge* build. The postmerge build compiles and packages the project, runs integration tests, applies a version tag to the Git repository and the artifact (a Docker image in our case), and delivers the artifact to Docker Hub. At the end of the build, Jenkins pushes the tag to GitHub to indicate which source code version was used to develop the Docker image with the matching tag:

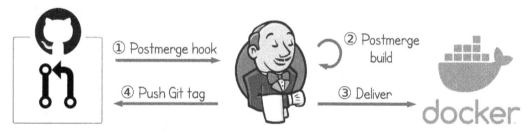

Figure 3.3 – Postmerge build workflow

Two projects are configured in Jenkins: adder and subtractor.

Five users are created, admin, adder-admin, adder-user, subtractor-admin, and subtractor-user, with the following permissions:

User	Permissions
admin	Global administrator.
adder-admin	Create, delete, view, and run the adder project.
adder-user	View and run the adder project.
subtractor-admin	Create, delete, view, and run the subtractor project.
subtractor-user	View and run the subtractor project.

We will examine GitHub build triggers for both AWS Jenkins and firewalled Jenkins where GitHub can reach the former but cannot reach the latter:

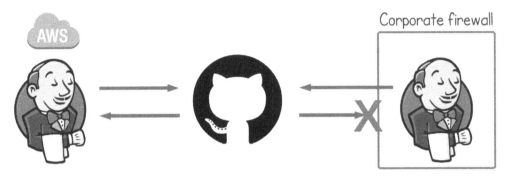

Figure 3.4 – Network access rules between GitHub, AWS Jenkins, and firewalled Jenkins

Let's get started.

Creating two sets of projects and users in Jenkins

We will create two projects in two folders:

1. Log in to Jenkins as the admin user, click New Item, choose Folder, and name it adder. Do this again, naming the folder subtractor:

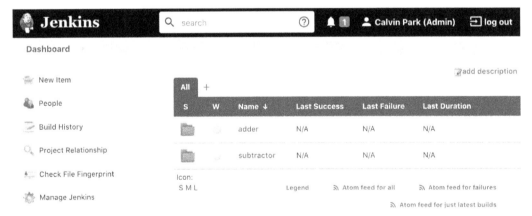

Figure 3.5 – The adder and subtractor folders created on Jenkins

2. Next, we will create the project admins and project users. Click Manage Jenkins | Manage Users | Create User. Fill out the form to create four users:

- `adder-admin`
- `adder-user`
- `subtractor-admin`
- `subtractor-user`

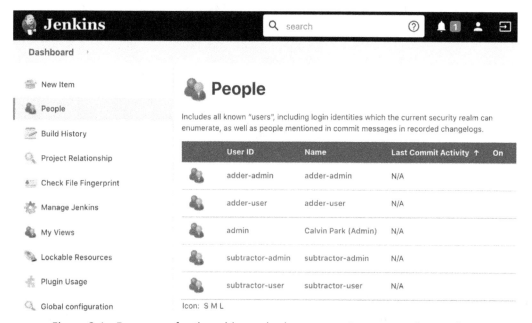

Figure 3.6 – Four users for the adder and subtractor projects created on Jenkins

3. Let's configure the adder project to have its own access rules. Go back to the home page of Jenkins and click the adder folder and then Configure. Under the Properties section, check the box for Enable project-based security.

4. Then, add adder-admin and adder-user and give correct permissions. Click Add user or group to add both users. For adder-admin, check all the boxes to give full permissions to the folder. For adder-user, check the following boxes:

- Credentials: View
- Job: Build, Cancel, Read, Workspace
- Run: Replay, Update
- View: Read
- SCM: Tag

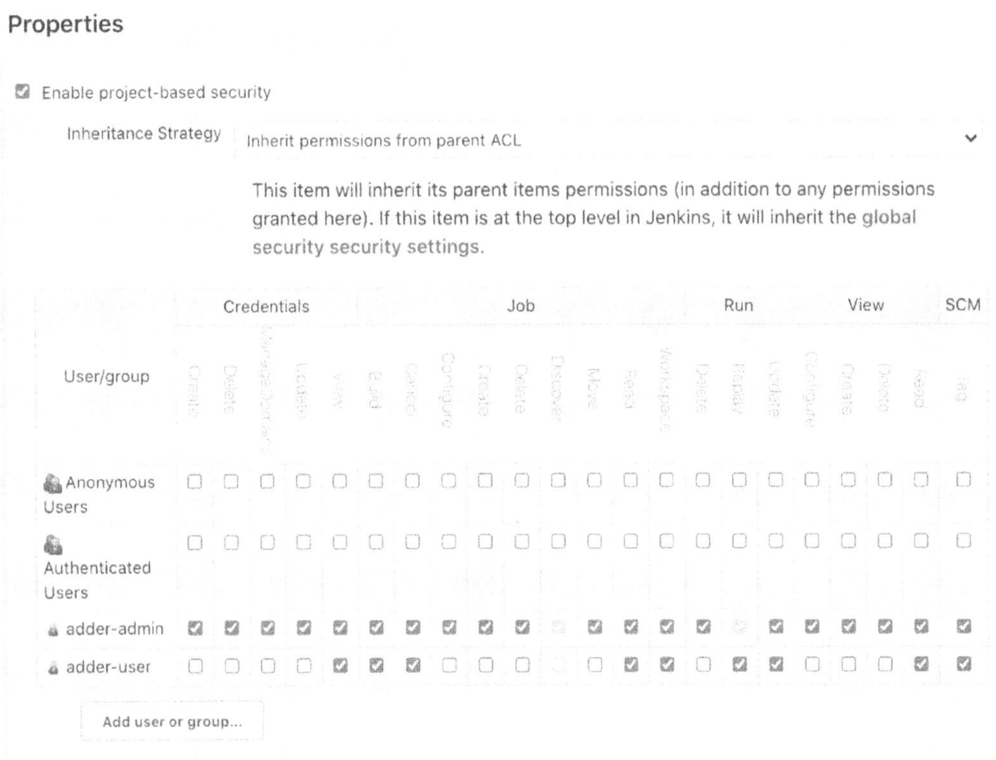

Properties

☑ Enable project-based security

Inheritance Strategy   Inherit permissions from parent ACL   ⌄

> This item will inherit its parent items permissions (in addition to any permissions granted here). If this item is at the top level in Jenkins, it will inherit the global security security settings.

User/group	Credentials					Job									Run			View				SCM
	Create	Delete	Manage Domains	Update	View	Build	Cancel	Configure	Create	Delete	Discover	Move	Read	Workspace	Delete	Replay	Update	Configure	Create	Delete	Read	Tag
Anonymous Users	☐	☐	☐	☐	☐	☐	☐	☐	☐	☐	☐	☐	☐	☐	☐	☐	☐	☐	☐	☐	☐	☐
Authenticated Users	☐	☐	☐	☐	☐	☐	☐	☐	☐	☐	☐	☐	☐	☐	☐	☐	☐	☐	☐	☐	☐	☐
adder-admin	☑	☑	☑	☑	☑	☑	☑	☑	☑	☑		☑	☑	☑	☑		☑	☑	☑	☑	☑	☑
adder-user	☐	☐	☐	☐	☑	☑	☑	☐	☐	☐		☐	☑	☑	☐	☑	☑	☐	☐	☑	☑

Add user or group...

Figure 3.7 – Project-based security permissions for adder-admin and adder-user

5. Go back to the home page of Jenkins and follow the same steps for the subtractor folder, this time with subtractor-admin and subtractor-user.

6. Now, log out of Jenkins and log in as adder-admin. We will see that only the adder folder is displayed, since adder-admin does not have the permission to see or do anything on the subtractor folder:

CREATING TWO SETS OF PROJECTS AND USERS IN JENKINS 87

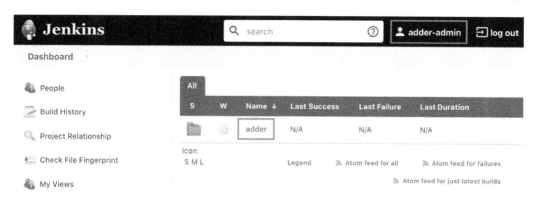

Figure 3.8 – adder-admin user can see the adder folder but can't see the subtractor folder

7. Click the adder folder, and note that the Configure, New Item, and Delete Folder options are available on the left:

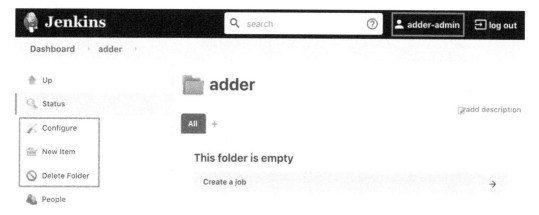

Figure 3.9 – adder-admin user can administrate the adder folder

8. Log out again and log in as adder-user. Much like adder-admin, we can see the adder folder but not the subtractor folder. Inside the adder folder, adder-user is not presented with the options for Configure, New Item, and Delete Folder:

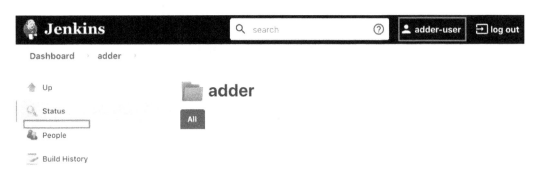

Figure 3.10 – adder-user can see the adder folder but can't administrate it

Folder-based role-based access control (RBAC) like this can be extended to use AD groups, which would allow multiple teams of engineers to have permission to see only their folders and projects.

Creating a static pipeline for build and unit tests

> **Required plugins**
> Docker Pipeline

Let's start building our CI pipelines, starting with some static pipelines.

First, go to github.com and create two repositories named adder and subtractor. Afterward, clone the repositories so that we can add files to them. The example command is for my repository – you should use your own:

```
~ $ git clone https://github.com/calvinpark/adder.git

Cloning into 'adder'...
remote: Enumerating objects: 5, done.
remote: Counting objects: 100% (5/5), done.
remote: Compressing objects: 100% (4/4), done.
remote: Total 5 (delta 0), reused 0 (delta 0), pack-reused 0
Unpacking objects: 100% (5/5), 5.45 KiB | 1.82 MiB/s, done.
```

In the adder repository, add the following adder.py file:

adder.py

```
import sys
import unittest

def adder(a: int, b: int) -> int:
    return int(a) + int(b)

def main(args):
    print(adder(args[1], args[2]))

class TestAdder(unittest.TestCase):
    def test_adder_success(self):
        self.assertEqual(adder(3, 5), 8)

    def test_adder_typeerror(self):
        self.assertRaises(TypeError, adder, 5)

    def test_adder_valueerror(self):
        self.assertRaises(ValueError, adder, 1, 'a')

    def test_main_indexerror(self):
        self.assertRaises(IndexError, main, [])

if __name__ == '__main__':
    main(sys.argv)
```

In the subtractor repository, add the following subtractor.py file:

subtractor.py

```python
import sys
import unittest

def subtractor(a: int, b: int) -> int:
    return int(a) - int(b)

def main(args):
    print(subtractor(args[1], args[2]))

class TestSubtractor(unittest.TestCase):
    def test_subtractor_success(self):
        self.assertEqual(subtractor(8, 5), 3)

    def test_subtractor_typeerror(self):
        self.assertRaises(TypeError, subtractor, 5)

    def test_subtractor_valueerror(self):
        self.assertRaises(ValueError, subtractor, 1, 'a')

    def test_main_indexerror(self):
        self.assertRaises(IndexError, main, [])

if __name__ == '__main__':
    main(sys.argv)
```

The scripts do what their names say – adder.py adds two numbers and subtractor.py subtracts two numbers:

```
~/adder $ python3 adder.py 1 2
3
~/subtractor $ python3 subtractor.py 3 1
2
```

Although it's a bit silly to compile Python scripts, here is how we can compile them. On your real project, this step can be substituted with a project compilation of your product code:

```
~/adder $ python3 -m compileall adder.py
Compiling 'adder.py'...
~/subtractor $ python3 -m compileall subtractor.py
Compiling 'subtractor.py'...
```

You might have noticed that the Python scripts have unit tests. Here is how we run them:

```
~/adder $ python3 -m unittest adder.py
....
----------------------------------------------------------------
Ran 4 tests in 0.000s
OK
~/subtractor $ python3 -m unittest subtractor.py
....
----------------------------------------------------------------
Ran 4 tests in 0.000s
OK
```

There, we have the fundamental building blocks of a software project: code, compilation, and unit tests. We will cover packaging, integration tests, deployment, and tagging in *Chapter 4, GitOps-Driven CD Pipeline with Docker Hub and More Jenkinsfile Features*.

Let's create a Jenkinsfile for each project. Here is the one for the adder:

Abridged syntax

A Jenkinsfile is made up of Groovy code, which has many closing curly braces. For the purpose of fitting the code onto a page, the closing curly braces are compressed into a single line. Correctly formatted files are in the book's GitHub repository listed at the beginning of the chapter.

---

adder/Jenkinsfile

```
pipeline {
    agent {
        docker {
            label 'docker'
            image 'python:3'
        }  }
    stages {
        stage('Compile') {
            steps {
                sh 'python3 -m compileall adder.py'
            }  }
        stage('Run') {
            steps {
                sh 'python3 adder.py 3 5'
            }  }
        stage('Unit test') {
            steps {
                sh 'python3 -m unittest adder.py'
}  }  }  }
```

Let's examine the elements of the Jenkinsfile:

- `pipeline`: This is the beginning of a declarative pipeline.

- `agent/docker`: This build uses a Docker container-based agent.

- `label`: This is the label of the node that will host the Docker container agent. In *Chapter 2, Jenkins with Docker on HTTPS on AWS and inside a Corporate Firewall*, we set docker as the label for the permanent agents, therefore a build will start a container in one of the permanent agents.

- `image`: This is the Docker image we're using for this build. We've chosen python:3 because our scripts are for Python 3.

- `stages`: Individual stages are listed inside the stages directive.

- `stage`: A specific stage of a build. A stage is a collection of related tasks. In this example, stage only contains steps, but later we will see that it can contain many more directives, including its own agent.

- `steps`: The list of actions to take in a stage.

- `sh`: We are running a shell command with python3 executable.

The pipeline will create a container from the python:3 image on one of the two permanent agents, compile the code, run it, then run unit tests. Let's add all the files to the project Git repository, make a commit, and push:

```
~/adder $ git add adder.py Jenkinsfile

~/adder $ git commit -m "Initial adder.py and Jenkinsfile"
[main 234e666] Initial adder.py and Jenkinsfile
 2 files changed, 49 insertions(+)
 create mode 100644 Jenkinsfile
 create mode 100644 adder.py

~/adder $ git push
Enumerating objects: 5, done.
Counting objects: 100% (5/5), done.
Delta compression using up to 8 threads
Compressing objects: 100% (4/4), done.
Writing objects: 100% (4/4), 822 bytes | 822.00 KiB/s, done.
Total 4 (delta 0), reused 0 (delta 0), pack-reused 0
To https://github.com/calvinpark/adder.git
   1f5f7d7..234e666  main -> main
```

On GitHub, we can see the newly added files:

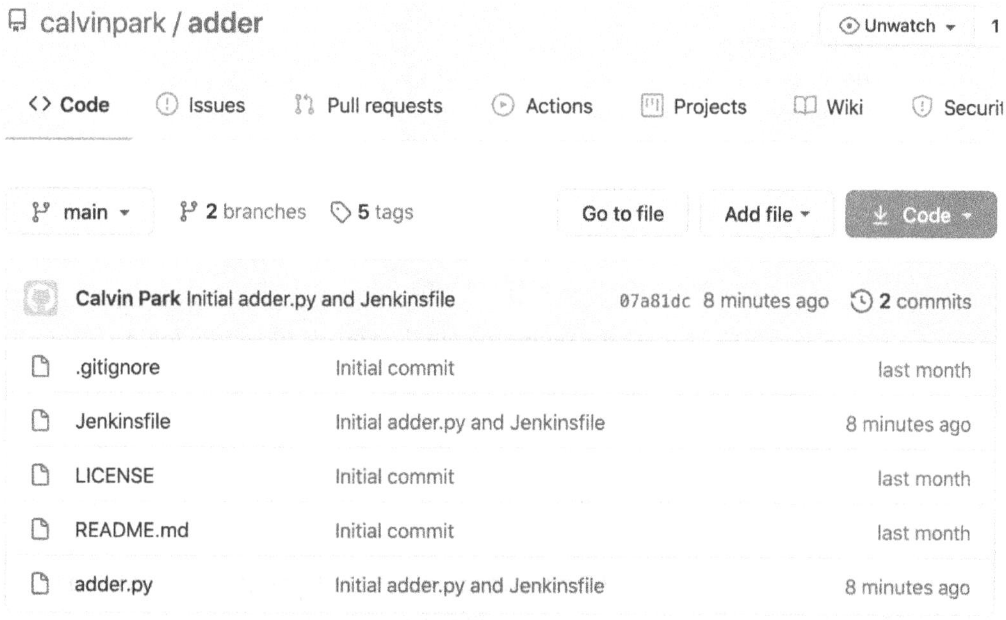

calvinpark / **adder**		⊙ Unwatch ▾ 1

<> Code	ⓘ Issues	⑃ Pull requests	⊙ Actions	Projects	Wiki	ⓘ Securit

⌥ main ▾	⌥ **2 branches**	◌ **5 tags**		Go to file	Add file ▾	⬇ Code ▾

	Calvin Park Initial adder.py and Jenkinsfile		07a81dc 8 minutes ago	◔ **2 commits**
▢	.gitignore	Initial commit		last month
▢	Jenkinsfile	Initial adder.py and Jenkinsfile		8 minutes ago
▢	LICENSE	Initial commit		last month
▢	README.md	Initial commit		last month
▢	adder.py	Initial adder.py and Jenkinsfile		8 minutes ago

Figure 3.11 – GitHub adder repository showing adder.py and the Jenkinsfile

It's time to make our first pipeline:

1. Log in to Jenkins as adder-admin, enter the adder folder, and click New Item. Name it premerge, choose Pipeline, and press OK:

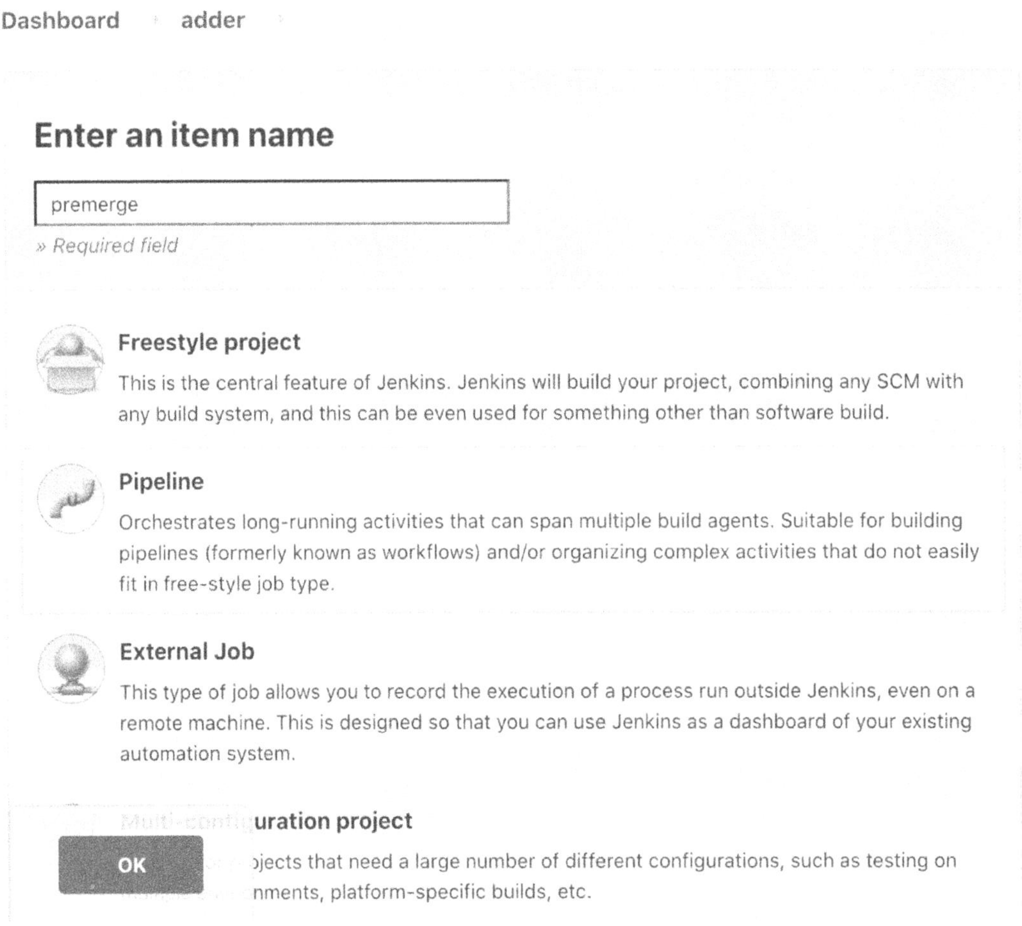

Dashboard   adder

Enter an item name

premerge

» *Required field*

Freestyle project

This is the central feature of Jenkins. Jenkins will build your project, combining any SCM with any build system, and this can be even used for something other than software build.

Pipeline

Orchestrates long-running activities that can span multiple build agents. Suitable for building pipelines (formerly known as workflows) and/or organizing complex activities that do not easily fit in free-style job type.

External Job

This type of job allows you to record the execution of a process run outside Jenkins, even on a remote machine. This is designed so that you can use Jenkins as a dashboard of your existing automation system.

 uration project

OK   ojects that need a large number of different configurations, such as testing on nments, platform-specific builds, etc.

Figure 3.12 – Create the premerge pipeline inside the adder folder

2. Now we're on the pipeline configuration page. Scroll down to the Pipeline **section**:

- Definition: `Pipeline script from SCM.`
- SCM: Git.
- Repository URL: `https://github.com/<GitHub Username>/adder.git.`
- Credentials: - none - (if our Git repository were private, we could create a credential with an SSH key and use it here to grant access).

- Branch Specifier: */main (take note that we're always building from the main branch, which is where our Python script and Jenkinsfile are).

Leave the rest unmodified and click Save to exit:

Figure 3.13 – The adder pipeline SCM configuration

3. Click Build Now, and we can see that the build ran successfully, indicated by the blue ball. Click #1 then Console Output to see the full console output of the build:

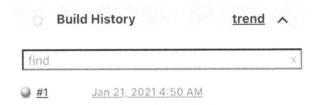

Figure 3.14 – Build history showing the first successful build

4. Now log back out, log in as adder-user, and see that we can't modify the pipeline as this user but we can still run the build.

Now we have a basic static pipeline configured. It runs the build as prescribed in the Jenkinsfile, but it's missing the bells and whistles. Let's continue to enhance the pipeline.

Displaying test results and a code coverage report

Required plugins
Cobertura, JUnit Plugin

We are already running unit tests, but we can't see the result unless we go into the console output and look for it. The test result can be exposed in a nicer format using the JUnit plugin. While we're at it, let's also add a code coverage report:

1. Open the Jenkinsfile and update the Unit test stage to generate two XML reports, junit.xml for the unit test results and coverage.xml for the code coverage report:

```
stage('Unit test') {
    steps {
        sh '''python3 -m pytest \
            -v --junitxml=junit.xml \
            --cov-report xml --cov adder adder.py
        '''
    } }
```

2. Also, add a post directive in the Jenkinsfile to pick up the two report files. The post directive should be on the same level as the stages directive:

```
pipeline {
    agent { ... }
    stages {
        stage('Compile') { ... }
        stage('Run') { ... }
        stage('Unit test') { ... }
    }
    post {
        always {
            junit 'junit.xml'
            cobertura coberturaReportFile: 'coverage.xml'
}   }   }
```

Running this Jenkinsfile will fail because the pytest Python module is not installed in the python:3 image. We can add a stage to install pip and pytest, but let's try something new this time.

3. Create a Dockerfile that extends Ubuntu 20.04 and installs python3, pip, and additional pip packages. This technique can be used to install or configure any additional tools that our build needs:

Dockerfile

```
FROM ubuntu:20.04

RUN apt-get -qq update && apt-get -qq -y install \
    python3 \
    python3-pip \
    && rm -rf /var/lib/apt/lists/*

RUN pip3 install \
    pytest \
    pytest-cov
```

4. Now, open the Jenkinsfile and update the agent directive to use the Dockerfile. The agent type is changed from docker to dockerfile. If your Dockerfile is not named Dockerfile, add filename '<Dockerfile name>' below label:

```
agent {
   dockerfile {
       label 'docker'
       //filename 'my.dockerfile'   // Uncomment and change
   }
}
```

label is, again, the label for the agent that will host the container for this build. Here is the complete Jenkinsfile:

Jenkinsfile

```
pipeline {
    agent {
        dockerfile {
            label 'docker'
    }   }
    stages {
        stage('Compile') {
            steps {
                sh 'python3 -m compileall adder.py'
        }   }
        stage('Run') {
            steps {
                sh 'python3 adder.py 3 5'
        }   }
        stage('Unit test') {
            steps {
                sh '''python3 -m pytest \
                    -v --junitxml=junit.xml \
                    --cov-report xml --cov adder adder.py
                '''
    }   }   }
    post {
        always {
            junit 'junit.xml'
            cobertura coberturaReportFile: 'coverage.xml'
}   }   }
```

5. Make a new commit with the changes and push:

```
~/adder $ git add Dockerfile Jenkinsfile
~/adder $ git commit -am "Add unit test and code coverage report"
~/adder $ git push
```

6. Run the build again. In the build log, we can see that a new Docker image was built from the `Dockerfile`. When we run the build again, the image cache will be used to avoid rebuilding the image. On the build summary page, we can see the test result and the code coverage report:

Build #4 (Jan 26, 2021 1:57:38 AM)

Started by user Calvin Park (Admin)

Revision: 10db7c7ea98c53d3ad9f762317e7681d2a9e75f7

* refs/remotes/origin/main

Test Result (no failures)

Cobertura Coverage Report

Packages: 100% **Files**: 100% **Classes**: 100% **Lines**: 94% **Conditionals**: 100%

Figure 3.15 – Build summary page with the test result and code coverage report

That covers the basics of the static pipeline. Let's turn it into a premerge CI pipeline.

Creating a premerge CI pipeline with GitHub PR hooks

> **Required plugins**
> GitHub Pull Request Builder

A static pipeline is nice, but what we really need is a premerge CI pipeline that is triggered when a PR is created or updated. GitHub triggers a premerge build when a PR is created or updated, Jenkins runs the build, and Jenkins reports the status back to the PR in GitHub:

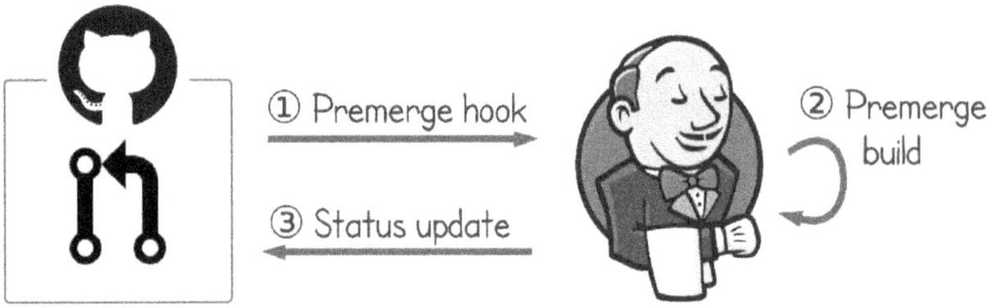

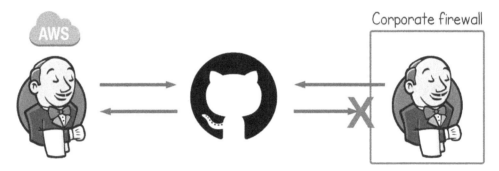

Figure 3.16 – Premerge build workflow

There are two ways to do this. For the AWS controller, GitHub can push the trigger directly since GitHub can reach the controller over the internet:

Figure 3.17 – Network access rules among GitHub, AWS Jenkins, and firewalled Jenkins

The firewalled controller, on the other hand, cannot be reached from GitHub, therefore we need a different way to trigger the build.

GitHub personal access token

Let's start by creating a personal access token from GitHub that is used by both the AWS controller and the firewalled controller:

1. Go to github.com | Settings (https://github.com/settings/profile) | Developer Settings | Personal access tokens | Generate new token. It should have the repo **scope:**

Figure 3.18 – GitHub personal access token with the repo scope

2. Save the GitHub personal access token as a credential in Jenkins. Go to Global Credentials **and click** Add Credentials. **Use the following configuration:**

- Kind: `Secret text`
- Scope: Global
- Secret: `<personal access token>`
- ID: `github-<GitHub Username>-pat`
- Description: `github-<GitHub Username>-pat`

3. Press OK **to save:**

Figure 3.19 – GitHub personal access token saved as a Jenkins credential

We will use this token in the coming sections.

> **GitHub App authentication is also available**
>
> GitHub App authentication is another way to manage the permissions of a robot account. Since the permissions are tied to a robot account rather than a user, the permissions can be reduced to just what is necessary. This can be useful in a large organization. Check it out if you find that using a user's personal access token isn't sufficient.

GitHub Pull Request Builder System Configuration

We can configure the common settings in System Configuration to avoid redundant work:

1. Go to System Configuration and find the GitHub Pull Request Builder section:

   - GitHub Server API URL: `https://api.github.com`

   - Credentials: `github-<GitHub Username>-pat`

   - Description: `<GitHub Username>`

   - Auto-manage webhooks: Uncheck

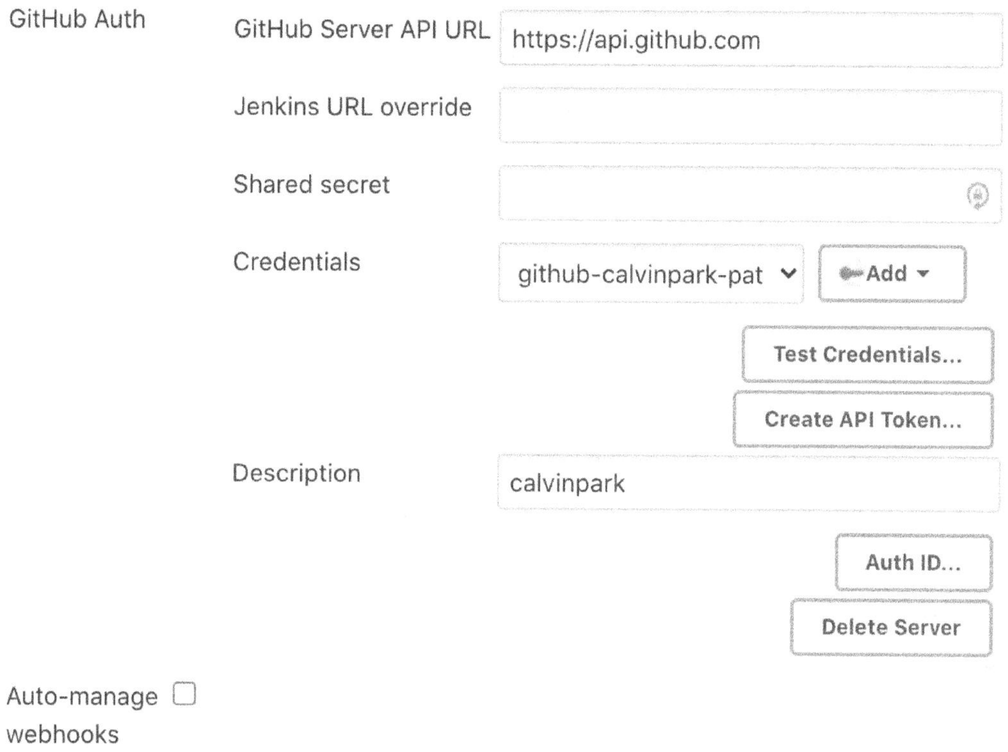

GitHub Auth		
	GitHub Server API URL	https://api.github.com
	Jenkins URL override	
	Shared secret	
	Credentials	github-calvinpark-pat ⌄ 🔑 Add ⌄
		Test Credentials...
		Create API Token...
	Description	calvinpark
		Auth ID...
		Delete Server

Auto-manage ☐
webhooks

*Figure 3.20 – GitHub Pull Request Builder – GitHub Auth and
Auto-manage webhooks configuration*

2. **Under** Application Setup, **add** Update commit status during build:

- Commit Status Context: `Jenkins CI`

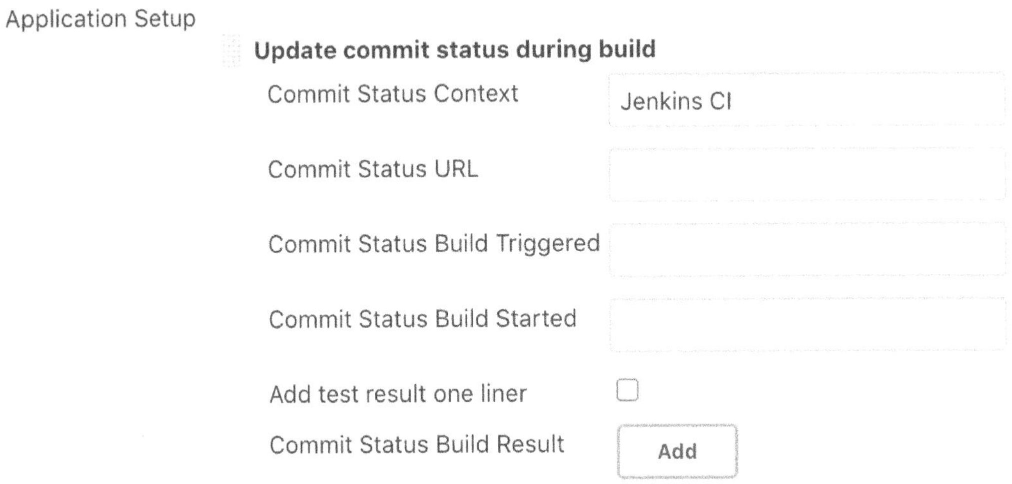

Application Setup

Update commit status during build

Commit Status Context	Jenkins CI
Commit Status URL	
Commit Status Build Triggered	
Commit Status Build Started	
Add test result one liner	☐
Commit Status Build Result	Add

Figure 3.21 – GitHub Pull Request Builder Application Setup configuration

This allows Jenkins to inspect the PR details when a hook is detected and update the commit with a build status after the build ends. We can use the Test Credentials button to validate the personal access token. Leave all the rest as the default and click Save.

Global configurations are now complete. Let's continue to set up the premerge trigger.

Configuring the premerge trigger

The premerge trigger for the AWS controller and the firewalled controller are slightly different. Let's take a look at the AWS controller first.

AWS controller

The AWS controller needs two configurations:

- GitHub must notify Jenkins when there's a PR event.
- Jenkins must listen for an event from GitHub and start a build.

Let's configure it:

1. Go to the adder repository in github.com, click Settings (this is the settings for the adder repository in https://github.com/<GitHub Username>/adder/ settings, not for our user account) | Webhooks | Add webhook:

    - Payload URL: https://jenkins-aws.lvin.ca/ghprbhook/ (the trailing slash is required)

- Let me select individual events:

  - **Check** Pull requests.

  - **Uncheck** Pushes.

This makes GitHub send a push hook to the URL when there is a PR change. Notice that it doesn't point to a specific pipeline. *All hooks go to one URL, and Jenkins distributes it to specific pipelines as needed.* **Click** Add webhook to save.

2. Now let's configure the adder premerge pipeline to listen to the push hook. Go to the pipeline and click Configure. **Find** GitHub project, check the checkbox, then enter the project URL:

- GitHub project | Project url: `<GitHub project URL>` (for example, `https://github.com/calvinpark/adder`)

Figure 3.22 – Jenkins adder premerge pipeline GitHub project configuration

3. Finally, let's configure the premerge pipeline to start a build when a push hook is detected. Find and check the box for GitHub Pull Request Builder:

- GitHub API credentials: **https://api.github.com** : `<GitHub Username>`.

- Admin list: `<GitHub Username>` (for example, `calvinpark`. This allows PRs from the specified GitHub user to start a build without additional checks. For information on allowing users not in the allow list to submit a PR and initiate a build, see the plugin home page at https://plugins.jenkins.io/ghprb/.)

- Use github hooks for build triggering: **Check**.

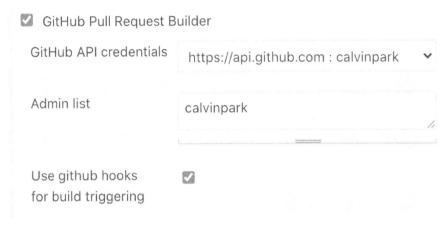

Figure 3.23 – AWS Jenkins adder premerge pipeline GitHub Pull Request Builder configuration

The premerge trigger is ready.

Let's take a look at how it's done for the firewalled controller.

Firewalled controller

The firewalled controller needs two configurations:

- Jenkins must poll GitHub at regular intervals to detect a PR event.

- Jenkins must listen for an event from GitHub and start a build.

Unlike the AWS controller, GitHub cannot send a push hook to the firewalled controller because the firewalled controller is, well, behind a firewall. We need to instead set up polling-based event detection:

1. Go to the adder premerge pipeline, click Configure, then find and check the box for GitHub Pull Request Builder:

   - GitHub API credentials: `https://api.github.com : <GitHub Username>`.

   - Admin list: `<GitHub Username>` (for example, `calvinpark`. This allows PRs from the specified GitHub user to start a build without additional checks. For information on allowing users not in the allow list to submit a PR and initiate a build, see the plugin home page at https://plugins.jenkins.io/ghprb/.)

   - Advanced | Crontab line: `H/2 * * * *` (this makes the plugin poll the status every 2 minutes. Every 2 minutes can be a bit too frequent for Jenkins, especially if there are hundreds of pipelines, but at the same time is a long time for someone to wait for a premerge CI build to start. It may make sense to make it every minute during the setup, then make it slower once it's established. Also, the H in this context stands for hash. This reduces the load spike on Jenkins by distributing the cron start times of pipelines – more on this can be found in the User Handbook (https://www.jenkins.io/doc/book/pipeline/syntax/#cron-syntax)):

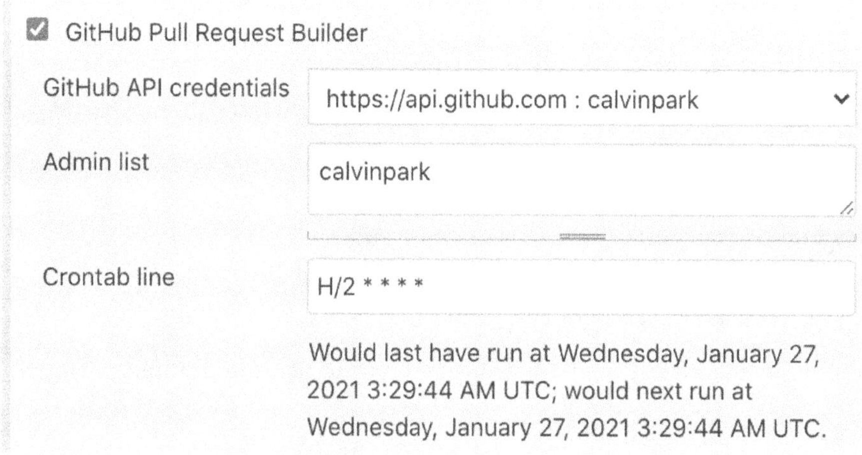

☑ GitHub Pull Request Builder

GitHub API credentials	https://api.github.com : calvinpark ⌄
Admin list	calvinpark
Crontab line	H/2 * * * *

Would last have run at Wednesday, January 27, 2021 3:29:44 AM UTC; would next run at Wednesday, January 27, 2021 3:29:44 AM UTC.

Figure 3.24 – Firewalled Jenkins adder premerge pipeline GitHub Pull Request Builder configuration. Unmodified fields were hidden for the screenshot

---

Build not triggering?

Due to a long-standing bug[1], H/1 * * * * runs every hour, not every minute.

Every minute should be written as * * * * * instead.

---

2. Now let's configure the adder premerge pipeline to start a build when a PR activity is detected. Find GitHub project, check the checkbox, then enter the project URL:

- GitHub project | Project url: `<GitHub project URL>` (for example, `https://github.com/calvinpark/adder`)

---

1 https://issues.jenkins.io/browse/JENKINS-22129

Figure 3.25 – Jenkins adder premerge pipeline GitHub project configuration

The premerge trigger is ready.

Let's try out the new triggers.

Testing the premerge trigger

Let's create a new PR in the adder Git repository to test the premerge trigger:

1. Add an echo to the Jenkinsfile:

Jenkinsfile

```
stage('Hello GitHub') {
    steps {
        echo "Hello GitHub!"
}   }
```

2. Make a new branch, commit, and push:

```
~/adder $ git checkout -b hello
~/adder $ git commit -am "Hello GitHub"
~/adder $ git push --set-upstream origin hello
[...]
remote: Create a pull request for 'hello' on GitHub by visiting:
remote:      https://github.com/calvinpark/adder/pull/new/hello
```

3. Follow the link on the console output to make a PR on GitHub, then wait for the adder premerge pipeline to start a build on its own. If a build doesn't start, examine the controller logs for error messages:

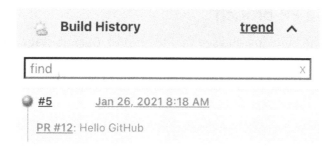

Figure 3.26 – Premerge build triggered by a PR creation. "PR #12" links to the GitHub PR page

The build is successful, and we can see on the GitHub PR page that it now has a successful check named Jenkins CI. If we click Details, it even takes us back to the Jenkins build:

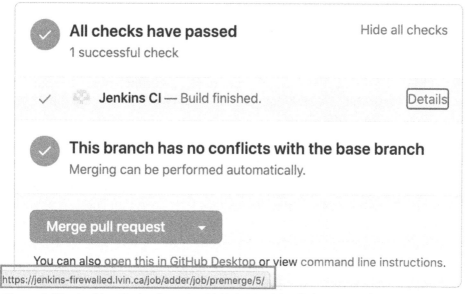

Figure 3.27 – Jenkins CI check on the GitHub PR with a link pointing back to the build

Success! Or is it?

Building the PR branch

Take a look at the build log and see that the Hello GitHub stage didn't run. Let's examine the build log to see why not:

```
GitHub pull request #1 of commit
69cc5fc003b114eda7abf46d72caa2c824a2b699, no merge conflicts.
[...]
 > git rev-parse refs/remotes/origin/main^{commit} # timeout=10
Checking out Revision a64306606c060bfad0891feca4e94ec06b7e5703
(refs/remotes/origin/main)
 > git config core.sparsecheckout # timeout=10
 > git checkout -f a64306606c060bfad0891feca4e94ec06b7e5703 #
timeout=10
```

The first line shows that the right PR (#1) was detected along with its commit sha1 (69cc5). However, a few lines below, it shows that the build looked for the main branch and checked out the sha1 for the main branch (a6430). Since the main branch doesn't have the Hello GitHub stage (our hello PR branch does), the build didn't run the Hello GitHub stage. What we need is to have the build use the PR branch rather than the main branch.

Go back to the pipeline and click Configure. Find the Pipeline | Definition | SCM section and apply the following configurations:

- Repositories | Advanced | Refspec (This is one long space-delimited line. See the book's GitHub repository for a text file with this line that you can copy): +refs/heads/*:refs/remotes/origin/* +refs/pull/${ghprbPullId}/*:refs/remotes/origin/pr/${ghprbPullId}/*
- Branches to build | Branch Specifier: ${ghprbActualCommit}
- Lightweight checkout: Uncheck

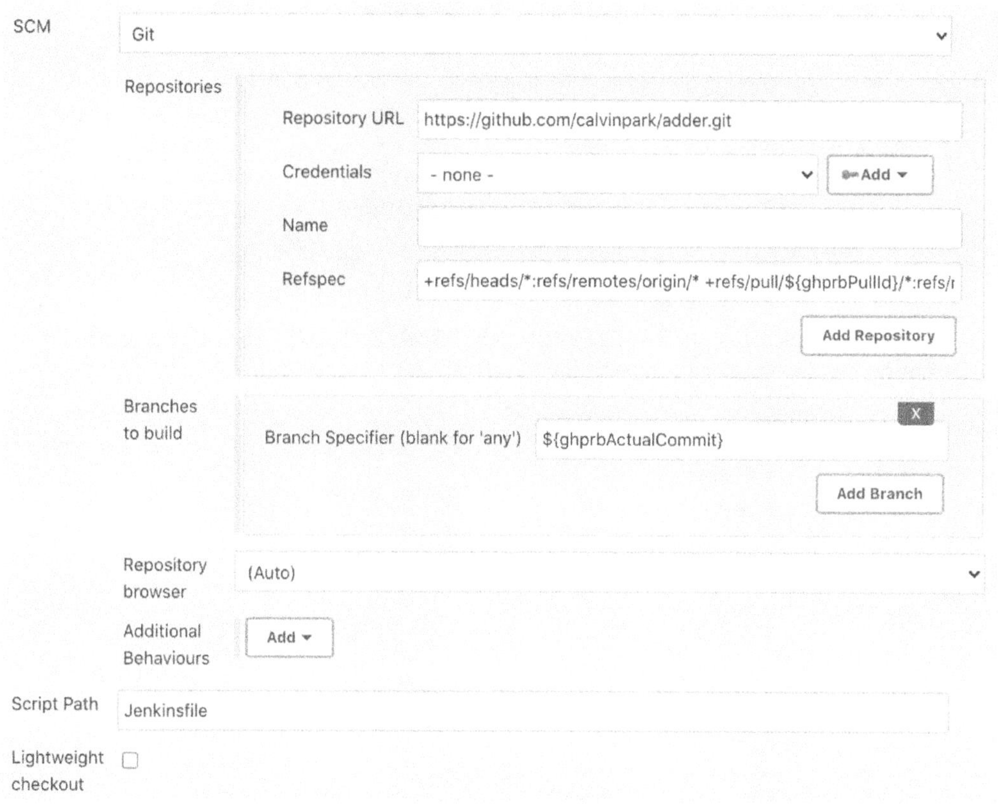

Figure 3.28 – SCM configuration to build the PR commit

This does a few things. First, it makes the build check out the sha1 contained in the ghprbActualCommit variable, which is populated by the GitHub Pull Request Builder plugin when a new hook is received. There is also an option to use a sha1 variable instead of ghprbActualCommit if we want to use GitHub's merge head rather than the actual commit that we've pushed. Second, it makes sure that the commit is fetched through the modified Refspec so that it's available to check out. Refspec is the set of rules that Git uses to determine which commits to download upon a clone. The change we've made ensures that our PR branch is downloaded. Finally, it makes sure that the commit is not pruned by a shallow clone. Save and rerun the build by pushing a new commit:

```
~/adder $ git commit -m "empty commit" --allow-empty
~/adder $ git push
```

This time, the `Hello GitHub` stage ran:

```
GitHub pull request #1 of commit
36077f946bc2562348f3c46b757554b0188d0d9a, no merge conflicts.
[...]
Checking out Revision 36077f946bc2562348f3c46b757554b0188d0d9a
(detached)
 > git config core.sparsecheckout # timeout=10
 > git checkout -f 36077f946bc2562348f3c46b757554b0188d0d9a #
timeout=10
Commit message: "empty commit"
[...]
[Pipeline] stage
[Pipeline] { (Hello GitHub)
[Pipeline] echo
Hello GitHub!
[...]
Finished: SUCCESS
```

Success! Or is it?

Building an arbitrary branch

Go back to the pipeline page and click Build Now to run without a PR. It will fail with the message that the `ghprbActualCommit` variable doesn't point to a Git commit:

```
 > git rev-parse ${ghprbActualCommit}^{commit} # timeout=10
ERROR: Couldn't find any revision to build. Verify the
repository and branch configuration for this job.
```

This means that while the pipeline works well as a PR validation, it doesn't allow us to run on an arbitrary branch for testing. In many cases this is fine, and in some cases, we want to actively prevent the `premerge` pipeline from being used outside of the PR validation.

If we want to use the premerge pipeline to build an *arbitrary branch or a commit*, we do the following:

1. Modify the Jenkinsfile and add a parameters directive. This should go on the same level as the agent and stages directives:

Jenkinsfile

```
pipeline {
    agent { dockerfile { label 'docker' } }

    parameters {
        string(name: 'REF', defaultValue:
'\${ghprbActualCommit}', description: 'Commit to build')
    }

    stages {
        stage('Hello GitHub') { ... }
[...]
```

2. Make a new commit, push, and *merge the PR.*

> **Be sure to merge!**
> If you skip this step, the pipeline gets into an in-between state with an invalid configuration.

3. Go to the pipeline, Configure, and under Definition | SCM | Branches to build | Branch Specifier, enter ${REF}:

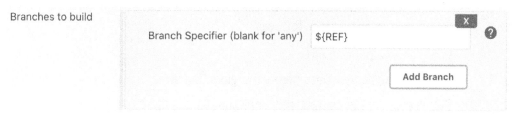

Figure 3.29 – Enter ${REF} in Branch Specifier to build the branch from the parameter

4. Running the pipeline will now ask us for a value for the REF parameter. Enter any branch name you want to build and click Build:

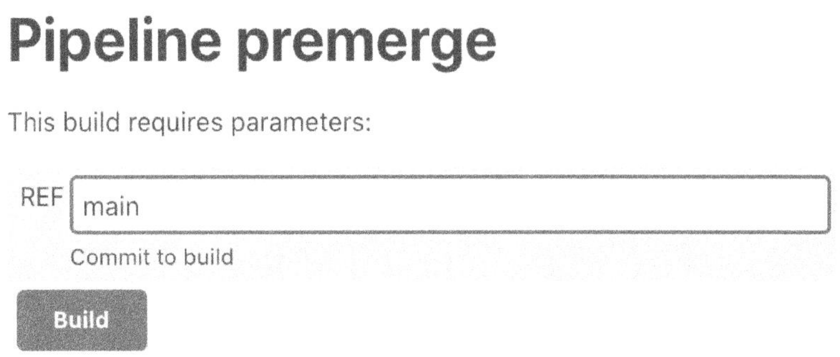

Pipeline premerge

This build requires parameters:

REF | main |

Commit to build

Build

Figure 3.30 – The REF build parameter for building an arbitrary branch

The pipeline will build the specified main branch.

The CI premerge pipeline is now complete. In addition to being triggered upon a PR activity, it can also run against an arbitrary branch. We have just one more configuration left.

Requiring a successful build for a merge

Let's finish off by making sure that merging a PR requires a successful premerge build. We want to check for possible mistakes and prevent shipping broken functionalities. Protected branches can be used to ensure that the code is thoroughly tested and reviewed:

1. Go to the **adder** repository in GitHub | Settings | Branches | Branch protection rules |Add rule. **Configure as follows:**

   - Branch name pattern: `main`

   - Require status check pass before merging: **Check**

   - Status checks that are required: `Jenkins CI`

2. Click Create **to save.**

3. Create a new PR to see that it now requires a successful build of Jenkins CI for a merge:

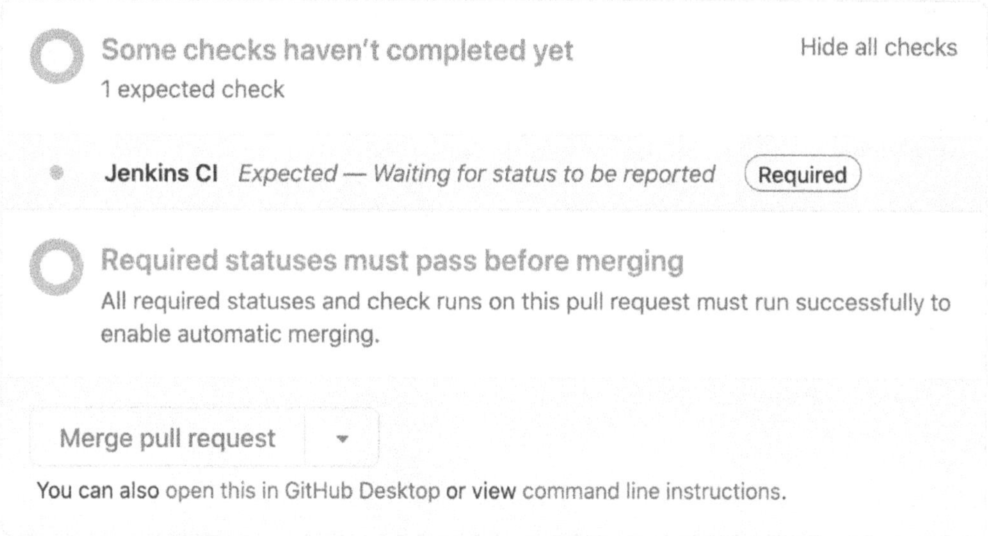

○ **Some checks haven't completed yet**   Hide all checks
   1 expected check

   •   **Jenkins CI** *Expected — Waiting for status to be reported*   (Required)

○ **Required statuses must pass before merging**
   All required statuses and check runs on this pull request must run successfully to
   enable automatic merging.

   [Merge pull request   ▾]

   You can also open this in GitHub Desktop **or view** command line instructions.

Figure 3.31 – PR waiting for a successful premerge build result before allowing a merge

Now our adder Git repository's main branch is protected with a validation
requirement that is set from automatically triggered premerge builds.

What about subtractor?

subtractor can be configured the same way as adder so we
have skipped the setup details. The most important learning from
subtractor is that adder users and subtractor users can't see
the projects outside of their permissions. This concept is revisited in
Chapter 10, Script Security, so keep the subtractor folder around
until you've gone through that chapter.

Summary

In this chapter, we've learned how to control user roles and permissions based on folders, which allows multiple teams to share a Jenkins instance. We've created a static pipeline equipped with unit tests and code coverage, and their outputs were nicely presented on the Jenkins UI. We've converted the static pipeline into a premerge CI pipeline that is triggered by GitHub PR activities, which allows us to validate the changes before a PR is merged. Finally, we've protected the Git repository's main branch by requiring a successful premerge build for a merge.

In the coming chapter, we will expand on the same idea and build a more advanced postmerge CD pipeline. Our work on advanced pipelines will feature practical activities such as versioning, Git tagging, Docker image creation, and pushing using Docker-out-of-Docker (DooD or DinD depending on the context).

4

GitOps-Driven CD Pipeline with Docker Hub and More Jenkinsfile Features

The chapter title is quite a mouthful. Let's unpack it to understand what we'll learn in this chapter.

We will create a postmerge continuous delivery (CD) pipeline that is triggered when a pull request (PR) is merged. The crux of it is the same as the premerge CI pipeline from *Chapter 3, GitOps-Driven CI Pipeline with GitHub* – listen for an event from GitHub and build accordingly.

Unlike the premerge pipeline, which was entirely self-contained, the postmerge pipeline will interact with external systems to resemble a pipeline for a real product more closely. It will generate a tag based on the list of current tags in the Git repository, build a Docker image as an artifact, log in to Docker Hub (the primary Docker registry at https://hub.docker.com), push the image to Docker Hub, and then push the tag back to GitHub.

In implementing the tasks, we will use additional features of a Jenkinsfile to explore various ways of handling common problems such as credential management, versioning, tagging, and Docker-outside-of-Docker (DooD). These features will get pretty complex, so get ready for it.

By the end of the chapter, we will have learned how to build a pipeline that can handle real-world product build and release scenarios.

In this chapter, we're going to cover the following main topics:

- Project overview
- Packaging the Docker image and running integration tests
- Versioning Git and Docker using Semantic Versioning
- Using more Jenkinsfile features with DooD and bare-metal agents
- Creating a static pipeline for packaging, integration tests, and delivery
- Creating a postmerge CD pipeline with a GitHub webhook and polling

Technical requirements

You need the GitHub account where you created the adder repository in *Chapter 3, GitOps-Driven CI Pipeline with GitHub*. This time, we'll be pushing tags back to the repository, so you need to create an SSH key and store it in your GitHub account. You also need a Docker Hub account where we'll be pushing adder Docker images.

Files in the chapter are available on GitHub at https://github.com/PacktPublishing/Jenkins-Administrators-Guide/blob/main/ch4.

Project overview

In this chapter, we will build a CD pipeline. Approving and merging a PR triggers a postmerge build in Jenkins to compile and package the project, run integration tests, apply version tags to the Git repositories and the artifacts, and deliver the artifacts (a Docker image in our case) to Docker Hub. At the end of the build, Jenkins pushes the Git tag to the GitHub repository:

Figure 4.1 – Postmerge CD build workflow

We will examine the GitHub build triggers for both AWS Jenkins, where GitHub has bidirectional communications, as well as firewalled Jenkins, where Jenkins can reach GitHub but GitHub cannot reach Jenkins:

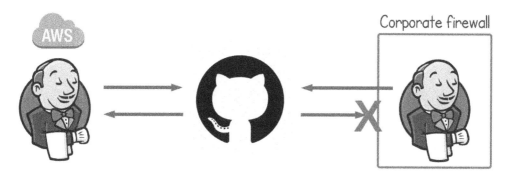

Figure 4.2 – Different access rules from GitHub to AWS Jenkins and firewalled Jenkins

In order to build this postmerge CD pipeline, we need to create and configure the following:

- Create build scripts and Dockerfiles that build, test, tag, and deploy the artifacts.

- Create a Jenkinsfile that calls the build scripts and Dockerfiles.

- Create a pipeline that uses the Jenkinsfile.

- Configure the pipeline to listen for PR merge events in GitHub, then start a build. AWS Jenkins and firewalled Jenkins will have different ways to listen.

- Configure GitHub repositories to notify the Jenkins pipeline.

In addition, we will learn about more complex syntax in Jenkinsfiles to allow a more versatile configuration, so that your actual product pipelines can handle more complex real-world scenarios.

The Jenkinsfile we'll develop uses many Dockerfiles and shell scripts across several stages, so it's easy to get lost in the process. Here's a *map* of the Jenkinsfile that shows all of its components, which you can reference as we progress through the chapter:

agent none

options / buildDiscarder

environment / credentials('dockerhub')

stages

    stage ('Get next version')

        agent label 'docker'

           git.dockerfile

               $ get-next-version.sh

    stage ('Get Docker group')

        agent label 'firewalled-firewalled-agent'

           $ get docker group ID

    stage ('Package, Test, & Deliver')

        agent label 'firewalled-firewalled-agent'

           cd-env.dockerfile

               $ package.sh builds adder.dockerfile

               $ integration-test.sh tests
                       calvinpark/adder Docker image

    stage ('Push Git tag')

        agent label 'docker'

           git.dockerfile

               $ git tag & push

post

    success

        Update build description

Figure 4.3 – Anatomy of postmerge.jenkinsfile

Are you excited? Let's get started!

Packaging the Docker image and running integration tests

In *Chapter 3, GitOps-Driven CI Pipeline with GitHub*, we learned how to run a CI build using Python. The CD builds will use shell scripts to learn a more generic way of running processes:

The files can be downloaded from the book's GitHub repository

There are many files in this chapter. Typing them out will help you learn better, but also consider downloading the files you already understand.

---

1. Let's start by creating a Dockerfile that defines our final deliverable Docker image. It's a simple Python 3 image with adder.py in it. entrypoint is overridden to run adder.py by default:

adder.dockerfile

```
FROM python:3
COPY adder.py /
ENTRYPOINT ["python", "/adder.py"]
```

2. Next, create a packaging shell script that builds the Docker image. The packaging script names the image <Docker Hub ID>/adder and uses a tag from the optional argument that defaults to latest.

   Replace calvinpark with your own Docker Hub ID:

package.sh

```
#!/bin/bash
set -e
VERSION=${1:-latest}
docker build -f adder.dockerfile -t calvinpark/adder:${VERSION} .
```

Run the script to build and tag the image:

```
~/adder $ ./package.sh
[...]
 => => naming to docker.io/calvinpark/adder:latest
~/adder $ ./package.sh customTag
[...]
 => => naming to docker.io/calvinpark/adder:customTag
```

3. Next, add an integration test shell script that runs the adder image with user input to validate the container:

integration-test.sh

```
#!/bin/bash
set -e

FIRST=$1
SECOND=$2
EXPECTED=$3
VERSION=${4:-latest}

SUM=$(docker run calvinpark/adder:${VERSION} ${FIRST} ${SECOND})

if [[ "${SUM}" == "${EXPECTED}" ]]; then
    echo "Integration test passed"
else
    echo "[ERROR] ${FIRST} + ${SECOND} returned ${SUM}, not
${EXPECTED}" >&2
    exit 1
fi
```

Jenkins treats a build process as a failure if the process doesn't return 0 as it exits. When a failed test exits with return code 1, Jenkins detects the error and marks the CD build as a failure:

```
~/adder $ ./integration-test.sh 13 42 55
Integration test passed

~/adder $ ./integration-test.sh 13 42 -1
[ERROR] 13 + 42 returned 55, not -1

$ echo $?
1
```

4. Finally, create a Dockerfile for an image that will be used as a CD build environment. It's an Ubuntu 20.04 image with Docker and curl installed.

We could have directly used the official docker image, but you can use this Dockerfile as a template for your actual project and customize it with additional tools:

cd-env.dockerfile

```
FROM ubuntu:20.04

RUN apt-get -qq update -y && \
    apt-get -qq install -y curl \
    && rm -rf /var/lib/apt/lists/*

RUN curl -fsSL https://get.docker.com | sh
```

The four files will be used shortly when we're putting together a CD pipeline. Before that, let's look at versioning.

Versioning Git and Docker using Semantic Versioning

We should version our deliverables so that a new version is differentiated from the old versions. Semantic Versioning (SemVer) (https://semver.org/) is a popular versioning format that looks like `major.minor.patch+build`, for example, `1.25.0+rc2`. We will version our deliverables by bumping the minor version on each PR merge:

1. Create a versioning script that inspects the current list of tags available in the Git repository and generates the next version by bumping the minor version. Run `chmod a+x gen-next-version.sh` afterward to make it executable:

get-next-version.sh

```
#!/bin/bash

set -e -o pipefail

next_version=$(git tag -l | egrep '^[0-9]+\.[0-9]+\.[0-9]+$' | sort
--version-sort | tail -1 | awk -F . '{print $1 "." $2+1 ".0"}')

if [ -z "${next_version}" ]; then
    next_version='0.1.0'
fi

echo "${next_version}"
```

Running this script on a Git repository without any SemVer tags yields 0.1.0. Once the repository is tagged with 0.1.0, running it again will yield 0.2.0:

```
~/adder $ ./get-next-version.sh
0.1.0
~/adder $ git tag 0.1.0
~/adder $ ./get-next-version.sh
0.2.0
```

There are many existing SemVer libraries, such as Mark Challoner's git-semver (https://github.com/markchalloner/git-semver), which supports more advanced use cases if you need a more robust and featureful solution. We will use the preceding simple script to learn the additional features of the Jenkinsfile in the next section.

2. Create a Dockerfile for an image based on Ubuntu 20.04 with git installed that defines the environment where the Git commands will run. Notice that in this image we are creating a user with a UID that matches the jenkins user's UID. This is because the git push command uses SSH, which requires a user to exist.

> **UID change alert**
> AWS Jenkins should change --uid 123 to --uid 1000, as well as the email address and the name.

Finally, the user.email and user.name variables are defined as they are also required by the git commit command:

git.dockerfile

```
FROM ubuntu:20.04

RUN apt-get -qq update -y && apt-get -qq install -y git \
    && rm -rf /var/lib/apt/lists/*

RUN adduser --disabled-password --gecos "" \
    --uid 123 --shell /bin/bash jenkins

RUN git config --system user.email robot_acct@example.com
RUN git config --system user.name "Robot Account"
```

We can build the image and test the scripts inside the container. By mounting the current directory and running the script in it, we can run git tag *inside* the container and have the tag be applied to the Git repository *outside* of the container. The following commands show that generating the next version and tagging inside the container resulted in new Git tags applied to the repository outside of the container:

```
~/adder $ docker build -f git.dockerfile . -t calvinpark/git
[...]
 => => naming to docker.io/calvinpark/git

          # Running inside the container
~/adder $ docker run -it -v $(pwd):$(pwd) -w $(pwd) \
             calvinpark/git ./get-next-version.sh
0.2.0

~/adder $ docker run -it -v $(pwd):$(pwd) -w $(pwd) \
             calvinpark/git git tag 0.2.0

~/adder $ docker run -it -v $(pwd):$(pwd) -w $(pwd) \
             calvinpark/git ./get-next-version.sh
0.3.0

          # Running outside the container
~/adder $ git tag -l
0.1.0
0.2.0
```

3. Create a new secret in GitHub and Jenkins so that we can push a Git tag during a postmerge build.

   Go to github.com and log in. Click the avatar image on the upper-right corner | Settings | SSH and GPG keys | New SSH key. Enter your SSH public key and click Add SSH key to save.

   Now, come back to Jenkins and create a credential in the adder folder. By creating this in the adder folder rather than the root folder, we're putting a restriction that only the adder pipelines can use this credential. Click the adder folder, then Credentials on the left, then (global) for the adder folder, not the parent Jenkins folder, and then finally Add Credentials:

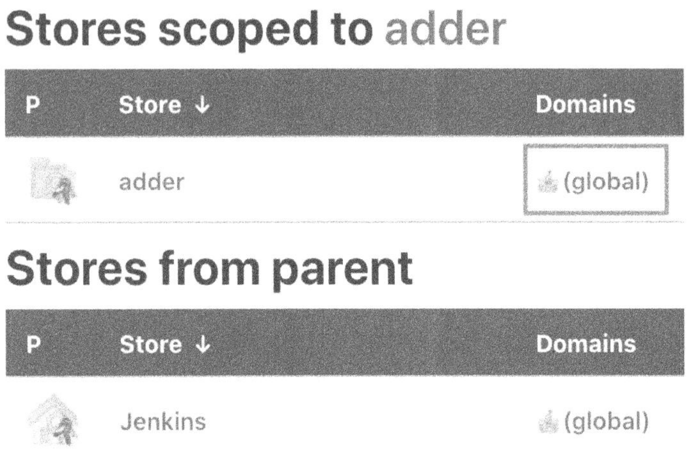

Stores scoped to adder

P	Store ↓	Domains
	adder	(global)

Stores from parent

P	Store ↓	Domains
	Jenkins	(global)

Figure 4.4 – Click (global) for the adder folder, not the parent Jenkins folder

Configure the credentials as follows:

- Kind: SSH Username with private key
- ID: `github-<GitHub Username>-priv` (for example, `github-calvinpark-priv`)
- Description: `github-<GitHub Username>-priv`
- Username: `<GitHub Username>` (for example, `calvinpark`)
- Private Key: SSH private key for the SSH public key that we saved in GitHub

4. Finally, create a new secret in Docker Hub and Jenkins so that we can push a Docker image during a postmerge build.

   Go to hub.docker.com and log in. Click the username on the upper-right corner | Account Settings | Security | New Access Token. Enter `Jenkins adder/postmerge` as a description, then click Create. Docker Hub shows us the new personal access token – we can use this in a `docker login` command. Copy and paste the token in a text editor and click Copy and Close.

   Come back to Jenkins and go to the adder folder's credential store just as we did in the previous step. Click Add Credentials and create a credential as follows:

   - Kind: Username with password
   - Username: `<Docker ID>` (for example, `calvinpark`)
   - Password: The Docker Hub personal access token that we just created
   - ID: `dockerhub-<Docker ID>-userpass` (for example, `dockerhub-calvinpark-userpass`)

- Description: `dockerhub-<Docker ID>-userpass`

We have now created all the build scripts and the credentials to be used in the postmerge CD pipeline. Let's continue to put them together using a Jenkinsfile.

Using more Jenkinsfile features with DooD and bare-metal agents

In the Jenkinsfile for the CD pipeline, we will explore additional Jenkinsfile syntax. We will examine the content of the postmerge `postmerge.jenkinsfile` pipeline in multiple small sections – the full file can be found in the book's GitHub repository. Let's start with the top-level Jenkinsfile directives.

agent none, buildDiscarder options, and credentials in environment variables

Let's look at the top-level `pipeline`, `agent`, `options`, `environment`, `stages`, and `post` directives.

`agent none` on the second line declares that each stage uses its own agent rather than using one agent for the entire build. It's also possible to specify a global agent here, and then override with specific agents on select stages:

```
pipeline {
    agent none
    options {
        buildDiscarder(logRotator(numToKeepStr: '1000'))
    }
    environment {
        DOCKER = credentials('dockerhub-calvinpark-userpass')
    }
    stages { [...] }
    post { [...] }
}
```

The `options/buildDiscarder` directive overrides the global default of 100-build history retention that we set in *Chapter 2, Jenkins with Docker on HTTPS on AWS and inside a Corporate Firewall*. Since the build logs for a postmerge CD pipeline can be a good auditing tool, we're keeping a longer build history.

> **Do not disable buildDiscarder**
>
> It's possible to disable `buildDiscarder` by setting `numToKeepStr` to `'-1'`. This is a bad idea because infinity is a very large number. Think of a very high retention count such as 10,000 builds. That's a ridiculously large number of builds, but infinity is much larger than 10,000. If there is a bug that submits a new build in a loop, the 10,000 retention count will stop the controller's disk from filling up, whereas an infinite retention count will fill the disk and break the controller. In practice, it's very rare to need an over-1,000-build *history*. Remember that this number doesn't restrict how frequently our builds can run – it merely restricts how many builds we can look back upon to check the logs. If you truly need a record of the history of every build ever, do not rely on Jenkins for history keeping, but instead archive the build information outside of Jenkins. Jenkins is not the right tool for keeping a record of an infinite number of builds.

The environment directive allows us to set the environment variables for the build, and also allows us to load Jenkins credentials into the variables. With `DOCKER = credentials('id')`, three environment variables are created:

- `DOCKER` contains `username:password`
- `DOCKER_USR` contains `username`
- `DOCKER_PSW` contains `password`

> **_USR and _PSW are always uppercase**
>
> Even with a variable name in lowercase, _USR and _PSW will be appended as uppercase, for example, docker, docker_USR, and docker_PSW.

This is only the beginning of how we can customize a pipeline using the `options` and `environment` directives. Take a look at the official Jenkins Handbook for the complete list of choices (https://www.jenkins.io/doc/book/pipeline/syntax/#options). Let's continue to the stages.

Using a custom Dockerfile for a dockerfile agent and running Groovy code in a script block

The first stage uses a dockerfile agent, which was covered in *Chapter 3, GitOps-Driven CI Pipeline with GitHub*, but this time with a custom filename. In a project with multiple Dockerfiles, we can organize and specify the Dockerfile using the filename parameter. Information on additional parameters such as buildArgs and registryUrl can be found in the official Jenkins Handbook (https://www.jenkins.io/doc/book/pipeline/syntax/#agent). Take note that we didn't have to mount any volumes or specify the work directory, unlike the commands we ran in the previous section. This is because Jenkins already handles both for us. It's quite convenient:

```
stage('Get next version') {
    agent {
        dockerfile {
            label 'docker'
            filename 'git.dockerfile'
        }   }
    steps {
        script {
            version_g = sh (
                script: 'sh get-next-version.sh',
                returnStdout: true
            ).trim()

            echo "Next version is ${version_g}"
            sh "echo -n ${version_g} > version_f"
}   }   }
```

In steps, we call get-next-version.sh to generate the next version number and store the result in a version_g Groovy variable so that it can be used in later stages. The access to the Groovy environment is made possible by the surrounding script directive. We also echo the version and write it in a file named version_f as another way to pass the value to later stages. This is a bit of a contrived example, but the code demonstrates how we can pick up the output from a shell command, save it as a Groovy variable, and use the Groovy variable in an echo pipeline step, then again in a new shell command. We'll see how the version_g Groovy variable and the version_f file are accessed from the later stages in the coming sections.

Let's continue to perhaps the most anticipated topic, DooD.

Docker-outside-of-Docker in Jenkins

Before jumping into DooD in Jenkins, let's go over how DooD permission works in general.

Understanding the Docker permissions requirements for DooD

Docker uses a client-server architecture, which means that it has two distinct software packages for the client and the server. The client software is the docker binary that we frequently use, and the server software is dockerd, which is the Docker daemon. When we issue a command such as docker ps, the client software docker binary contacts the server software dockerd and requests to run ps. The /var/run/docker.sock Docker socket is the contact point where the Docker daemon listens, thus the socket is also the location where the Docker client contacts to make the request.

An important fact is that the *Docker socket is a file*, just like everything else in Linux. This means that *running the* docker run *command* is actually using the docker binary to *write the* run *command to the Docker socket file*.

Writing to a file requires that the user who's initiating the write has the write permission to the destination file. In the case of the Docker socket, /var/run/docker.sock has srw-rw---- permissions and is owned by the root user and the docker group. This means that the user who's initiating the write must either be root or a member of the docker group:

```
agent:~$ ls -la /var/run/docker.sock
srw-rw---- 1 root docker 0 Mar 12 05:33 /var/run/docker.sock
agent:~$ id
uid=123(robot_acct) gid=30(dip) groups=30(dip),998(docker)
```

We made our VMs in *Chapter 1, Jenkins Infrastructure with TLS/SSL and Reverse Proxy*, such that the agent user (ubuntu or robot_acct) is a part of the docker group, so that the Jenkins agent can write to the Docker socket and use the Docker client to create containers. The permission rules are pretty straightforward for normal docker uses such as this:

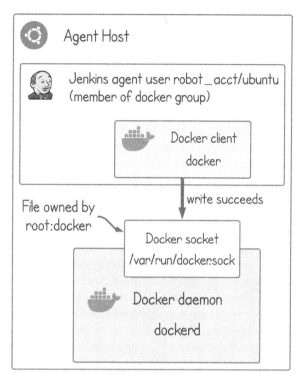

*Figure 4.5 – The Jenkins agent uses the Docker client to reach
the Docker daemon through the Docker socket*

The complexity comes when we try to run Docker commands inside a Docker container. In order for a container to run Docker commands, the container must have access to the Docker socket. We provide the socket to the container by bind mounting the socket from the host into the container, and that allows the container to use the Docker client to write to the socket. This is the essence of DooD.

The trouble is that when a container is created, the user inside the container is not the agent user. Since the container user is not the agent user, the container user is also not a member of the docker group. When the container user runs docker ps, the Docker client attempts to write to the Docker socket, but the write fails because the container user doesn't have the write permission to the socket:

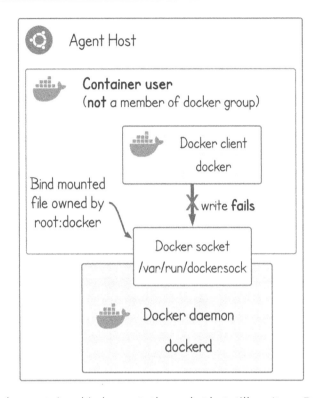

Figure 4.6 – Docker container bind mounts the socket but still can't run Docker commands because the container user doesn't have permission to write to the socket

The solution to this problem is rather simple – modify the container user's GID or group membership during docker run so that it gains write permission to the socket. Let's see how we can do this.

Start by finding the GID of the Docker socket. In our case, the GID is 998. We will make sure that the container user has a GID of 998 or a group membership of 998:

```
agent:~$ stat -c '%g' /var/run/docker.sock
998
```

There are two possible approaches. The first way is using the --user switch to specify the UID and GID of the container user. Running docker run --user $(id -u):998 docker id sets the UID and the GID of the container user as the current user's UID and 998:

```
agent:~$ docker run -it --user $(id -u):998 docker id
uid=1000 gid=998
```

The second way is using the `--group-add` switch to add group membership to the user inside the container. Unlike the previous command, `docker run --group-add 998 docker id` keeps the existing UID and GID and simply adds 998 to the list of groups. In the following output, you can see that the GID remained as 0, unlike the previous output, which changed it to 998. This approach is preferred because it's less intrusive than the previous approach:

```
agent:~$ docker run -it --group-add 998 docker id
uid=0(root) gid=0(root) groups=0(root),1(bin),2(daemon),3(sys),
4(adm),6(disk),10(wheel),11(floppy),20(dialout),26(tape),
27(video),998
```

Either way, the essence of the solution is that we're enabling the container user to have a GID or a group membership that grants the write permission to the bind-mounted Docker socket:

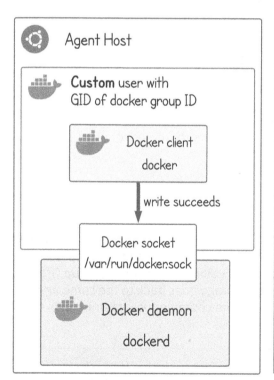
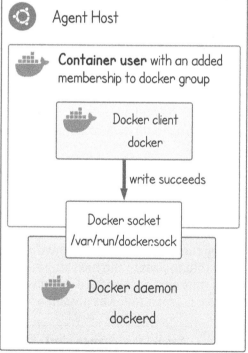

Figure 4.7 – Docker container running as a user with either a GID of the docker group ID or added membership to the docker group can run Docker commands because the container user has permission to write to the socket

If this is difficult to follow, just keep this in mind: *the user inside a DooD container must have write permissions to the /var/run/docker.sock file.*

Now, let's take a look at how this applies to Jenkins.

As we discussed before, we need to make sure that the user inside the DooD container has permission to write to the Docker socket. This has to be done in two parts: first, figure out which group owns the Docker socket, and then use the --group-add switch to add the group membership to the user inside the container:

```
stage('Get Docker group') {
    agent { label 'firewalled-firewalled-agent' }
    steps {
        script {
            docker_group = sh (
                script: "stat -c '%g' /var/run/docker.sock",
                returnStdout: true
            ).trim()
}    }   }
stage('Package, Test, & Deliver') {
    agent {
        dockerfile {
            label 'firewalled-firewalled-agent'
            filename 'cd-env.dockerfile'
            args "-v /var/run/docker.sock:/var/run/docker.sock:rw
    --group-add ${docker_group}"
        }   }
    steps { [...] }
}
```

The two stages in the preceding snippet perform the two parts. The first stage picks a specific agent (firewalled-firewalled-agent in this case) and looks up the group ID by running a shell script, and then stores the value in the docker_group Groovy variable to pass to the next stage. The second stage goes back to the same agent and runs on a dockerfile agent while giving the container user membership to the docker_group that was passed from the previous stage.

Agent label change alert
AWS Jenkins should change firewalled-firewalled-agent to aws-aws-agent.

There are two disadvantages of this approach:

- The stages are pinned to use a specific agent.
- Two stages are required.

There are three ways to mitigate this:

- If all agents in our Jenkins have the same group ID for the docker group, we can skip the entire first stage and hardcode the group ID on the second. This way, we don't need two stages and don't need to pin them to a specific agent.

- Another option is to change the group that owns the Docker socket to the primary GID of the agent user (1000 for AWS Jenkins and 30 for firewalled Jenkins). To be specific, we would run chgrp $(id -g) /var/run/docker.sock on all agents. In Jenkins DooD, the container user has the same GID as the agent user by default, so if the socket is owned by the agent user's GID, the user inside the container has write permissions to the socket natively. This way, you can skip the first stage, don't pin to a specific agent, and even omit the --group-add flag entirely.

- The final option is to change the primary GID of the agent user to the docker group's ID (998). This is the same idea as the second option except this time, we're going the opposite way by modifying the agent user's GID rather than the Docker socket's group ID. This option is the most difficult to implement because it requires that all agents in our Jenkins have the same group ID for the docker group. In addition, it must be possible to modify the primary GID of the agent user – if the robot account's GID was set by Active Directory, we won't be able to modify it.

> **Alternate DooD and DinD methods**
>
> There are even more options such as using a proper Docker-in-Docker (DinD) through a Sysbox container runtime if mounting the Docker socket is considered a security issue. DinD and DooD techniques are still evolving, so search online for more information.

That covers the intricacies and pitfalls of running DooD on Jenkins. It's not a very polished process, but it's very powerful so it'll be difficult to ignore.

Let's continue to learn how to log in to Docker Hub and push an image.

Variable handling, Docker Hub login, and docker push

Once we're inside the Docker container created from cd-env.dockerfile, we run several shell commands. By surrounding the commands with triple quotes (''' or """), each line gets interpreted as one shell command, as if we're on a shell typing one line at a time and pressing *Enter*:

```
steps {
    sh """
        export version_s=$(cat version_f)

        ./package.sh \${version_s}
        ./integration-test.sh 11 8 3 ${version_g}

        echo \${DOCKER_PSW} > docker-password
        export HOME=${WORKSPACE}
        cat docker-password | docker login --username
\${DOCKER_USR} --password-stdin

        docker push calvinpark/subtractor:${version_g}
    """
}
```

The variables in this step are a mix of shell variables and Groovy variables. Unfortunately, both Groovy and shell use the same ${var_name} syntax to reference a variable, and determining whether a variable is a shell variable or a Groovy variable can get quite confusing. The exact rules can be found in the official documentation (https://www.jenkins.io/doc/book/pipeline/jenkinsfile/#string-interpolation). In the first line, we read the version_f file and store the value in a version_s shell variable. The next line calls package.sh with the version_s shell variable to create the deliverable calvinpark/adder:${version_s} container. The next line calls integration-test.sh with the version_g Groovy variable, which was set in the very first stage. integration-test.sh validates that the calvinpark/adder:${version_g} container can calculate *3+8=11*. The Docker image is validated and we're ready to push it to Docker Hub.

The next block follows Docker's recommended way[1] of logging in non-interactively by writing the password into a file and piping the content into the docker login command. Before calling docker login, HOME is redefined as ${WORKSPACE} because docker login creates a file in the ${HOME}/.docker/ directory. The container we're running in doesn't have the jenkins user created, therefore ${HOME} doesn't point to a valid directory. By redefining HOME as ${WORKSPACE}, docker login is able to create the file to store the credentials and the login succeeds. Finally, the deliverable Docker image is pushed to Docker Hub.

---

1 https://docs.docker.com/engine/reference/commandline/login/#provide-a-password-using-stdin

> **The example code is not the most optimal way**
>
> We have taken a roundabout approach in order to learn the different ways we can write a Jenkinsfile. This example code can be written more cleanly with the withRegistry() and image.push() methods, so check them out in the official documentation (https://www.jenkins.io/doc/book/pipeline/docker/#custom-registry).

We've learned about the ways to handle shell and Groovy variables, as well as logging in to Docker Hub and pushing an image. Let's now continue to learn even more Jenkinsfile tricks.

Bare-metal agents, Groovy language features, and alternate ways to run Docker and handle credentials

The final stage uses a bare-metal agent – it runs directly on the agent, rather than encapsulating our processes inside a predefined container. This is usually a bad idea because it relies on the environment defined by the agent. For example, this bare-metal agent lacks git, which is required for this stage. We will soon see a way to overcome this:

> **The bare-metal agent is for accessing the hardware**
>
> The bare-metal agent is most useful when we need direct access to the hardware. If we need specially configured USB devices attached to the agent (for example, testing a phone or a tablet), it is often far simpler to use a bare-metal agent rather than trying to bring the hardware into a container.

```
stage('Push Git tag') {
  agent { label 'docker' }
  steps { script {
    try {
      container = docker.build("git", "-f git.dockerfile .")
      container.inside {
        withCredentials([sshUserPrivateKey(
            credentialsId: 'github-calvinpark-priv',
            keyFileVariable: 'KEYFILE')]) {
          withEnv(['GIT_SSH_COMMAND=ssh -o
StrictHostKeyChecking=no -i ${KEYFILE}']) {
            sh "git tag ${version_g}"
            sh "git push origin ${version_g}"
      } } } }
    catch (Exception e) {
      sh "git tag -d ${version_g} || true"
      throw e
} }}}
```

In the steps directive, we start with the script directive to enter the Groovy environment and use a try/catch block. This try/catch block is not some special Jenkins-made directive, but instead a normal Groovy try/catch block. Inside the script directive, we have access to the full Groovy environment. The try/catch block is just one example of using Groovy's language features – you can use conditionals, loops, classes, closures, or any other language features of Groovy. It's powerful to a fault, which tempts users to embed complex code into Jenkinsfiles when the best practice is to move the business logic out to separate scripts.

Do not write complex code in Jenkinsfiles

The first item in the official Jenkins documentation's Best Practices list is *Making sure to use Groovy code in Pipelines as glue*, and the second item is *Avoiding complex Groovy code in Pipelines* (https://www.jenkins.io/doc/book/pipeline/pipeline-best-practices/). Jenkinsfiles should be a way to execute your build script, not the build script itself. Jenkinsfiles can't be debugged outside of Jenkins, which means we can't write unit tests that will catch bugs in the offline coding environment – pushing incremental changes to a Git repository and then attempting to execute the Jenkinsfile to see where it breaks gets tiresome very quickly.

Inside the `try` block, we learn yet another way of creating a container and using it as an environment. The `container = docker.build()` line builds a Docker image using `git.dockerfile`. With the use of the `container.inside` block, we create a container from the image and run the enclosed commands inside the container environment. This works as if we had specified `git.dockerfile` as the source of a dockerfile agent. Just like the dockerfile agent, the workspace directory is automatically mounted.

Inside the container, we learn a new way of using a secret from the Jenkins credential store as well as a new way of setting an environment variable. Using the `withCredentials` directive, we fetch the private key stored in the `github-calvinpark-priv` secret and save it in a file. The location of the file is stored in the `KEYFILE` variable, and this variable is used in the `withEnv` directive to specify the private key to be used during the subsequent Git operations. Finally, with the Docker container, credentials, and environment variable defined, we run the `git tag` and `git push` commands.

> **withCredentials can handle various credential types**
>
> The `withCredentials` directive has different ways of handling the various credential types (file, text, user/pass, pub/priv, and so on) and the full syntax can be found in the official documentation (https://www.jenkins.io/doc/book/pipeline/jenkinsfile/#handling-credentials).

In the catch block, we prepare for the case where `git push` fails, perhaps due to a network issue or a credential rotation. We clean up the tag that's not pushed, then throw the error again to signal to Jenkins that there has been an error.

We've learned when and how to use a bare-metal agent, about access to the powerful Groovy language features, and new ways to start a container and use a credential. Let's continue to the final section of the Jenkinsfile.

post

In the post section, we expose the version number onto the Jenkins UI so that it's easy to find the build log of a specific version. In addition to the `always` condition, there are other conditions, including `success`, `unsuccessful`, and `cleanup`. `unsuccessful` is often useful for sending an email to alert the developers, while `cleanup` is used for a resource release in case there are locks applied during a build:

```
post {
    always {
        script {
            currentBuild.description = version_g
}   }   }
```

Whew! That was a pretty complex Jenkinsfile. The file is too big to fit onto a page, but you can compare your Jenkinsfile with the one from the book's GitHub repository (https://github.com/PacktPublishing/Jenkins-Administrators-Guide/blob/main/ch4/postmerge.jenkinsfile).

We have created a Jenkinsfile for a postmerge CD pipeline to identify the next version number, build a Docker image, run integration tests, log in to Docker Hub, push the Docker image, tag the Git repository, push the Git tag, and then finally update the build description. We've used various agent types and DooD, learned several ways to use the credentials, several ways to pass the variables across the pipeline stages, and a way to override the buildDiscarder options, and also looked at how to use Groovy's native language features. Let's save it to our Git repository.

Saving the files, making a PR, and merging

Since we locked the main branch of the adder GitHub repository in *Chapter 2, Jenkins with Docker on HTTPS on AWS and inside a Corporate Firewall*, we can't simply make a new commit and push to the main branch. Instead, make a new branch, commit, push, and make a PR:

```
~/adder $ git checkout -b postmerge origin/main
Branch 'postmerge' set up to track remote branch 'main' from
'origin'.
Switched to a new branch 'postmerge'

~/adder $ chmod 755 *.sh  # Make the shell scripts executable

~/adder $ git add .

~/adder $ git commit -m "postmerge"

~/adder $ git push origin -u postmerge
```

Once the premerge build passes, we can merge the PR.

With the files ready in the Git repository, let's continue to make a static pipeline.

Creating a static pipeline for packaging, integration tests, and delivery

Let's make our postmerge pipeline. The steps are similar to the premerge pipeline. Log in to Jenkins as adder-admin, enter the adder folder, and click New Item. Enter postmerge as the name, choose Pipeline, and press OK:

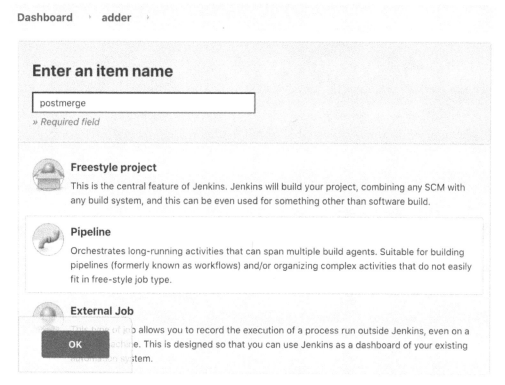

Figure 4.8 – Creating a postmerge pipeline in the adder folder

Now we're on the pipeline configuration page. Scroll down to the Pipeline section and enter the following:

- Definition: Pipeline script from SCM

- SCM: Git

- Repository URL: git@github.com:<GitHub Username>/adder.git (for example, git@github.com:calvinpark/adder.git. This time we're using the SSH remote instead of the HTTPS remote because we need to push to it)

- Credentials: `<GitHub Username>` (github-`<GitHub Username>`-priv) (for example, `calvinpark` (github-calvinpark-priv))
- Branch Specifier: `*/main` (take note that we're always building from the main branch, which is where our postmerge build scripts are)
- Script Path: `postmerge.jenkinsfile`

Leave the rest unmodified and click Save to exit:

Figure 4.9 – The adder postmerge pipeline SCM configuration with the SSH GitHub URL

Click Build Now, and you can see that the build ran successfully, indicated by the blue ball. Notice that the version number 0.1.0 is shown on the Build History widget:

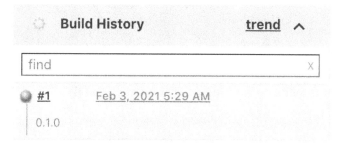

Figure 4.10 – Build History showing a successful build for version 0.1.0

Click #1 and then Console Output to see the full console output of the build. In the build log, we can see that all of our steps worked. Let's go through the build log output to identify the steps that ran:

1. dockerfile agent worked to provide an isolated environment with the preconfigured tools:

```
firewalled-firewalled-agent does not seem to be running inside
a container
$ docker run -t -d -u 123:30 ...
```

2. get-next-version.sh generated 0.1.0 as the version:

```
+ sh get-next-version.sh
[Pipeline] echo
Next version is 0.1.0
```

3. DooD worked without a permission issue:

```
$ docker run -t -d -u 123:30 -v /var/run/docker.sock:/var/run/
docker.sock:rw --group-add 998 ...
```

4. The credentials we've set up using environment worked to allow a successful Docker login:

```
+ docker login --username calvinpark --password-stdin
[...]
Login Succeeded
```

5. A Docker image with version 0.1.0 was built and pushed:

```
+ docker push calvinpark/adder:0.1.0
```
```
The push refers to repository [docker.io/calvinpark/adder]
```

6. A bare-metal agent with a container generated with withDockerContainer worked to provide an isolated environment on the fly:

```
Running on firewalled-firewalled-agent in /home/robot_acct/
firewalled-firewalled-agent/workspace/adder/postmerge
```
```
[...]
```
```
[Pipeline] withDockerContainer
```
```
firewalled-firewalled-agent does not seem to be running inside
a container
```
```
$ docker run -t -d -u 123:30 ... git cat
```

7. The credentials we've set up using withCredentials worked to allow a successful Git push:

```
# Credentials from withCredentials
```
```
[Pipeline] withCredentials
```
```
[...]
```
```
# Git tag & push
```
```
+ git tag 0.1.0
```
```
+ git push origin 0.1.0
```
```
Warning: Permanently added 'github.com,192.30.255.113' (RSA) to
the list of known hosts.
```
```
To github.com:calvinpark/adder.git
```
```
 * [new tag]  0.1.0 -> 0.1.0
```

We can go to GitHub and Docker Hub to check that the 0.1.0 tag and the image exist:

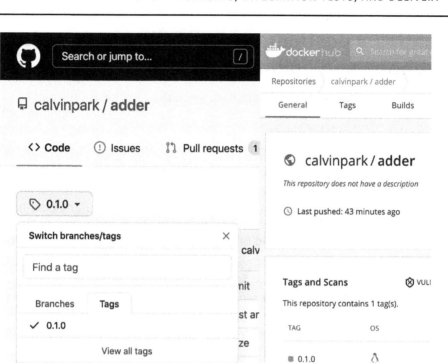

Figure 4.11 – GitHub and Docker Hub showing the 0.1.0 tag and image

8. Finally, the build description was set to 0.1.0 (although it's a bit hard to tell from the logs):

```
[Pipeline] stage
[Pipeline] { (Declarative: Post Actions)
[Pipeline] script
[Pipeline] {
[Pipeline] }
[Pipeline] // script
[Pipeline] }
[Pipeline] // stage
```

Whew! That was a lot to take in, wasn't it? We've gone through the entire build log to identify the various tasks we've configured on the Jenkinsfile and confirmed that they all worked correctly. By checking GitHub and Docker Hub, we were able to see the artifacts of our build. Let's continue to turn it into a CD pipeline.

Creating a postmerge CD pipeline with a GitHub webhook and polling

As we saw with the premerge CI pipeline, the static pipeline is only half of the challenge. Let's convert the static pipeline into a postmerge CD pipeline that is triggered by a GitHub PR merge.

The premerge pipeline runs when a PR is created or updated with a new commit. The postmerge pipeline runs when a PR is merged *to a specific branch*. Our postmerge pipeline bumps the version number and deploys an artifact, so it should only run when a PR is merged to the main branch, not other branches. If the postmerge pipeline ran when a PR is merged to a branch other than main, the version would be bumped and a new artifact would be deployed even though there are no changes to the artifact.

Create multiple postmerge pipelines if there are multiple active branches

A Git repository may have more than one branch from where the artifacts are versioned and deployed. These branches are sometimes called feature branches. For example, one branch may handle 1.x.x versions, while another handles 2.x.x versions. In such a case, we can create two postmerge pipelines for the two branches, so that each pipeline uses its own versioning and deployment rules. It is possible to combine them into one – a premerge pipeline often serves all PRs regardless of the target branch. However, this is not recommended for a postmerge pipeline because of the different nature of the two. The premerge pipeline build history is no longer useful once a PR is merged – they are effectively disposable one-time validation tasks. Postmerge pipelines are different in that they are useful even after the build and deployment are complete – the build history serves as an auditing tool for generating and deploying artifacts. If a deployed container is misbehaving, the build log for that version is often the first place to look. Therefore, it's often useful to have a clean build history for the postmerge pipelines. Separating each feature branch into its own postmerge pipeline helps with that.

In *Chapter 3, GitOps-Driven CI Pipeline with GitHub*, the GitHub Pull Request Builder (GHPRB) plugin did an excellent job of monitoring for PR changes and triggering the premerge builds. Unfortunately, the GHPRB does not handle postmerge triggers. Thankfully, the native Jenkins features can handle the postmerge triggering for both AWS Jenkins and firewalled Jenkins. Let's continue to see how they work.

Configuring the postmerge trigger

The postmerge triggers are configured differently on an AWS controller than a firewalled controller. Let's look at the AWS controller first, since it's the simpler of the two.

AWS controller

We will start by adding a new webhook to send a notification to Jenkins when there is a push event. Go to the GitHub adder repository | Settings | Webhooks | Add webhook, as follows:

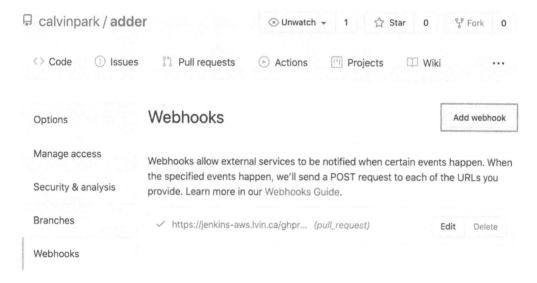

Figure 4.12 – Adding a new webhook from the GitHub adder repository | Settings | Webhooks page

The payload URL is https://<JENKINS URL>/github-webhook/ (for example, https://jenkins-aws.lvin.ca/github-webhook/). The trailing slash is important! Not ending the URL with a slash will cause the build triggers to fail[2]. Keep everything else as the default since it already sends push events. Click Update webhook to save and enable the webhook. This is what it should look like:

---

2 https://issues.jenkins.io/browse/JENKINS-17185

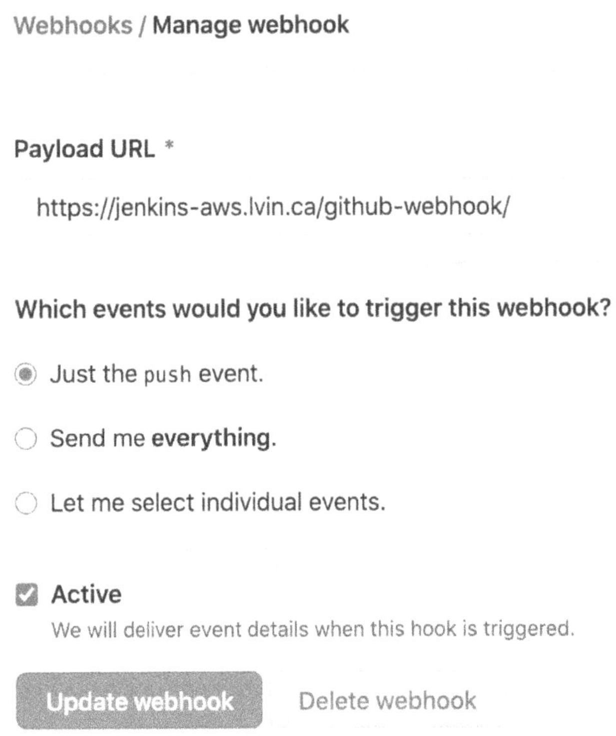

Webhooks / **Manage webhook**

Payload URL *

https://jenkins-aws.lvin.ca/github-webhook/

Which events would you like to trigger this webhook?

◉ Just the push event.

○ Send me **everything**.

○ Let me select individual events.

☑ **Active**
  We will deliver event details when this hook is triggered.

Update webhook Delete webhook

Figure 4.13 – GitHub webhook configured for push events.
Unmodified fields were hidden for the screenshot

GitHub is now configured to push a hook when a branch is updated. Let's configure the postmerge pipeline to listen to the hook.

On Jenkins, go to the postmerge pipeline and click Configure. There are three changes we need to make.

First, check GitHub project and add the GitHub adder repository URL (for example, `https://github.com/calvinpark/adder/`) for Project url so that the postmerge pipeline listens to the hooks for the GitHub adder repository.

Second, under Build Triggers, check GitHub hook trigger for GITScm polling so that a build is triggered when a hook is received. I've hidden the unmodified fields, but this is what the changed fields should look like:

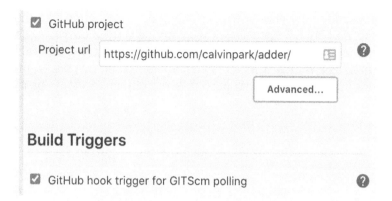

Figure 4.14 – Jenkins adder postmerge configured to listen to GitHub webhooks. Unmodified fields were hidden for the screenshot

But wait a minute. Why are we configuring *polling* when there is already a push hook? Aren't those two mutually exclusive? This strangeness is due to the way the GitHub plugin works in order to protect Jenkins from attacks.

Recall that we configured GitHub to send the push hooks to https://jenkins-aws.lvin. ca/github-webhook/. That HTTP endpoint on our Jenkins is wide open to the internet, and there is no guarantee that a POST request to the endpoint is a legitimate push hook from GitHub.

When a push hook is received, the GitHub plugin does not process the hook and start a build because the request might be fake. Instead, *the GitHub plugin asks Jenkins to go to GitHub directly to see whether there really is a change.* For this, the GitHub plugin triggers *one-time polling* on the Git plugin. The word *polling* makes us think of a continuous loop, but in this case the polling is just a way to trigger a lookup.

With that understanding, let's read the option text again. GitHub hook trigger for GITScm polling means that when a push hook is received, this checkbox determines whether one-time polling should ask the Git plugin to go to GitHub and look for a change, then start a build if there is one. Thank you Sebass van Boxel and Sam Gleske for helping me understand this mysterious option. Let's continue.

The last change is a bit interesting. Under Pipeline | Definition | SCM | Branches to build, change Branch Specifier from */main to main. By default, Jenkins prepends the branch name with */ so that a branch from any Git origin can be used. This is a harmless default that casts a wide net, which is quite effective. The trouble is that the GitHub plugin uses this value to determine the branch to monitor. We want the postmerge pipeline to run when a PR is merged into the main branch, but not when a side branch is created to open a PR. In the eyes of GitHub, both activities are push events and they both push webhooks to Jenkins. It is the Jenkins GitHub plugin's job to read the webhook details and trigger only the right pipeline with the matching branch name, and Branch Specifier is the value the GitHub plugin uses to match the branch name. When the value includes */, it ends up matching all branches because * is interpreted as a wildcard. As a result, if you don't remove the */, the postmerge pipeline ends up running on every push to every branch. It should be changed to look like this:

Figure 4.15 – Specifying main as the branch name so that the hook triggers a build only when a change is pushed to the main branch and no other branches

Click Save to exit. The postmerge trigger is ready. Now let's see how it's done for the firewalled controller.

Firewalled controller

GitHub can't push a hook to the firewalled controller, so we must use polling to identify the merges. Go to the adder/postmerge pipeline and click Configure. Under Build Triggers, check Poll SCM and schedule to poll every 2 minutes. This can be adjusted as needed the same way the polling schedule was adjusted for the premerge pipeline. The Ignore post-commit hooks checkbox doesn't apply to the firewalled controller because it can't receive any hooks from GitHub:

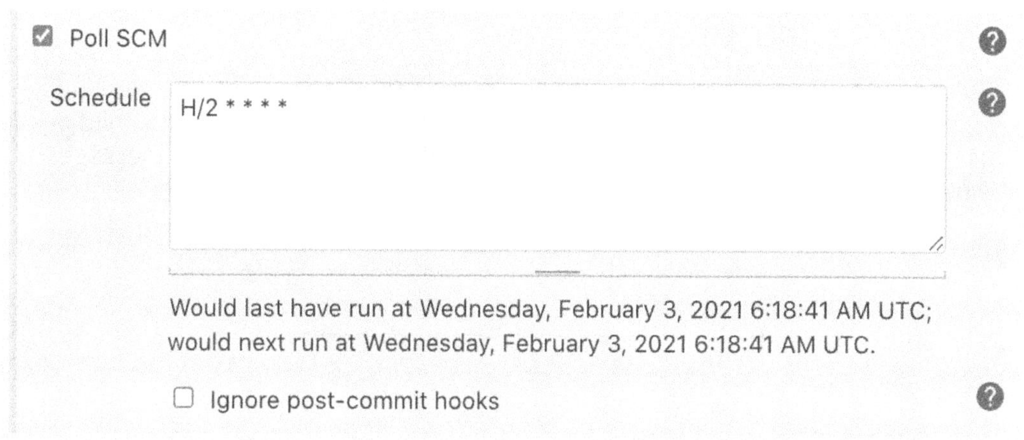

☑ Poll SCM   ?

Schedule   H/2 * * * *   ?

Would last have run at Wednesday, February 3, 2021 6:18:41 AM UTC;
would next run at Wednesday, February 3, 2021 6:18:41 AM UTC.

☐ Ignore post-commit hooks   ?

Figure 4.16 – Configuring Poll SCM to monitor for changes in the GitHub repository

Thankfully, Jenkins is smart enough to monitor only for the changes in the `main` branch without any further configurations.

> **Build not triggering?**
>
> Due to a long-standing bug[3], H/1 * * * * runs every hour, not every minute.
>
> Every minute should be written as * * * * * instead.

One major shortcoming of this configuration is that a build doesn't trigger for every merge that happens between the polls. For example, suppose that a pipeline polls every 5 minutes. If there were three merges within the 5-minute window, only one postmerge build is triggered with all three changes in it, rather than three postmerge builds each with one change. Guaranteeing a postmerge build for each merge requires Jenkins to receive push hooks, and there are two solutions to this problem:

- If you're using GitHub Enterprise that's deployed within the corporate firewall, GitHub Enterprise *can* push hooks to firewalled Jenkins. In this case, follow the instructions for AWS controller configuration and build on each hook received.

- There are solutions, such as localtunnel or ngrok, that allow you to run an agent on a machine inside the firewall and relay the push hook sent to their server to firewalled Jenkins. While this is an effective solution, keep in mind that this is tunneling the traffic across the corporate firewall, which may be against your company's IT policy.

The postmerge trigger is ready. Let's trigger it.

---

3   https://issues.jenkins.io/browse/JENKINS-22129

Testing the postmerge trigger

Create a branch with a minor change and make a PR. Once the premerge build succeeds, merge the PR, then wait for the postmerge pipeline to start a build on its own. If a build doesn't start, examine the Jenkins logs for the error messages (https://jenkins-firewalled.lvin.ca/log/all):

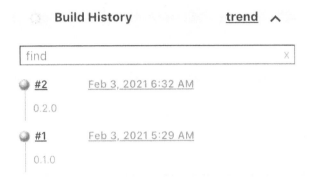

Figure 4.17 – Postmerge build triggered by a PR merge

Building the PR branch or an arbitrary branch

In *Chapter 3, GitOps-Driven CI Pipeline with GitHub*, we modified the premerge pipeline to build the PR branch and also to allow the building of an arbitrary branch. The target branch of a postmerge build is already known (main in our case), so there's no need to build the PR's target branch. Also, since the build generates tags and artifacts, building an arbitrary branch on a postmerge pipeline is usually a bad idea. This is especially true if we intend to use the build history as an auditing tool for bugs in the artifacts.

And that's all! We've created a postmerge CD pipeline that is triggered by a GitHub PR merge. It builds, tests, and deploys a Docker image, and tags the Git repo.

Summary

We have learned how to create a postmerge CD pipeline using various Jenkinsfile syntax options. We've learned three different ways of launching a Docker container (docker, dockerfile, and bare-metal agent), so that we can always run a build in an isolated environment in any situation. We've learned two ways of handling credentials so that we can interact with external systems. We've learned how to get around the user requirements of using Git inside a container. Finally, we've learned how to configure a postmerge CD pipeline so that it is triggered when a PR is merged.

By now, you should have a production-grade Jenkins with premerge and postmerge pipelines set up for your product. This is already good enough to carry the software development life cycle (SDLC) of one or two teams. In the second part of the book, we will learn how to scale Jenkins to serve a larger team, an organization, or even a company. Before jumping to the second part of the book, the next chapter will go through the detailed steps on how to set up Jenkins on AWS. Be sure to check it out if you are unclear on some of the AWS tasks.

5
Headfirst AWS for Jenkins

In this chapter, we will learn how to create and update the AWS services required for running Jenkins on AWS. We will learn how to create EC2 instances, security groups, application load balancers, IAM users for Let's Encrypt Certbot, Elastic IPs (EIPs), and DNS records in Route 53. If you were able to create the services by yourself, then feel free to skip this chapter. If you are unsure, if you are stuck, or if you have never used AWS before, then you have come to the right place. No steps have been skipped and no prior knowledge assumptions have been made. The structure of this chapter aligns with the first two chapters to help you navigate easily.

In this chapter, we are going to cover the following topics with click-by-click instructions and plenty of screenshots:

- Logging in to AWS
- Navigating the AWS console
- EC2 instances and EIPs
- Let's Encrypt
- Setting up an application ELB for the AWS Jenkins controller

Technical requirements

All the links in this chapter can be found in the README.md file inside the chapter 5 folder in the GitHub repository (https://github.com/PacktPublishing/Jenkins-Administrators-Guide/tree/main/ch5). Look for the heading you are currently reading, and you will find the links under the same heading in the README file.

Before you can begin, you will require an Amazon Web Services (AWS) account. If you don't have one, follow the instructions available at https://aws.amazon.com/premiumsupport/knowledge-center/create-and-activate-aws-account/ to create one. If you are using a personal credit card for AWS, set up a budget and alerts as explained here, https://docs.aws.amazon.com/awsaccountbilling/latest/aboutv2/budgets-create.html, to ensure that there are no unexpected charges on your card.

Logging in to AWS

First, let's learn how to log in to our AWS account and select the correct region. To log in to our AWS account, let's open our browser and navigate to https://console.aws.amazon.com. AWS caches the region we were using last, and it will automatically select that region for us. It is always a good idea to check the region as soon as we log in to avoid creating resources in the wrong region. So, let's make sure that we are in the correct AWS Region by looking at the navigation bar at the top of the page as shown. The region I will be using is N. Virginia. Every region is associated with a code name. The code name for N. Virginia is us-east-1. You can pick the region that is closest to you:

Figure 5.1 – Top navigation bar showing the selected region

If we are not in the correct region, let's select the region we would like to use by clicking on the drop-down arrow next to the region and selecting the region in the list that appears.

Now that we have learned how to log in to AWS, let's learn how we can find our way around the AWS console.

Navigating the AWS console

AWS offers a lot of services, such as EC2 (compute service), RDS (database service), EKS (Kubernetes service), and so on. Each of these services has a dedicated dashboard. To create, view, modify, or delete these services, we must first navigate to the service's dashboard. Let's now see how we can do that.

After logging in to the AWS console and verifying the region, we need to search for the resource using the search bar located at the top of the screen. As we start typing in the name of the resource, AWS will start suggesting the services available, and when we find what we are looking for in the suggestion box, we can click on it to go to the service's dashboard.

As an example, say we want to navigate to the EC2 dashboard. To do this, enter ec2 in the search bar, as shown in the following screenshot:

Figure 5.2 – Searching for EC2 in the search bar

Once the suggestions load, click on EC2 in the suggestion box to go to the EC2 dashboard.

Now that we know how to log in to AWS and how to navigate the dashboard, let's look at some important points we need to keep in mind while reading the chapter.

Important notes

- *Use the new EC2 experience*: When you are in the EC2 dashboard, make sure that New EC2 Experience is selected. You should be able to find the toggle button for this just below the AWS logo on the upper left-hand side of the window.

- *The color and location of the buttons might change*: AWS constantly updates the console to provide the best possible experience for us. As part of this, they may change the color and placement of the buttons, but they most likely will not change the text in the buttons. So, when looking for buttons, make sure to match the text and not the color and location.

Now that we know how to log in, find our way through the AWS console, and have looked at some important notes to keep in mind, let's now create our EC2 instances for our controller, agent, and Docker cloud host.

EC2 instances and EIPs

To create the necessary Elastic Compute Cloud (EC2) instance and attach an EIP to them, we need to perform the following steps:

1. Create an SSH key pair.

2. Create a security group.

3. Create an EC2 instance.

4. Create and attach an EIP.

Let's now look at how we perform each of these steps.

Step 1 – Create an SSH key pair

An SSH key pair consists of a public key and a private key that is used for authenticating with Linux hosts. The virtual machine will store our public key and we will store the private key. The private key will be used when we connect to the virtual machine via SSH. If we lose our private key, we will not be able to connect to our virtual machines.

The private key is like our password!

The private key is like our password! We should not upload it to GitHub, Pastebin, or any other publicly accessible service, nor should we share it with anyone.

To create a key pair in AWS, we must first navigate to the EC2 dashboard. To do that, search for ec2 in the search bar at the top and click on the appropriate search result as described in the *Navigating the AWS console* section earlier.

We now need to navigate to the key pairs portion of the EC2 dashboard. To do that, click on Key Pairs under Network & Security in the left navigation pane, as shown in the following screenshot. Depending on the size of the screen, we may need to scroll down to reveal this option:

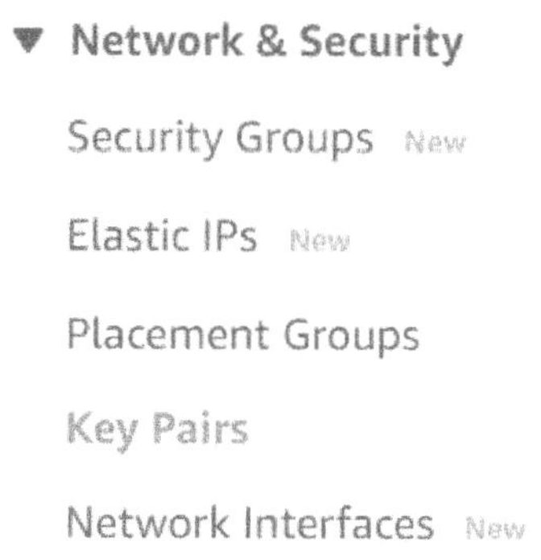

▼ **Network & Security**

Security Groups New

Elastic IPs New

Placement Groups

Key Pairs

Network Interfaces New

Figure 5.3 – Left navigation pane showing the Key Pairs option

To create a new key pair, click on the Create key pair button.

Every key pair must have a unique name and it must be either a pem type or ppk type. We will be using the pem type as it is supported by the SSH command on Linux, Mac, and Windows 10. Let's fill in this information in the create key pair form:

- Name: jenkins-keypair.
- File format: **Select** pem:

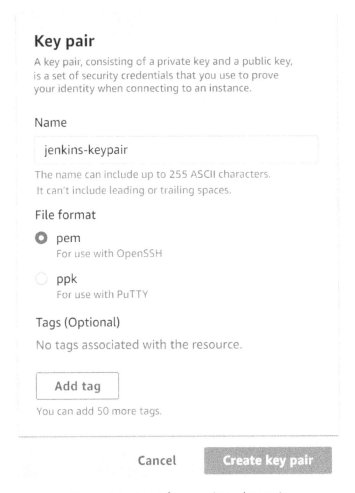

Figure 5.4 – Form for creating a key pair

When done, click on the Create key pair button to create the key pair. Our private key will automatically download to our PC. The public key will be stored for us by AWS. Once this process is complete, we will see a green bar at the top of the window that says successfully created key pair.

Now that we have a key pair, let's create our security groups.

Step 2 – Create a security group

A security group is a virtual firewall that filters packets based on the origin, port, and destination just before or just after it is sent or received by our virtual machines. Security groups work at a virtual machine level and not at a network level, in other words, different virtual machines in the same network can have different security groups. We can attach up to five security groups for a single network interface.

To create a security group, we must first navigate to the EC2 dashboard. To do that, search for ec2 in the search bar at the top and click on the appropriate search result as described in the *Navigating the AWS console* section earlier.

Now, navigate to the security groups portion of the EC2 dashboard by clicking on Security Groups under Network & Security in the left navigation pane, as shown in the following screenshot. Depending on the size of our screen, we may need to scroll down to reveal this option:

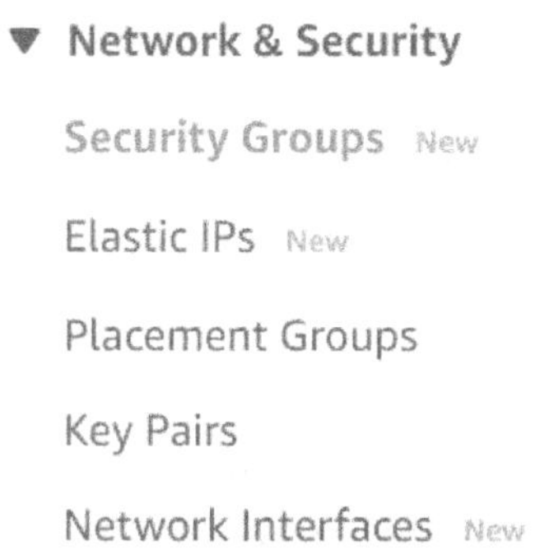

▼ **Network & Security**

  Security Groups   New

  Elastic IPs   New

  Placement Groups

  Key Pairs

  Network Interfaces   New

Figure 5.5 – Left navigation pane showing the Security Groups option

Let's now create the security group that we will attach to our virtual machines. To do this, click on the Create security group button. The form for creating a security group is broken down into three sections on the same page. Let's look at each of them one by one.

Basic Details

In the Basic Details section, we need to enter the name of our security group, a brief description that explains what it is for, and the VPC it belongs to. Enter the following information in the Basic Details section:

- Name: `jenkins-vm`.
- Description: `Allow access for debugging and inbound agent`.
- VPC: Select default VPC:

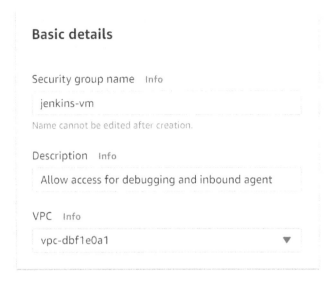

Figure 5.6 – Form for Basic Details of the security group

Now that we are done with the Basic Details section, let's look at the Inbound Rules section.

Inbound rules

In the Inbound Rules section, we need to specify a list of IPs (using Classless Inter-Domain Routing (CIDR) notation) from where the connections to the resource will originate and what port on the virtual machine these communications will land on.

For the jenkins-vm security group, we want to allow connections originating from our PC for SSH and HTTP on ports 22 and 8080, respectively. We also want to allow connections originating from the IP of the inbound agent on port 50000.

To add a rule to this security group, click on the Add rule button. A row of input fields should appear. Before we fill in the input fields, let's first learn a little bit about what each field signifies and how the input fields behave:

- The Type column denotes the type of connection, in other words, SSH, HTTP, and so on. In cases when we select a type that has a well-known port range and protocol, the port range and protocol will be auto-filled in for us. For example, when we select the type as SSH, port 22 will be filled in for us under the port range column, and TCP will be filled in under the protocol column.

- The Source column denotes the CIDRs from where the connections will originate. The source column has two input fields under it, a dropdown and a text field. This text field is a bit different from other text fields we see on the screen. It is different because when we start typing a CIDR, AWS will start autocompleting. When we find the CIDR we are looking for in the autocomplete box, we need to click on it to select it. Once clicked, it will appear in a blue box below the text field. This means that the CIDR address has been selected and connections originating from this CIDR will be allowed in. Depending on what we select in the dropdown, the CIDRs may be autoselected for us. If we select Anywhere, then 0.0.0.0/0 will be selected for us and it will appear in a blue box below the text field. If we select My IP, then AWS will determine the public IP address of our PC and it will select the CIDR for that IP.

Now that we have understood the fields and how they behave, let's add an SSH inbound rule that allows connections from our PC. To do that, enter the following information in the row of input fields:

- Type: SSH.

- Source: Select My IP in the dropdown.

- Description: SSH access from Lalit IP. Replace my name with your name/AD username:

Figure 5.7 – Form for adding Inbound Rules in the security group

Add two more rules using the same method described previously using the following information:

Type	Protocol	Port range	Source	Description
Custom TCP	TCP	8080	My IP.	Web access from Lalit IP for debugging
Custom TCP	TCP	50000	Custom – Egress IPs for the machine that will be used as an inbound agent. If your organization has multiple egress CIDRs, add a new rule for each of them.	TCP access for inbound agent

The final list of Inbound Rules should appear as shown in the following screenshot. The values for My IP will be different for you:

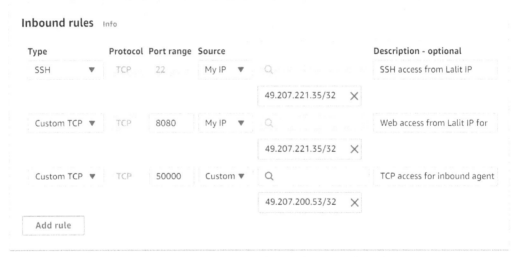

Figure 5.8 – Table showing all the Inbound Rules added

> **Important note about using "My IP"**
>
> As mentioned previously, when you select My IP, AWS determines the public IP of our PC and selects that for the rule. Depending on how our internet connection is set up, our IP might change when our router restarts or when something changes on our ISP's end. When this happens, AWS will not automatically update the inbound rules. We will have to edit them manually. To learn how to edit your inbound rules manually, look at the next sub-section.

We do not have to add any outbound rules as we want our virtual machines to be able to communicate with the internet. So, leave the outbound rules section unchanged and create the security group. To create the security group, scroll to the bottom of the screen and then click on the Create security group button. In a couple of seconds, we will see a green message at the top of our window saying that the security group was created successfully.

Now that we have created a security group, let's take a quick detour and look at how we can edit an inbound rule in a security group.

Editing inbound rules for a security group in AWS

There are some cases when we may want to edit our security group. One such case is when our IP changes and we are no longer able to reach our AWS resources. In this section, we are going to see how we can edit the inbound rules for our security group. Navigate to the security groups dashboard using the steps mentioned earlier.

In the upper portion of the screen, make sure the jenkins-vm security group is selected (when selected, a tick mark will appear in the first column of the table), click on the Actions button, and, in the dropdown that appears, select Edit inbound rules as shown:

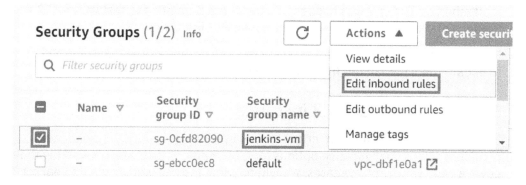

Figure 5.9 – Editing the inbound rules

In the screen that appears, note that all rules have the source as Custom and none of them have My IP as the source. This is expected. Behind the scenes, AWS updates My IP to Custom for the source. This is one of the reasons why it is preferable to include a detailed description for every inbound rule. We added our name/AD username as part of the description of our inbound rules that allow connections only from our PC. Using the description, we should be able to easily determine the rules that need to be updated (*hint: it is the rules that have the port ranges 8080 and 22*). To update the CIDRs to the IP address of our PC, select My IP in the source column and AWS will automatically update the CIDR for that rule as shown:

Inbound rules Info

Type	Protocol	Port range	Source		Description - optional
Custom TCP ▼	TCP	8080	My IP ▼	🔍	Web access from Lalit IP for
				49.207.213.49/32 ✕	
SSH ▼	TCP	22	My IP ▼	🔍	SSH access from Lalit IP
				49.207.213.49/32 ✕	
Custom TCP ▼	TCP	50000	Custom ▼	🔍	TCP access for inbound agent
				49.207.200.53/32 ✕	

Figure 5.10 – Updating the source to "My IP"

When done, let's save our changes by clicking on the Save rules button. In a few seconds, we will see a green bar at the top of our window that says that our security rules were successfully modified. Once the green bar is displayed, this means that the changes have taken effect for all the EC2 instances that this security group is attached to. The changes happen live and there is no need to restart the EC2 instance or reattach the security group to the EC2 instances.

Now that we have created the necessary security groups and we have learned how to update them in case our IP changes, let's create our EC2 instances.

Step 3 – Create an EC2 instance

EC2 instances are nothing but virtual machines running on the AWS cloud. AWS provides various configuration options for CPU, memory, disk, and networking that cover a wide variety of use cases. We will be using the key pair we created earlier to access these virtual machines.

Afterwards, we must first navigate to the EC2 dashboard. To do that, search for ec2 in the search bar at the top and click on the appropriate search result as described in the *Navigating the AWS console* section earlier.

To create an EC2 instance, we must navigate to the instances portion of the EC2 dashboard. To do that, click on Instances under Instances in the left navigation pane as shown:

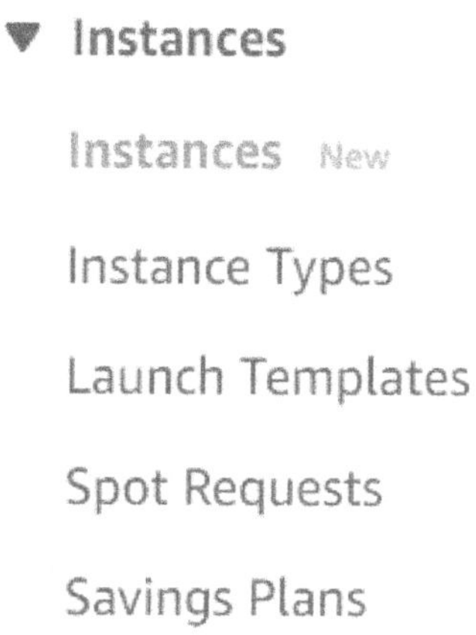

▼ **Instances**

Instances New

Instance Types

Launch Templates

Spot Requests

Savings Plans

Figure 5.11 – Left navigation pane showing the Instances option

Let's now create a virtual machine for our Jenkins controller. To do that, click on the Launch instances button. A multi-page form will appear. Let's now look at how to fill in this form, page by page.

Page 1 – OS

On the first page of the form, we need to select the Operating System (OS) for our virtual machine. We are going to be using Ubuntu Server 20, so search for Ubuntu Server 20 and click on the Select button next to the Ubuntu Server 20.04 LTS (HVM), SSD Volume Type listing as shown:

Figure 5.12 – Selecting the AMI for the instance

As soon as we click on the select button, we will be taken to the next page. Let's now see how we can fill in the second page.

Page 2 – Instance size

On the second page, we need to select the size of our virtual machine, in other words, the number of CPU cores, the amount of RAM, the networking and storage capabilities, and so on. If we are creating a dev/test instance, we can go with the t2.micro instance type. If we are building a production instance, we need to select the instance type based on the scale at which we will be operating Jenkins. Let's select the instance type of our choice as shown:

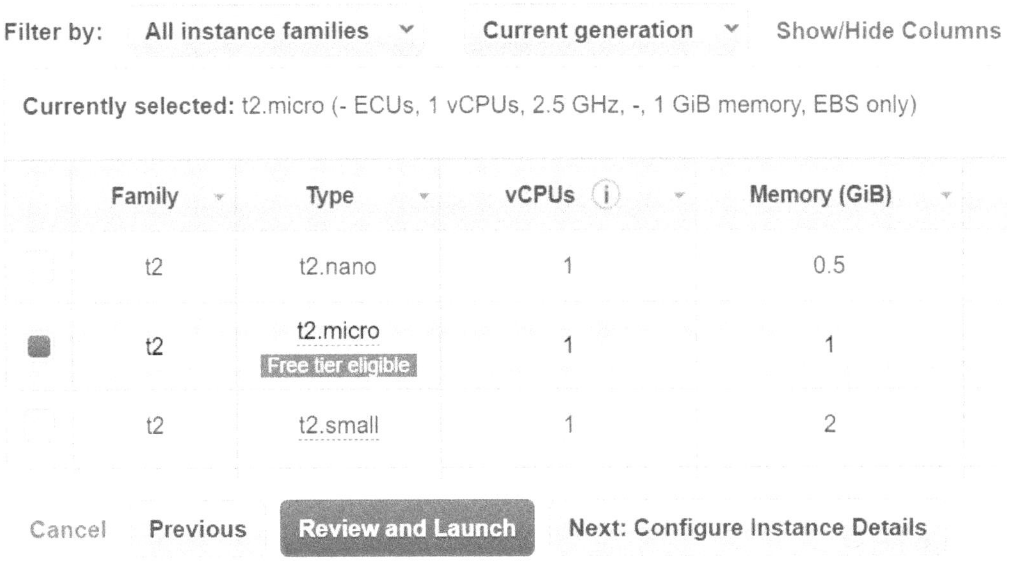

Filter by: All instance families   ⌄   Current generation   ⌄   Show/Hide Columns

Currently selected: t2.micro (- ECUs, 1 vCPUs, 2.5 GHz, -, 1 GiB memory, EBS only)

	Family	Type	vCPUs ⓘ	Memory (GiB)
	t2	t2.nano	1	0.5
■	t2	t2.micro Free tier eligible	1	1
	t2	t2.small	1	2

Cancel   Previous   **Review and Launch**   Next: Configure Instance Details

Figure 5.13 – Selecting the instance family

When we are done, click on the Next: Configure Instance Details **button to go to the** next page.

Page 3 – Instance details

On the third screen, we need to fill in some details about our instance, such as the VPC the instance needs to be launched in and the IAM role. We can leave most of these as the default settings. Let's just fill in the following information:

- Number of instances: 1.
- Network: **Default VPC.**

- Subnet: **Pick any subnet.**

- Auto-assign Public IP: Disable. **We pick** Disable **here because we will be attaching an EIP to the instance in the next section:**

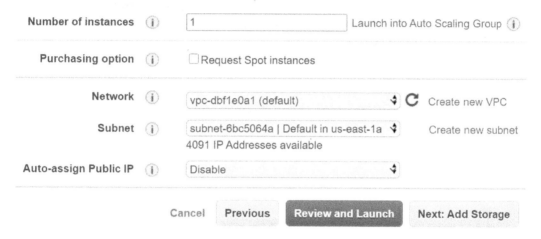

Figure 5.14 – Form for the Basic Details of the instance

When done, click on the Next: Add Storage button to go to the next page.

Page 4 – Storage

On the fourth page, we need to specify our storage requirements. If we are creating a dev/test instance, we can make use of an 8 GB disk size with a gp2 volume. If we are creating a production instance, we need to pick the disk size and volume type based on the scale at which we plan to run Jenkins. Let's fill in the information based on our requirements:

- Size: **8**

- Volume Type: General Purpose SSD (gp2):

Figure 5.15 – Selecting the storage for the instance

When done, click on the Next: Add Tags button to go to the next page.

Page 5 – Tags

On the fifth page, we need to add tags. It is always a good practice to add tags to all the resources that we create. We need to add one tag with a key as a name so that it is easy to search for and identify the purpose of the instance. To do that, click on the Add another tag button. A row of input fields will appear. In the row that appears, add a tag with Key as Name and Value as Jenkins-controller, as shown in the following screenshot:

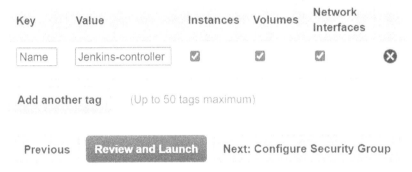

Figure 5.16 – Adding tags for the instance

When done, click on the Next: Configure Security Group button to go to the next page.

Page 6 – Security groups

On the sixth page, we need to select the security groups that must be attached to our instance. We have already created the security groups that we would like to attach to our instance, so we need to select the security group we created for the virtual machines and the default security group. We select the default security group to allow all our virtual machines to talk to each other. To do that, select the radio button with the label Select an existing security group and, in the table that appears, select the default security group, as well as the jenkins-vm security group, as shown:

Assign a security group: ○ Create a **new** security group

◉ Select an **existing** security group

Security Group ID	Name	Description
▣ sg-ebcc0ec8	default	default VPC security group
▣ sg-0cfd8209	jenkins-vm	Allow access for debugging and inbound agent

Figure 5.17 – Adding security groups for the instance

When done, click on the Review and Launch button to go to the final page.

Page 7 – Review

On the seventh and final page, review the settings that we have selected and then click on the Launch button when ready. If we need to make any changes, we can click on the Previous button to go back to the previous screen.

When we click on the launch button, a dialog box will pop up asking us to select a key pair. Since we have already created a key pair, select Choose an existing key pair in the first dropdown and, in the second dropdown, select the key pair that we created earlier. Check the checkbox acknowledging that you have the private key file and then click on the Launch Instances button:

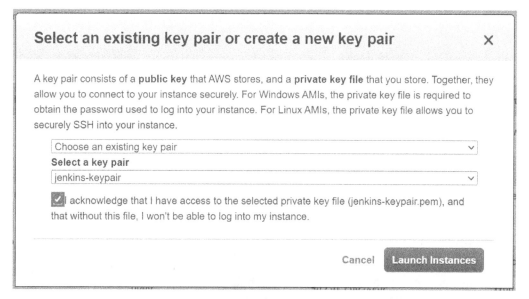

Figure 5.18 – Dialog for selecting the key pair for the instance

Review the information shown on screen and then click on the View Instances button. This will take us back to the EC2 dashboard. Create two more instances (one for the agent and another for the Docker cloud host) using the same procedure. Be sure to give an appropriate name when creating the tag on the fifth page.

We will not be able to SSH to our instances until we create and attach EIPs to the instances. Let's now create and attach EIPs to our instances.

Step 4 – Create and attach an EIP

An EIP is a static public IP that will not change unless it is deleted. We will be attaching EIPs to our virtual machines so that we can reach them from anywhere and these IPs will never change irrespective of what happens to the virtual machine. Even if we have to recreate the virtual machine from scratch, we can retain the same IPs.

To create an EIP, we must first navigate to the EC2 dashboard. To do that, search for ec2 in the search bar at the top and click on the appropriate search result as described in the *Navigating the AWS console* section earlier.

Let's now create an EIP for our Jenkins controller. The EIP will ensure that the public IP of the controller will not change even if we stop and start the virtual machine.

To create an EIP, we must navigate to the EIP portion of the EC2 dashboard. To do that, click on Elastic IPs under Network & Security in the left navigation pane, as shown. Depending on the size of the screen, we may need to scroll down to reveal this option:

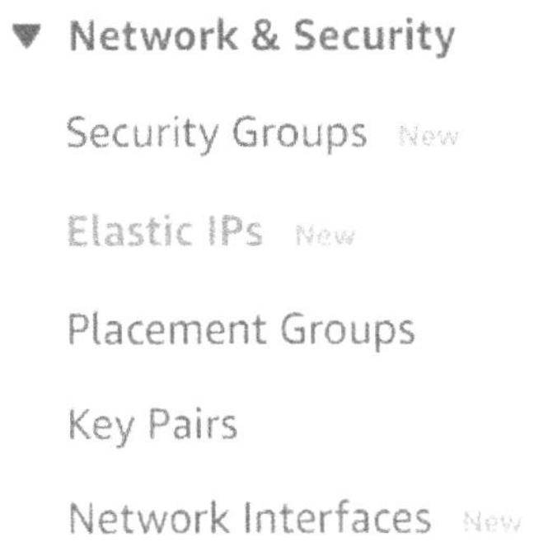

▼ **Network & Security**

    Security Groups   New

    Elastic IPs   New

    Placement Groups

    Key Pairs

    Network Interfaces   New

Figure 5.19 – Elastic IPs in the left navigation bar

To create a new EIP, click on the Allocate Elastic IP address button.

We only need to enter the tags for our EIP and leave everything else as their default settings. We need to add a name that will help us identify what this EIP will be used for. To add the name tag for our EIP, scroll down to the Tags – optional section. Add the name tag by clicking on the Add new tag button. In the row that appears, enter Name in the Key field and Jenkins Controller EIP in the Value field, as shown:

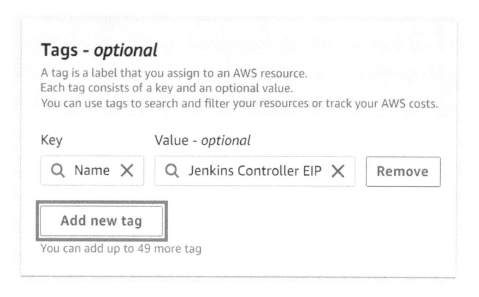

Figure 5.20 – Adding a tag for the EIP

When done, scroll to the bottom of the screen and click on the Allocate button to create our EIP. The allocation should take a couple of seconds. When the allocation is complete, we will see a green bar at the top of our window telling us that our public IP has been successfully allocated. In this green bar, click on the Associate this Elastic IP address button.

In the form that appears, select Instance under Resource Type and, in the textbox for Instance, start typing in Jenkins controller. As we start typing, AWS will start suggesting the EC2 instances in our account. As soon as we see the virtual machine we created for the Jenkins controller, click on it to select it, as shown in the following screenshot. As soon as you click on the name of the instance, AWS will replace the name with the instance ID. AWS does this because the instance ID is a unique ID across all of AWS, but the instance name does not have to be unique across all of AWS. Please note that the instance ID that you see on screen will be different from mine:

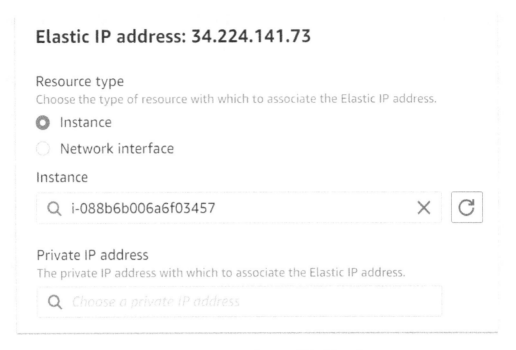

Figure 5.21 – Form to associate an EIP with an instance

Let's leave the remainder as their defaults and click the Associate button at the bottom of the screen to associate this EIP with the Jenkins controller instance. After a few seconds, we will see a green bar at the top of our window that says EIP associated successfully. We will now be able to SSH to our instance using the public IP.

Let's now create and attach two more EIPs for our Jenkins agent and Docker cloud host using the preceding steps.

Now that we have all our instances created and accessible to us, we can go back to the steps in *Chapter 1, Jenkins Infrastructure with TLS/SSL and Reverse Proxy,* to install Docker.

Let's Encrypt

Let's Encrypt is a non-profit certificate authority that provides X.509 certificates for TLS communication at no charge.

When we request a certificate from Let's Encrypt, we need to provide proof of ownership for our domain. This can be done in one of two ways. We can either manually add a TXT record to our DNS provider or we can allow Certbot to automatically create a TXT record for us in Route 53. The latter method will only work if we are managing our DNS records in Amazon Route 53. The former method can be used for any DNS provider.

Manual verification

When we start the Certbot Docker container with the manual verification flag, we will see the following block of text:

```
- - - - - - - - - - - - - - - - - - - - - - - - - - - - - - - - - - - -
Please deploy a DNS TXT record under the name

_acme-challenge.jenkins-firewalled.lvin.ca with the following
value:

kLmhtIfqI5PZFuk-lXna13Z4_oIYDmaJoPd6RaFgwqQ

Before continuing, verify the record is deployed.
- - - - - - - - - - - - - - - - - - - - - - - - - - - - - - - - - - - -

Press Enter to Continue
```

In this block of text, we can see that there are two pieces of information, namely, record name and record value. In my case, the record name is _acme-challenge. jenkins-firewalled.lvin.ca and the value is kLmhtIfqI5PZFuk-lXna13Z4_ oIYDmaJoPd6RaFgwqQ. These values will be different for you. Please use the values that are displayed on your screen and do not use the values from the book.

In this section, let's see how we can add a TXT record. If you are not using Route 53 to manage your DNS, then use the links in the *Other DNS providers* section (*see* page 203) to add a TXT record to your DNS provider. If you are using Route 53, continue reading to learn how to add a TXT record to Route 53.

To add a TXT record, first, navigate to the Route 53 dashboard by searching for route53 in the search bar at the top, and then click on the appropriate search result as described in the *Navigating the AWS console* section earlier.

To add a TXT record, we need to first navigate to our hosted zone. A hosted zone is a container for all our DNS records, and it can only be associated with a single domain. The name of the hosted zone will be the same as our domain name.

To navigate to the hosted zone dashboard, click on the Hosted zone link under DNS management, as shown:

Route 53 Dashboard Info

Figure 5.22 – Link to the hosted zones

In the hosted zone dashboard, we will see a list of all our hosted zones. To add a DNS record to our specific hosted zone, we need to navigate to the dashboard for that hosted zone. To do that, click on your domain name. In my case, it is lvin.ca:

Figure 5.23 – Clicking on the hosted zone link

The dashboard for our hosted zone will show us the list of DNS entries that we have added previously. To add a new record, click on the Create record button. In the form that appears, fill in the following information:

- Record name: Enter the record name that is displayed by Let's Encrypt. In my case, it is _acme-challenge.jenkins-firewalled. Note that you need to remove the domain name before you enter the value. Let's Encrypt will display the record name as _acme-challenge.jenkins-firewalled.lvin.ca, but you only need to enter _acme-challenge.jenkins-firewalled.

- Record type: **Select** TXT.

- Value: Enter the record value that is displayed by Let's Encrypt. In my case, it is kLmhtIfqI5PZFuk-lXna13Z4_oIYDmaJoPd6RaFgwqQ:

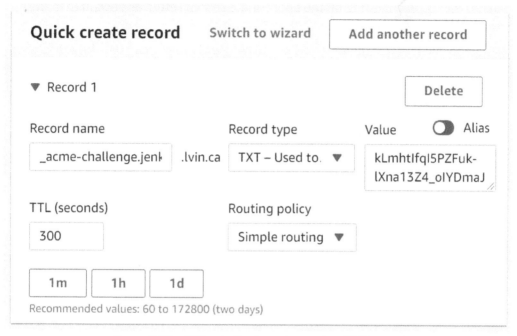

Figure 5.24 – Filling in the form to create a TXT record

Once we have entered all the details, create a record by clicking on the Create records button.

After a few seconds, we will see a green bar at the top of our windows that says the record was successfully created. Wait for about 15 minutes for the DNS propagation to complete and then continue with the Let's Encrypt process.

Now, let's look at how to use the automated verification method.

Automated verification for AWS Route 53

To use the automated verification process, we need to create an IAM user for Certbot with a policy that allows it to modify DNS entries in a Route 53-hosted zone. To use the automated verification method, we need to perform the following steps:

1. Find our hosted zone ID.
2. Create an IAM policy.
3. Create an IAM user.

Let's now look at how we perform each of these steps.

Step 1 – Find our hosted zone ID

Let's navigate to the Route 53 dashboard by searching for `route 53` in the search bar at the top and clicking on the appropriate search result, as described in the *Navigating the AWS console* section earlier.

A hosted zone is a container for all our DNS records, and it can only be associated with a single domain. The name of the hosted zone will be the same as our domain name. To navigate to the hosted zone dashboard, click on the Hosted zone link under DNS management, **as shown:**

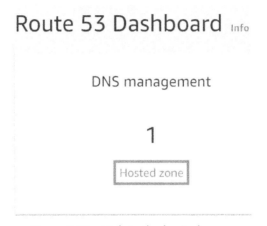

Figure 5.25 – Link to the hosted zones

In the hosted zone dashboard, we will see a list of all our hosted zones along with the hosted zone ID. Let's note down the hosted zone ID that is displayed under the Hosted zone ID column in the same row as our domain name, as shown:

Figure 5.26 – Finding the hosted zone ID

Now that we have the hosted zone ID, let's create an IAM policy.

Step 2 – Create an IAM policy

To create a policy, we must first navigate to the IAM dashboard. Navigate to the IAM dashboard by searching for IAM in the search bar at the top and clicking on the appropriate search result as described in the *Navigating the AWS console* section earlier.

To navigate to the policies section of the IAM dashboard, click on Policies in the left navigation pane:

▼ **Access management**

User groups

Users

Roles

Policies

Identity providers

Account settings

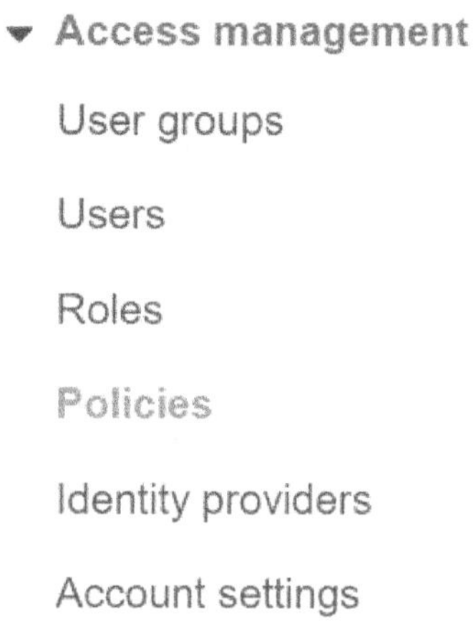

Figure 5.27 – Clicking on Policies in the left navigation pane

Now that we are in the policies section of the IAM dashboard, let's create a new policy that will allow Certbot to do the following:

- Get the status of a request to Route 53.
- List hosted zones in Route 53.
- Add record sets to our hosted zone in Route 53.

To do that, click on the Create policy button. We will be presented with a multi-page form. Let's look at how we can fill in this form, page by page.

Page 1 – Policy document

On this page, click on the JSON tab and then paste the following JSON document into the editor as shown. Please replace YOURHOSTEDZONEID with the hosted zone ID you copied in the previous section. *You can copy this JSON document from the README file located inside the folder for this chapter in the GitHub repository:*

```
{
    "Version": "2012-10-17",
    "Statement": [
        {
            "Effect": "Allow",
            "Action": [
                "route53:GetChange",
                "route53:ListHostedZones"
            ],
            "Resource": "*"
        },
        {
            "Effect": "Allow",
            "Action": "route53:ChangeResourceRecordSets",
            "Resource": "arn:aws:route53:::hostedzone/YOURHOSTEDZONEID"
        }
    ]
}
```

When done, our screen should look like the following screenshot. Notice that I have replaced YOURHOSTEDZONEID with my hosted zone ID on line 15:

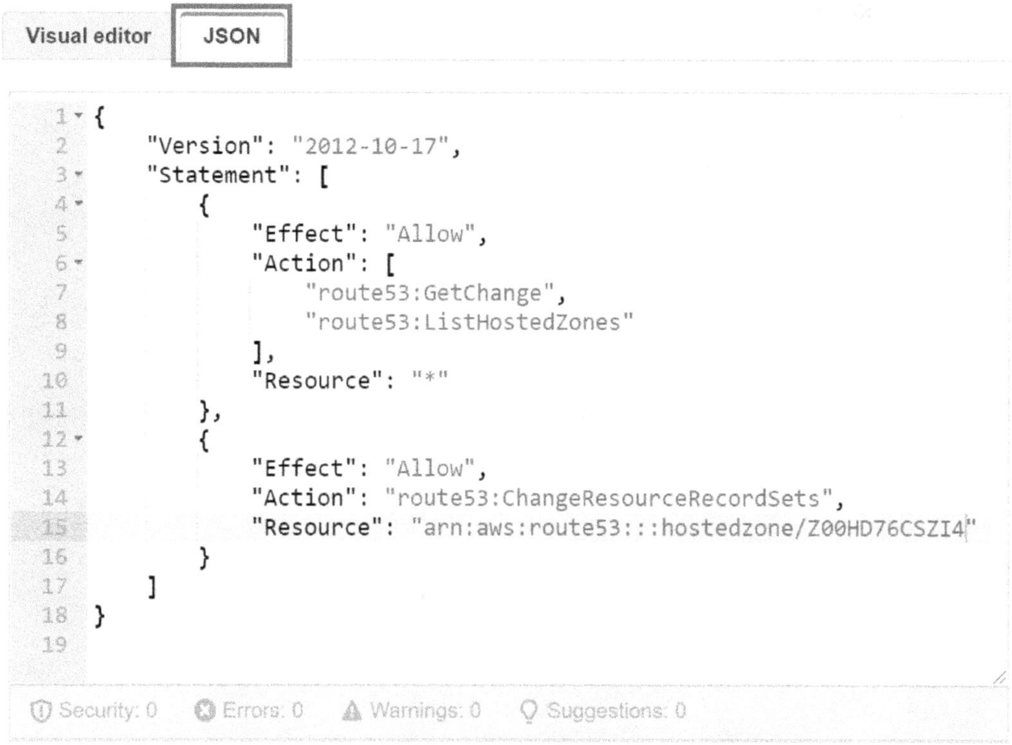

```
 1 ▾ {
 2       "Version": "2012-10-17",
 3 ▾     "Statement": [
 4 ▾         {
 5               "Effect": "Allow",
 6 ▾             "Action": [
 7                   "route53:GetChange",
 8                   "route53:ListHostedZones"
 9               ],
10               "Resource": "*"
11           },
12 ▾         {
13               "Effect": "Allow",
14               "Action": "route53:ChangeResourceRecordSets",
15               "Resource": "arn:aws:route53:::hostedzone/Z00HD76CSZI4"
16           }
17       ]
18   }
19
```

Figure 5.28 – Adding the policy document

When done, click on the Next: Tags button to go to the next page.

Page 2 – Tags

We do not need to add a name tag, unlike the other resources, because in the next screen, AWS gives us the option to add a name and description. So, click on the Next: Review button to go to the last page.

Page 3 – Review

On this page, enter the following information as shown:

- Name: `certbot-policy`
- Description: This policy allows the automated verification by the Lets Encrypt certbot.

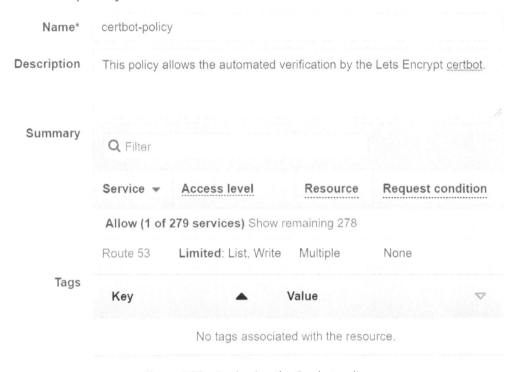

Review policy

Name*   certbot-policy

Description   This policy allows the automated verification by the Lets Encrypt certbot.

Summary

Q Filter

Service ▼	Access level	Resource	Request condition
Allow (1 of 279 services) Show remaining 278			
Route 53	**Limited**: List, Write	Multiple	None

Tags

Key ▲	Value ▽

No tags associated with the resource.

Figure 5.29 – Reviewing the Certbot policy

Once we are done entering the information, click on the Create policy button. In a few seconds, we should see a green banner at the top of our windows that says `certbot-policy` has been created.

Now that we have created the policy, let's look at how we can create an IAM user that will make use of this policy.

Step 3 – Create an IAM user

To create an IAM user, we must first navigate to the users portion of the IAM dashboard. To do that, click on Users in the left navigation pane as shown:

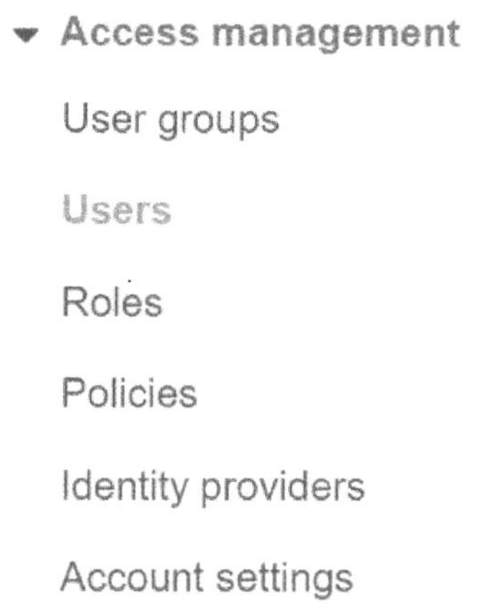

▾ **Access management**

User groups

Users

Roles

Policies

Identity providers

Account settings

Figure 5.30 – Clicking on Users in the left navigation pane

In the dashboard, click on the Add user button. We will be presented with a multi-page form. Let's now see how we fill in this form, page by page.

Page 1 – User details

On this page, we need to enter the name of the user and the type of access we want to grant it. We need to select programmatic access because Certbot will make use of the AWS APIs for Route 53 and it does not use the AWS Management Console. Fill in the following information as shown:

- User name: `jenkins-certbot`.
- Access type: **Select** Programmatic access:

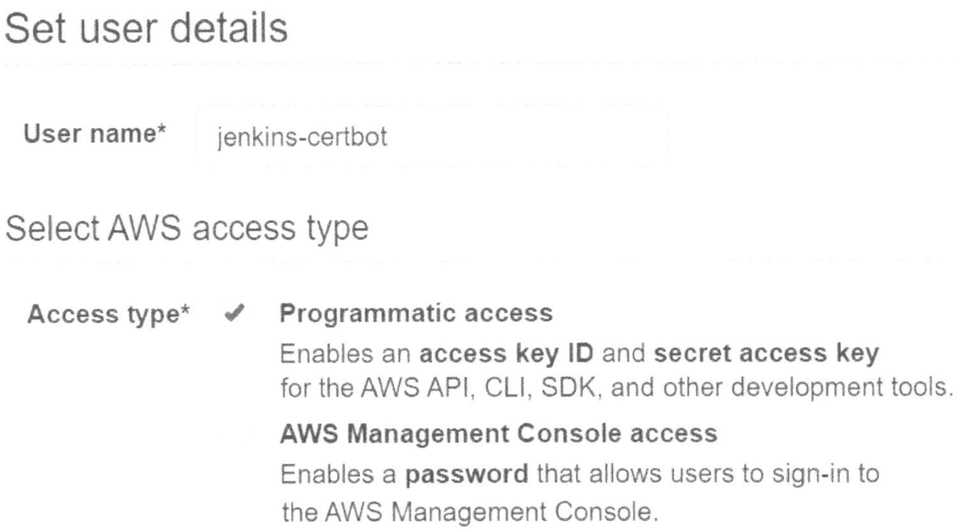

Set user details

User name*   jenkins-certbot

Select AWS access type

Access type* ✔ **Programmatic access**

Enables an **access key ID** and **secret access key**
for the AWS API, CLI, SDK, and other development tools.

AWS Management Console access

Enables a **password** that allows users to sign-in to
the AWS Management Console.

Figure 5.31 – Filling in the user information and type of access

When done, click on the Next: Permissions **button to go to the next page.**

Page 2 – Permissions

On this page, we will have to select the permissions for our user account. To do that,
click on Attach existing policies directly, and, in the search bar, search for cer tbot.
As we start typing, AWS will filter the list of results. Click on the checkbox next to
certbot-policy **as shown:**

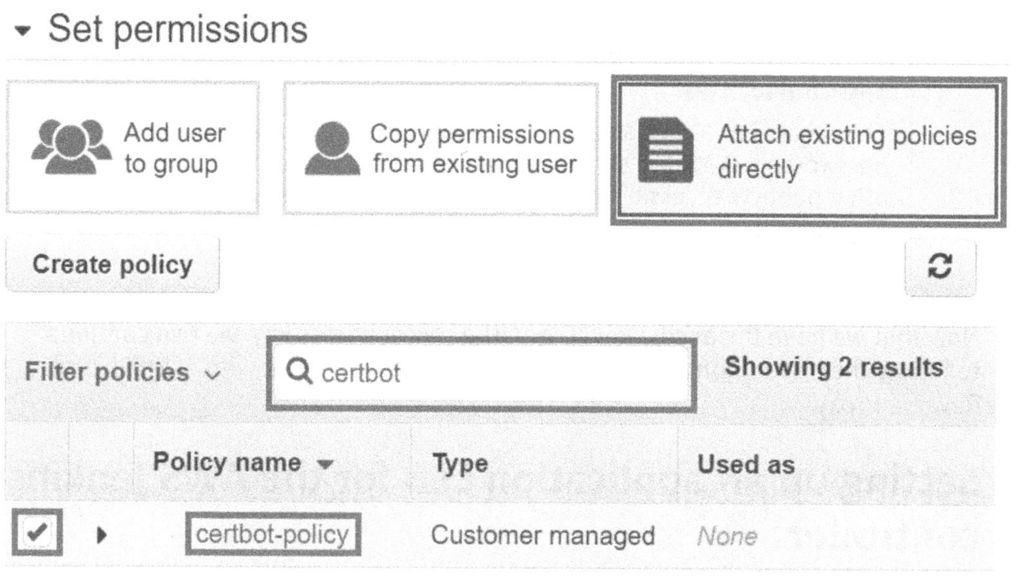

▾ Set permissions

| | Add user to group | | Copy permissions from existing user | | Attach existing policies directly |

Create policy   ↻

| Filter policies ⌄ | Q certbot | | Showing 2 results |

	Policy name ▾	Type	Used as
✓ ▸	certbot-policy	Customer managed	*None*

Figure 5.32 – Attaching a policy to the user

When done, click on the Next: Tags button to go to the next page.

Page 3 – Tags

We do not need to add any tags for this user as we have already given the user a name and we will be able to identify the user based on the name. Now, click on the Next: Review button to go to the next page.

Page 4 – Review

Review the information shown on screen and, if it looks good, click on the Create user button to go to the last page.

Page 5 – User credentials

In a few seconds, we will see a message that says Success, and below that message, we will see a Download csv button. Click on that button to download a CSV file that contains the access key ID and the secret access key. When we open this file in Microsoft Excel or any other text editor, we will see one row with the access key ID in the third column and the secret access key in the fourth column.

> **The access key ID and secret access key are like our username and password for AWS!**
>
> The access key ID and secret access key are like our username and password for AWS! We should not upload it to GitHub, Pastebin, or any other publicly accessible service, nor should we share it with anyone. Anyone with these credentials can access our AWS account and they will be able to blow through our wallet!

Now that we have the access key ID and the secret access key, we can continue to follow the instructions in *Chapter 1, Jenkins Infrastructure with TLS/SSL and Reverse Proxy*.

Setting up an application ELB for the AWS Jenkins controller

To set up an application ELB for the AWS Jenkins controller, we need to perform the following steps:

1. Create a TLS certificate in AWS Certificate Manager.
2. Create a security group.
3. Create an ALB.

Let's now look at how we perform each of these steps.

Step 1 – Create a TLS certificate in AWS Certificate Manager

A certificate manager is an application that handles the complexity of creating, storing, and renewing TLS certificates. AWS provides a managed service called Amazon Certificate Manager (ACM) for this purpose.

To create a certificate in ACM, you first need to navigate to the ACM dashboard. To do that, search for acm in the search bar at the top and then click on the appropriate search result as described in the *Navigating the AWS console* section earlier.

Depending on whether you have used ACM before or not, the screen that will be displayed will be different.

If you have not used ACM before, you will be presented with a screen with the heading AWS Certificate Manager in a big font at the center of the screen with two

Get started buttons. If you see this screen, click on the Get started button located below the Provision certificates sub-heading, as shown:

Provide the name of your site, establish your identity, and let ACM do the rest. ACM manages renewal of SSL/TLS certificates issued by Amazon or by your own private Certificate Authority.

Get started

Figure 5.33 – Button for provisioning certificates

If you have used ACM before, you will be presented with a dashboard like the EC2 dashboard. If you see this dashboard, click on the Request a certificate button.

The next screen that you will be presented with will be a multi-page form and this will be the same irrespective of the previous screen. Let's now look at how you fill in this form, page by page.

Page 1 – Certificate type

On this screen, we need to select whether we require a public or a private certificate. A public certificate will be trusted by all the browsers, whereas a private certificate will only be trusted by PCs that have the CA installed. Since we need the certificate to be trusted by all the browsers, let's select Request a public certificate, as shown:

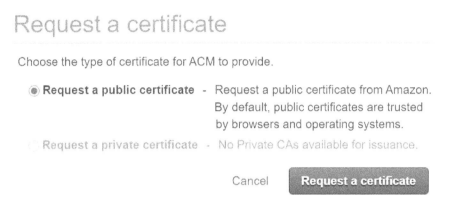

Figure 5.34 – Form for requesting a certificate

When done, click on the Request a certificate **button to go to the next page.**

Page 2 – Add domain names

On this page, we need to enter the domain name for our Jenkins instance. This domain name will be added to the certificate's common name. Let's enter the domain name that we want to use for Jenkins, as shown:

Figure 5.35 – Form for entering the domain name

When done, click on the Next button to go to the next page.

Page 3 – Select validation method

On this page, we need to select how we want AWS to validate ownership of our domain. The easiest way to prove ownership of our domain to AWS is DNS validation. As part of DNS validation, we need to add a TXT record to our domain. If we are using AWS Route 53, then AWS will automatically update Route 53 for us! Therefore, let's select DNS validation as shown:

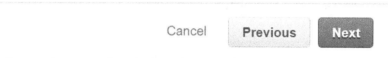

⦿ **DNS validation**

Choose this option if you have or can obtain permission to modify the DNS configuration for the domains in your certificate request. Learn more.

◯ **Email validation**

Choose this option if you do not have permission or cannot obtain permission to modify the DNS configuration for the domains in your certificate request. Learn more.

Cancel   **Previous**   **Next**

Figure 5.36 – Form for selecting the type of validation

When done, click on the Next button to go to the next page.

Page 4 – Add tags

On this page, we need to add tags. It is always a good practice to add tags to all the resources that we create. We need to add one tag with a key as the name so that it is easy to search for and identify the purpose of the certificate. Enter Name under Tag Name and enter Jenkins-certificate under Value, as shown:

Tag Name   Value

Name   Jenkins-certificate

Add Tag

Cancel   **Previous**   **Review**

Figure 5.37 – Form for adding tags

When done, click on the Review button to go to the next page.

Page 5 – Review

On this page, review the information on the screen and then click on the Confirm and request button to go to the next page.

Page 6 – Validation

This page will tell us all the information that we need in order to add a CNAME record. To view all the details, click on the arrow next to the name of your domain. Clicking it will expand to show you the DNS entry that you need to add, as shown in the following screenshot:

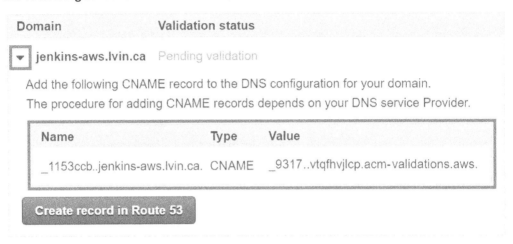

Figure 5.38 – Screen showing the DNS record set that must be added

If you are managing your DNS entries using Route 53, click on the Create record in Route 53 button. A dialog box will open showing the changes that will be made in your hosted zone; review them, and then click on the Create button.

If you are not managing your DNS using Route 53, open a new tab, log in to your DNS provider, and add the CNAME record. The instructions for adding a record vary significantly depending on the vendor you are using. Use the links in the *Other DNS providers* section (*see* page 203) to add the DNS record.

After adding the record set manually or AWS updating Route 53 automatically, you may have to wait for up to 15 minutes for the DNS propagation to complete. When complete, you will see the validation status change to Success, as shown in the following figure:

> **Important note**
> You do not have to refresh this page! AWS will automatically update the status in real time.

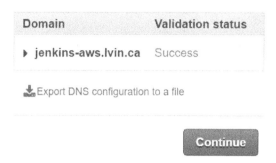

Figure 5.39 – Screen after successfully adding the record set

Once the validation status has changed to Success, click on the Continue button and you will be able to see your certificate in the dashboard.

Now that we have our TLS certificate, let's create a security group for our load balancer.

Step 2 – Create a security group

For the Application Load Balancer (ALB), we need to open ports 80 and 443 to the world. To do that, create a new security group as we did previously using the following information:

- Name: jenkins-lb.
- Description: Allow inbound for HTTP and HTTPS.
- VPC: Select default VPC.
- Inbound rules: Refer to the following table:

Type	Protocol	Port range	Source	Description
HTTP	TCP	80	Anywhere	Allow HTTP traffic
HTTPS	TCP	443	Anywhere	Allow HTTPS traffic

Now that we have a security group, let's create our application load balancer.

Step 3 – Create an ALB

An elastic ALB distributes incoming traffic among a set of targets. It also performs TLS termination, DDoS prevention, and path- and port-based routing. We need to create a load balancer that will redirect requests on port 80 to 443, perform TLS termination, and will forward requests on port 443 to the Jenkins controller instance on port 8080. To do this, we need to perform the following steps:

1. Create the ALB.

2. Update the ALB rules for redirecting requests on ports 80 to 443.

3. Add a CNAME DNS record to point to the ALB.

The reason why we need to perform *step 2* after we are done creating the ALB is that, as part of the load balancer creation form, AWS does not provide a way to update the load balancer rules.

Let's now look at how to create the ALB.

Step 3.1 – Create the ALB

To create an ALB, we need to first navigate to the EC2 dashboard. To do that, navigate to the EC2 dashboard by searching for ec2 in the search bar at the top and click on the appropriate search result as described in the *Navigating the AWS console* section earlier.

To navigate to the load balancers portion of the EC2 dashboard, click on Load Balancers under the Load Balancing heading in the left navigation pane, as shown in the following screenshot. We may need to scroll down to reveal this option depending on the screen size:

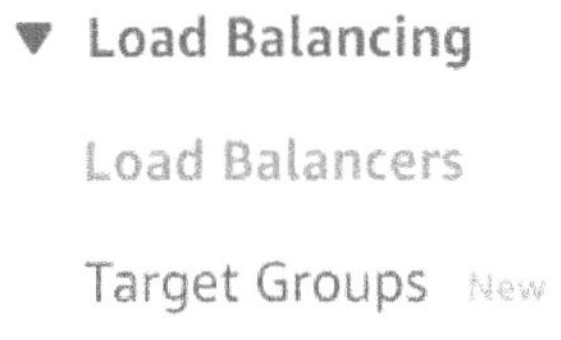

▼ **Load Balancing**

Load Balancers

Target Groups   New

Figure 5.40 – Left navigation pane showing the Load Balancers option

Let's now create an ALB that will redirect requests on ports 80 to 443, perform TLS termination, and forward requests on port 443 to the Jenkins controller instance on port 8080. To do that, click on the Create load balancer button. In the screen that appears, click on the Create button under the Application Load Balancer heading, as shown:

Figure 5.41 – Button for creating the ALB

A multi-page form will appear. Let's now look at how to fill in this form page by page.

Page 1 – Configure the Load Balancer

On this page, we need to fill in the basic information and some networking information. Fill in the following information under basic information:

- Name: Jenkins-lb
- Scheme: Internet facing

In the Listeners section, we need to enter the ports that we want the ALB to listen to and accept connections on. By default, the ALB will listen to and accept connections on port 80 as indicated by the first row in the table. For our use case, we need the ALB to accept connections on port 443 as well. To do that, click on the Add listener button. In the row that appears, select HTTPS (Secure HTTP) under the Load Balancer Protocol column, as shown:

Listeners

A listener is a process that checks for connection requests, using the protocol and port that you configured.

Load Balancer Protocol	Load Balancer Port	
HTTP ∨	80	⊗
HTTPS (Secure HTTP) ∨	443	⊗

Add listener

Figure 5.42 – Form showing the listeners for ALB

In the Availability Zones section, we need to select the subnets where we want to deploy the load balancer. We need to select at least two for high availability purposes. Select the default VPC in the VPC dropdown and then select the subnet that contains the Jenkins controller instance and one more subnet, as shown in the following screenshot:

Figure 5.43 – Form for picking the availability zones for ALB

Let's leave all the other settings as their default settings and then click on the Next: Configure Security Settings button at the bottom of the screen to go to the next page.

Page 2 – Configure Security Settings

On this page, we need to choose the certificate that we want the ALB to use while performing TLS termination. Since we created a certificate in ACM in the previous section, select the Choose a certificate from ACM radio button, and in the dropdown for Certificate name, select the certificate we created earlier:

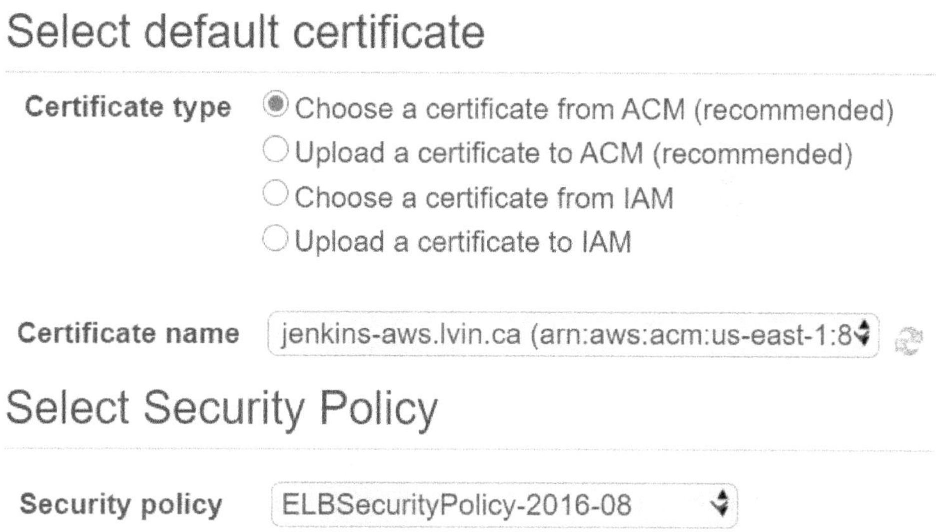

Select default certificate

Certificate type ⦿ Choose a certificate from ACM (recommended)
○ Upload a certificate to ACM (recommended)
○ Choose a certificate from IAM
○ Upload a certificate to IAM

Certificate name | jenkins-aws.lvin.ca (arn:aws:acm:us-east-1:8⇕ | ⟳

Select Security Policy

Security policy | ELBSecurityPolicy-2016-08   ⇕ |

Figure 5.44 – Form for selecting the default certificate

When done, leave all the other options as their defaults and click on the Next: Configure Security Groups button to go to the next page.

Page 3 – Configure Security Groups

On this page, we need to attach security groups to the ALB. We need to attach two security groups to the ALB, namely, the default security group and the security group we created for the load balancer earlier in this section. The default security group will enable the ALB to talk to the instances, and the jenkins-lb security group will allow the external requests to reach the ALB on ports 80 and 443. To do this, select Select an existing security group, and, in the list of security groups that appear, select the security group we created for the load balancer (jenkins-lb) and the default security group, as shown:

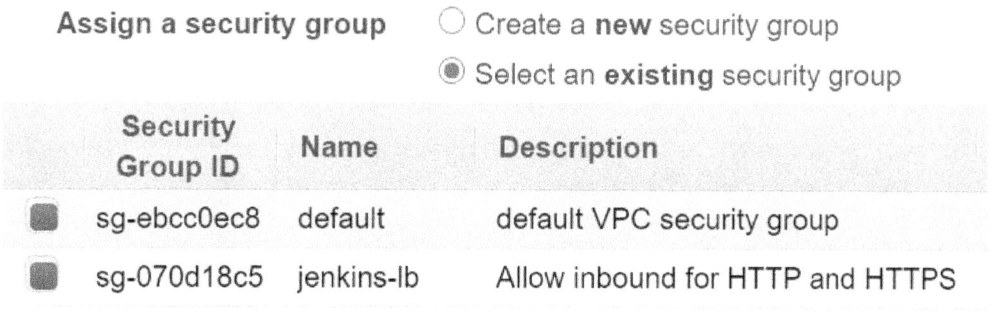

Assign a security group   ○ Create a **new** security group

⦿ Select an **existing** security group

	Security Group ID	Name	Description
■	sg-ebcc0ec8	default	default VPC security group
■	sg-070d18c5	jenkins-lb	Allow inbound for HTTP and HTTPS

Figure 5.45 – Form for selecting the security groups for ALB

When done, click on the Next: Configure Routing button to go to the next page.

Page 4 – Configure Routing

On this page, we need to create a target group that will hold all the instances that the ALB will load balance requests across. To create a target group, fill in the following information under the Target group section:

- Name: jenkins-tg
- Target type: Instance
- Protocol: HTTP
- Port: 8080

We need to change the health check path to /login because requests to / will fail with an HTTP unauthorized error code. When this happens, the ALB will register the node as unhealthy. This is a false positive. To prevent this, change Path to /login under the Health checks section, as shown:

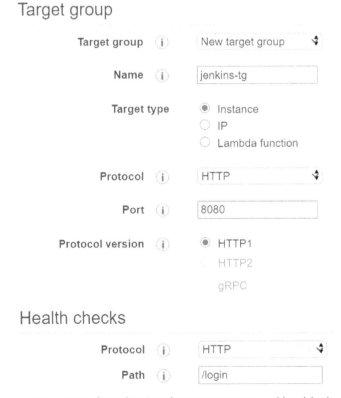

Figure 5.46 – Form for selecting the target group and health checks

When done, click on the Next: Register Targets button to go to the next page.

Page 5 – Register Targets

On this page, we need to select the instance to add to the target group. Since we need the ALB to send the requests to our Jenkins controller, select Jenkins-controller under Instances. **Next to the** Add to registered **button, enter 8080 in the** on port **text field and then click on the** Add to registered **button:**

Figure 5.47 – Table showing the list of instances for registration

When done, click on the Next: Review button to go to the last page.

Page 6 – Review

Let's review the information shown on screen and then click on the Create button at the bottom of the screen. After a few seconds, we will see a message that says Successfully created load balancer. Click on the Close button and it will take us back to the load balancer dashboard.

Now that we have created a load balancer, we need to enable the HTTP to HTTPS redirect. To do that, we need to update the load balancer rules. Let's now see how these can be updated.

Step 3.2 – Update the ALB rules for redirecting requests on ports 80 to 443

Let's select the Listeners tab in the bottom portion of the screen. In the table, select the checkbox to the left of HTTP : 80 and then click on the Edit button, as shown:

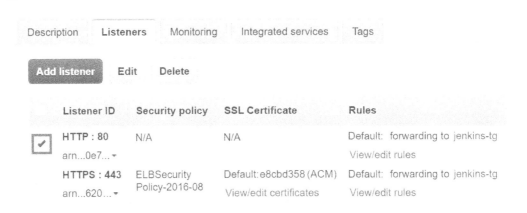

Figure 5.48 – Tab showing the listeners

Under the default action, click on the trash can icon to delete the default rule, as shown:

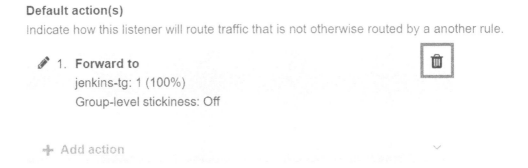

Figure 5.49 – Default actions for HTTP 80 listener

Click on the Add action button and select Redirect to..., as shown:

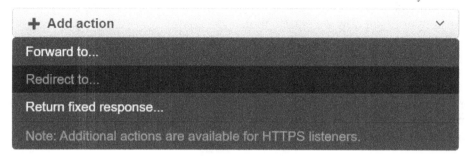

Figure 5.50 – Adding an action for ALB

Enter 443 for the port and then click on the blue tick mark, as shown:

Figure 5.51 – Adding HTTP to HTTPS redirect

When done, click on the Update button to save your changes.

In a couple of seconds, we will see a message that says successfully modified listener on port 80 appear next to the button. We now have a load balancer that can redirect requests on ports 80 to 443, perform TLS termination, and can forward requests on port 443 to the Jenkins controller instance on port 8080.

We have one last step remaining, in other words, to add a DNS record so that when we enter jenkins-aws.lvin.ca in our browser, we will be able to see our Jenkins instance. Let's now see how we can do that.

Step 3.3 – Add a CNAME DNS record to point to the ALB

We must first find the DNS name of our load balancer. To do that, click on the back button in our browser and it will take us back to the load balancer dashboard. In the bottom portion of the screen, click on the Description tab and copy the DNS name for our load balancer, as shown:

Figure 5.52 – Screen using the DNS name for the ALB

Log in to your DNS provider and add a CNAME record pointing to this DNS name. The name for the CNAME should be the domain name you want to use for Jenkins.

If you are not managing your DNS using Route 53, then open a new tab and log in to your DNS provider and add the CNAME record. The instructions for adding a record vary significantly depending on the vendor you are using. Use the links in the *Other DNS providers* section (*see page 203*) to add the DNS record.

If you are managing your DNS records using Route 53, navigate to the Route 53 dashboard by searching for route 53 in the search bar at the top and click on the appropriate search result, as described in the *Navigating the AWS console* section earlier.

To add a CNAME record, we first need to navigate to our hosted zone. A hosted zone is a container for all our DNS records, and it can only be associated with a single domain. The name of the hosted zone will be the same as our domain name.

To navigate to the hosted zone dashboard, click on the Hosted zone link under DNS management, as shown:

Figure 5.53 – Link to the hosted zones

In the Hosted zones dashboard, we will see a list of all of our hosted zones. To add a DNS record to our specific hosted zone, we need to navigate to the dashboard for our hosted zone. To do that, click on your domain name. In my case, it is `lvin.ca`:

Figure 5.54 – Clicking on the hosted zone link

The dashboard for our hosted zone will show the list of DNS entries that we have added previously. To add a new record, click on the Create record button. In the form that appears, fill in the following information:

- Record name: `jenkins-aws`.
- Record type: **Select** CNAME.

- Value: Enter the record value that you copied previously. In my case, it is `jenkins-lb-2058402629.us-eadt-1.elb.amazonaws.com`:

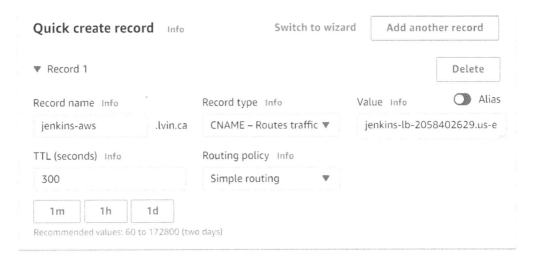

Figure 5.55 – Filling in the form to create a CNAME record

Leave the rest of the fields as their default settings and click on the Create records button. After a few seconds, we will see a green bar at the top of our windows that says Record jenkins-aws.lvin.ca was successfully created.

Wait up to 15 minutes for the DNS propagation to complete. Then open another tab in your browser, navigate to the domain name for your Jenkins instance, and you should see the home page of Jenkins load.

We now have our application ready and working. We can now continue with the remainder of the steps in *Chapter 2, Jenkins with Docker on HTTPS on AWS and inside a Corporate Firewall.*

Other DNS providers

If you are using a DNS provider other than Route 53, use the following links to learn how to add a new record:

- GoDaddy (CNAME record): https://in.godaddy.com/help/add-a-cname-record-19236

- GoDaddy (TXT record): https://in.godaddy.com/help/add-a-txt-record-19232

- Azure: https://docs.microsoft.com/en-us/azure/dns/dns-operations-recordsets-portal

- WordPress: https://wordpress.com/support/domains/custom-dns/

- Google Cloud: https://cloud.google.com/dns/docs/records

- Cloudflare: https://support.cloudflare.com/hc/en-us/articles/360019093151-Managing-DNS-records-in-Cloudflare

- Akami: https://learn.akamai.com/en-us/webhelp/edge-dns/edge-dns-user-guide/GUID-686C0A1A-511E-4C50-97A8-54B23CE68FFD.html

- Alibaba: https://partners-intl.aliyun.com/help/doc-detail/27144.htm

- Oracle Cloud: https://docs.oracle.com/en-us/iaas/Content/DNS/Tasks/managingdnszones.htm#Managing_DNS_Service_Zones

- IBM Cloud: https://cloud.ibm.com/docs/dns-svcs?topic=dns-svcs-managing-dns-records

That brings us to the end of this chapter. Let's now look back at what we have learned.

Summary

In this chapter, we saw click-by-click instructions for how to create and update the necessary services for running Jenkins on AWS. We learned how to create EC2 instances, security groups, ALBs, elastic IPs, and DNS records in Route 53. We also learned how to create the necessary resources on AWS for automated and manual verification for Let's Encrypt. We also learned how to update the security groups in case our IP changes. Using the skills we have learned in this chapter, we can easily and effortlessly spin up a Jenkins controller along with an agent and Docker cloud host on the AWS cloud.

Congratulations! You have successfully completed Part 1 of this book! You should have a production-grade Jenkins instance with premerge and postmerge pipelines for your product. In the second part of this book, we will learn how to scale Jenkins to serve a larger team, an organization, or even a company. In the next chapter, we will learn about Jenkins Configuration as Code (JCasC), which will allow us to codify the steps that we have performed so far to set up Jenkins.

Part 2

Jenkins Administration

In this section, you will learn how to manage Jenkins configuration as code using the Jenkins Configuration as Code (JCasC) plugin. You will also learn how to back up and restore Jenkins, along with preparing for disaster recovery. Finally, you will learn how to upgrade Jenkins to the next version.

This part of the book comprises the following chapters:

- *Chapter 6, Jenkins Configuration as Code (JCasC)*

- *Chapter 7, Backup and Restore and Disaster Recovery*

- *Chapter 8, Upgrading the Jenkins Controller, Agents, and Plugins*

6

Jenkins Configuration as Code (JCasC)

We've spent *Part 1* of the book setting up Jenkins by hand, one button click at a time. Wouldn't it be nice if all of this could be codified? It would allow us to download a configuration file and run it to have fully configured Jenkins out of the box.

Jenkins Configuration as Code (JCasC), officially introduced in *DevOps World - Jenkins World 2018*, is a solution heading in that direction. It provides a way to declaratively apply a configuration to the Jenkins controller, agents, and plugins. Its admirable stated goal is to *Configure ALL Jenkins initial setup* with *no hands on keyboard* and *no click on UI* (https://www.jenkins.io/projects/jcasc/), which fully resonates with my aspiration to automate all things.

It's not perfect of course. Most notably, it doesn't support importing credentials (although there are advanced workarounds for determined admins). This means that any configuration item that requires a secret, such as an agent, cannot be fully configured with JCasC. Agent is one of the most fundamental components of Jenkins, so I think it's safe to say that JCasC hasn't met its stated goal, at least at this time.

With the limitations in mind, let's see how far we can codify Jenkins using JCasC.

> **Progress!**
>
> In the short time between writing this chapter and publishing the book, the GitHub Pull Request Builder (GHPRB) plugin has added full support for JCasC. This is a long awaited feature, so kudos to GHPRB community for developing this feature. This means that, as you go through the chapter, you can ignore my wishing for the full JCasC support from GHPRB.

In this chapter, we're going to cover the following main topics:

- Downloading and understanding the current configuration
- Converting controller configuration to JCasC
- Converting agent configuration to JCasC
- Converting Docker cloud configuration to JCasC
- Converting pipeline configuration to JCasC
- Redeploying Jenkins using JCasC
- Reverting back to the original Jenkins
- Retrospective
- Advanced: JCasC Plugin – Groovy Scripting Extension

Technical requirements

You need the same VMs that were used in *Part 1*, as we will build a new Jenkins instance on them. The new instance will be configured by JCasC.

Files in the chapter are available in GitHub: https://github.com/PacktPublishing/Jenkins-Administrators-Guide/blob/main/ch6.

Downloading and understanding the current configuration

> **Required plugins**
> Configuration as Code

JCasC uses a YAML file to store its data. The structure of the YAML file is something we'll have to examine and learn about, but one of the best features of JCasC is that it allows us to download the current configuration in the JCasC configuration format. This significantly lowers the barrier to entry – if we can't figure out the JCasC syntax for a configuration, we can configure it manually then look up how JCasC codified it. Let's download and examine our current configuration:

1. Go to Manage Jenkins, Configuration as Code, and click View Configuration to see the current configuration. You should be able to identify various components, such as users, credentials, agents, Docker cloud, and other plugins. Some configurations, such as Jenkins URL, should look familiar.

2. Go back to Configuration as Code and click Download Configuration to save the current configuration as a file. It should be named `jenkins.yaml`. Open the file in a text editor to see that it contains the same YAML file that we saw earlier through View Configuration.

Can we set up a new Jenkins and apply this file to configure everything through the magic of JCasC? As the warning message says on the Configuration as Code page, it doesn't quite work that way:

> ⚠ Export is not intended to offer a directly usable jenkins.yaml configuration. It can be used for inspiration writing your own, be aware export can be partial, or fail for some components.

Figure 6.1 – Sad reality

Let's examine why.

User passwords aren't codified

Open the `jenkins.yaml` file and take a look at the `jenkins/securityRealm/local/users` key:

```
jenkins:
  securityRealm:
    local:
      allowsSignup: false
      enableCaptcha: false

      users:

      - id: "admin"
        name: "Calvin Park (Admin)"

        properties:
        - "apiToken"
        - mailer:
            emailAddress: calvinspark@gmail.com
        - "myView"
        - preferredProvider:
            providerId: "default"
        - "timezone"
        - userConfiguration:
            displayForReadOnlyPermission: true
```

Each user has the id, name, and several other keys, but the password key is missing. If this configuration file were applied on a new Jenkins where these users don't already exist, JCasC would create the users without a password, *disabling login.* Since all users are in the configuration file, *all users can't log in, including the admin.* This is obviously a problem. Jenkins running inside a corporate firewall is often configured with AD or LDAP for user login, so it's often simpler to just delete the entire users section and let it be recreated through new logins. Jenkins that uses the Jenkins' own user database is a bit more complex to manage. It's possible to manually add the missing password key to the configuration file and pre-populate it in clear text – users can log in using the pre-populated password then change the password. However, when this configuration file is applied again in the future to configure new plugins, the changed passwords will revert to the pre-populated value. Depending on the number of users, this may or may not be acceptable, let alone the security issues of having a clear text password written in a file. For Jenkins using Jenkins' own user database, I am yet to find an effective way to codify the user passwords.

Let's continue to examine perhaps the biggest gap in JCasC.

Secrets aren't portable

Take a look at the credentials key:

```
credentials:
  system:
    domainCredentials:
    - credentials:

      - usernamePassword:
          description: "dockerhub-calvinpark-userpass"
          id: "dockerhub-calvinpark-userpass"
          password: "{AQAAABAAAAAQ...xfQKUbLhLVlY1dg=}"
          scope: GLOBAL
          username: "calvinpark"
```

A secret is stored as an encrypted value; however, the key to decrypt it exists *only in the original Jenkins where the secret was generated from.* If this configuration file were applied on a new Jenkins, the credentials would be generated but their secrets would be incorrect because the new Jenkins wouldn't have the decryption key from the old Jenkins.

As a quick demonstration, let's decrypt a credential to see its secrets. Go to Manage Jenkins and click Script Console.

> **Be very careful with the Script Console**
>
> The Script Console is like a secret lair of super-admins with far too much power over the entire Jenkins. It gives you a textbox where you can run any Groovy code to manage Jenkins. Delete all jobs? Sure. Decrypt all secrets? Yessir. Send an email to all users with a picture of a cat? With pleasure. The code will execute as an admin *without any validations* and *the changes are immediate and irreversible*. It's a powerful tool so be *very* careful using it.

Type the following code into the textbox and click Run to decrypt the private key in the github-calvinpark-priv credential:

```
import com.cloudbees.plugins.credentials.CredentialsProvider as cp

def secrets = cp.lookupCredentials(com.cloudbees.plugins.
credentials.common.StandardCredentials.class)

for (secret in secrets)
  if (secret.id == "github-calvinpark-priv")
    println(secret.privateKey)
```

In the original Jenkins where the secret was created, the decrypted secret shows a correct value for the SSH private key:

 # Script Console

```
1  // Original Jenkins
2  import com.cloudbees.plugins.credentials.CredentialsProvider as cp
3
4  def secrets = cp.lookupCredentials(
5    com.cloudbees.plugins.credentials.common.StandardCredentials.class)
6
7  for (secret in secrets)
8    if (secret.id == "github-calvinpark-priv")
9      println(secret.privateKey)
10
11
```

Run

Result

```
-----BEGIN OPENSSH PRIVATE KEY-----
b3BlbnNzaC1rZXktdjEAAAAABG5vbmUAAAAEbm9uZQAAAAAAAAABAAABlwAAAdzc2gtcn
NhAAAAAwEAAQAAAYEAsiyBvl85LQrWypMSBsXbEGGaGFLOUOp6LkIIoEkZgDF+zrK9zbNF
```

Figure 6.2 – SSH private key correctly decrypted on Jenkins where the secret was created

In contrast, when the secret is exported using JCasC and imported to a different Jenkins, the decrypted secret shows an incorrect value:

 # Script Console

```
1  // New Jenkins
2  import com.cloudbees.plugins.credentials.CredentialsProvider as cp
3
4  def secrets = cp.lookupCredentials(
5    com.cloudbees.plugins.credentials.common.StandardCredentials.class)
6
7  for (secret in secrets)
8    if (secret.id == "github-calvinpark-priv")
9      println(secret.privateKey)
10
```

Run

Result

{AQAAABAAAAowlZeZqs9Ukr0RAn56CC2qD5PAlXVbxgSfF71YQhFmdpWOeXDFgZLeOKKB9gy2neUxj
YttXQcElmraZ+RiMdhLFZl/jcOVSjkY5FX98PgQL88rmfgXr7rHR789odbAAphP+bXQU4WCNcUHtvB
rdf2gLL4aZplSWEuNwdDPDpOakY+FmtvISa5dUe46jsQEv2z1dTbjQKUvSz2krXr7xpy0+Iv2FQQ6L

Figure 6.3 – A secret can't be decrypted on a new Jenkins where the secret was imported

Using this secret to push a Git tag will of course fail, which shows that the credentials can't be correctly configured through JCasC. This is a gap that the Jenkins developers are actively working to fill[1].

Don't try to transfer the decryption key from one Jenkins to another

Some determined people have managed to transfer the decryption key from the old Jenkins to the new, but it's not officially supported. I strongly advise you not to try this since it could corrupt even existing secrets, resulting in disabled login, broken agent connections, and broken pipelines. If you're thinking *challenge accepted!*, make a sandbox instance dedicated for the experiment.

Let's continue to see the final reason why we shouldn't use the existing configuration file without modifying it first.

---

1 https://github.com/jenkinsci/configuration-as-code-plugin/issues/1141

Most entries are auto-generated defaults

Step back to realize that most of the keys in the configuration file are autogenerated defaults. For example, there's a giant map of color names and hex codes under unclassified/ansiColorBuildWrapper/colorMaps. There is no reason for us to track this as a change, since installing the ansicolor plugin creates this entry. Nothing stops us from reapplying the defaults, but it's generally best practice to specify only the changes we need. In the coming sections, we'll identify the keys we've modified and build a concise JCasC configuration file.

> **Workarounds exist – just understand the limitations**
>
> You may have noticed that the password and secret decryption concerns are only applicable to building a new Jenkins from an existing configuration. We can even work around the secret decryption issue by first creating a blank Jenkins and backing up the blank Jenkins with its decryption key, then building a new Jenkins from the backed-up blank Jenkins. My goal in highlighting these issues is not to dissuade you from using JCasC but instead to give you ample warnings about the limitations as you're building a production-scale Jenkins for your company relying on JCasC. Many people are trying to solve the same problems as you are, so find the workaround that works for your situation.

In the coming sections, we will pick through the entries in jenkins.yaml to find the ones we need, then collect them to a new file named jcasc.yaml. We'll do this by first remembering a configuration that we set from *Chapter 1, Jenkins Infrastructure with TLS/SSL and Reverse Proxy*, through *Chapter 4, GitOps-Driven CD Pipeline with Docker Hub and More Jenkinsfile Features*, looking for that configuration in jenkins.yaml, then copying it to jcasc.yaml, where we keep just the entries we need.

For example, let's say you see a notation like this:

- Set # of executors to 0:

```
jenkins:
  numExecutors: 0
```

It means that we have "set # of executors to 0" sometime in the earlier chapters, and JCasC has recorded this change as jenkins/numExecutors: 0 in jenkins.yaml. Search for that entry in jenkins.yaml to confirm that you have the same entry, copy the entry to jcasc.yaml, then continue to the next item.

Let's start building a JCasC configuration file, starting with the controller configurations. Create a new file for `jcasc.yaml` and open it in a text editor to follow along.

Converting controller configuration to JCasC

Let's go through all the changes we made to the controller, then find out how to apply the same changes using JCasC by looking up the relevant section from `jenkins.yaml`:

- Install plugins

  When you look through the configuration file you've downloaded, you will find that there's no section for specifying the plugins that we want to install. This is because JCasC does not install plugins[2]. Thankfully, the `jenkins/jenkins` Docker image has a built-in plugin installer that we can use. Let's update our Dockerfile for Jenkins to install the plugins:

jenkins.dockerfile

```
FROM jenkins/jenkins:2.263.1-lts

RUN jenkins-plugin-cli --plugins \
    active-directory \
    ansicolor \
    cobertura \
    configuration-as-code \
    docker-plugin \
    docker-workflow \
    ghprb \
    junit

# From chapter 2
USER root
RUN  usermod -u 123 -g 30 jenkins
USER jenkins
```

  We will use this image to redeploy Jenkins once the JCasC configuration file is ready. Let's continue to the next item, which is our first real JCasC configuration.

---

2 https://github.com/jenkinsci/configuration-as-code-plugin#installing-plugins

- Set # of executors to 0:

```
jenkins:
  numExecutors: 0
```

As we discussed earlier, this means that we have set # of executors to 0 in our Jenkins, and JCasC recorded it as jenkins/numExecutors: 0. Look for this entry in your jenkins.yaml, confirm that you have the same entry, then copy it over to jcasc.yaml. Let's continue.

- Set Jenkins URL to https://jenkins-aws.lvin.ca/:

```
unclassified:
  location:
    url: https://jenkins-firewalled.lvin.ca/
```

The same as before, understand that this change is for setting the Jenkins URL. Look for this entry, confirm that you have it, copy it to jcasc.yaml, then continue.

- Set Pipeline Default Speed/Durability Level to Performance-optimized:

```
unclassified:
  globalDefaultFlowDurabilityLevel:
    durabilityHint: PERFORMANCE_OPTIMIZED
```

Understand that this change is for setting Pipeline Default Speed/Durability Level. Look for this entry in jenkins.yaml, confirm that you have it, and copy it over to jcasc.yaml.

Did you notice that the top-level entry is unclassified, which is the same as the previous item? jcasc.yaml is a YAML file, which means it shouldn't have a duplicate entry on the same level. When you copy this entry, make sure you merge this entry with the previous entry by putting globalDefaultFlowDurabilityLevel on the same level as location.

Your jcasc.yaml file should look like this at this time, with just one unclassified entry:

jcasc.yaml

```
jenkins:
  numExecutors: 0

unclassified:
  location:
    url: "https://jenkins-firewalled.lvin.ca/"

  globalDefaultFlowDurabilityLevel:
    durabilityHint: PERFORMANCE_OPTIMIZED
```

Let's continue.

- Configure Global Build Discarders to use the Log Rotation strategy and keep at most 100 builds:

```
unclassified:
  buildDiscarders:
    configuredBuildDiscarders:
    - simpleBuildDiscarder:
        discarder:
          logRotator:
            numToKeepStr: "100"
```

It's the same drill. Understand that this is for configuring Global Build Discarders, look for the entry, copy it, and merge it with the existing entries in jcasc.yaml.

- Set Authorization Strategy to Project-based Matrix Authorization Strategy. Give Job Discover permission to Anonymous Users, give three permissions to Authenticated Users, and give Administrator permission to the admin user:

```
jenkins:
  authorizationStrategy:
    projectMatrix:
      permissions:
      - "Credentials/View:authenticated"
      - "Job/Discover:anonymous"
      - "Overall/Administer:admin"
      - "Overall/Read:authenticated"
      - "View/Read:authenticated"
```

It's the same idea again, but notice that the top-level entry is jenkins not unclassified. This should be merged with the entry from our first entry like this:

jcasc.yaml snippet

```
jenkins:

  numExecutors: 0

  authorizationStrategy:
    projectMatrix:
      permissions:
      - "Credentials/View:authenticated"
      - "Job/Discover:anonymous"
      - "Overall/Administer:admin"
      - "Overall/Read:authenticated"
      - "View/Read:authenticated"
```

We're almost done.

- **Change** Markup Formatter **to** Safe HTML:

```
jenkins:
  markupFormatter:
    rawHtml:
      disableSyntaxHighlighting: false
```

The same as the previous entry, be sure to merge the entry with the existing top-level jenkins entry.

- **And finally,** in CSRF Protection, Enable proxy compatibility:

```
jenkins:
  crumbIssuer:
    standard:
      excludeClientIPFromCrumb: true
```

Again, merge this entry with the existing top-level jenkins entry.

Congratulations, you have codified the controller configurations! You can compare it with jcasc.controller.yaml from the book's GitHub repository (https://github.com/PacktPublishing/Jenkins-Administrators-Guide/blob/main/ch6/jcasc.controller.yaml). The order of the entries doesn't matter in a YAML file but the hierarchy does.

Let's continue to codify agent configuration.

Converting agent configuration to JCasC

Similar to the controller configuration, let's identify all the changes we've made for agent configuration, then find out how to apply the same changes using JCasC by looking up the relevant section of jenkins.yaml.

This time I won't repeat the same steps, but instead discuss the meaning of the entries more in depth:

- Save the SSH private key as a credential in the Jenkins credentials store:

```
redentials:
  system:
    domainCredentials:
    - credentials:
      - basicSSHUserPrivateKey:
          description: "firewalled-agent-robot_acct-priv"
          id: "firewalled-agent-robot_acct-priv"
          privateKeySource:
            directEntry:
              privateKey: "{AQAAABAu4YD...UN8IhfZyc5nQi}"
          scope: SYSTEM
          username: "robot_acct"
```

As discussed before, privateKey is stored as an encrypted value, and only the current Jenkins has the decryption key. In fact, you can see in the book's GitHub repository that I've posted the encrypted value for my private key, because I know that it can't be decrypted by your Jenkins.

That doesn't mean that this key is useless. We can use JCasC not only to make a new Jenkins but also to manage the configuration of an existing Jenkins. For example, storing this key as an encrypted value in the YAML file will allow us to recover it using the same Jenkins instance in case we lose the key. If we regularly save jenkins.yaml from JCasC in Git, we would even be able to tell when the secret has changed.

Even on a new Jenkins where we know that the decryption will fail, we still need to make the credential so that the pipelines relying on it won't point to a credential that doesn't exist. We'll correct the credentials once the new Jenkins is running.

- Create an SSH agent with the SSH private key secret and SSH host key:

```
jenkins:
  nodes:
  - permanent:
      labelString: "docker"
      launcher:
        ssh:
          credentialsId: "firewalled-agent-robot_acct-priv"
          host: "192.168.1.158"
          port: 22
          sshHostKeyVerificationStrategy:
            manuallyProvidedKeyVerificationStrategy:
              key: "ssh-rsa AAAABzaC1yc2...EAAAn6WLUDFE="
      name: "firewalled-firewalled-agent"
      numExecutors: 10
      remoteFS: "/home/robot_acct/firewalled-firewalled-agent"
      retentionStrategy: "always"
```

- Create a stub for the inbound agent:

```
jenkins:
  nodes:
  - permanent:
      labelString: "docker"
      launcher:
        jnlp:
          tunnel: "35.85.99.221:50000"
          workDirSettings:
            disabled: false
            failIfWorkDirIsMissing: false
            internalDir: "remoting"
      name: "inbound-agent"
      numExecutors: 10
      remoteFS: "/home/robot_acct/inbound-agent"
      retentionStrategy: "always"
```

Notice that the inbound agent connection can't be fully configured in JCasC – in addition to this stub, we'd need to SSH into the agent and initiate an inbound connection to the controller. The inbound connection originates from outside of Jenkins (that is, from the agent), therefore JCasC cannot configure it. That is to say that JCasC can only configure the items inside Jenkins but not outside. In contrast, the SSH agent connection originates from Jenkins, therefore it can be fully configured in JCasC as we saw earlier.

Merge the values into the same jcasc.yaml file. We have now codified the controller and the agent configurations. You can see the values you should have at this point in jcasc.controller.agent.yaml from the book's GitHub repository (https://github. com/PacktPublishing/Jenkins-Administrators-Guideblob/main/ch6/jcasc.controller. agent.yaml).

Let's continue to configure Docker cloud.

Converting Docker cloud configuration to JCasC

Let's identify all the changes we've made for Docker cloud, then find out how to apply the same changes using JCasC by looking up the relevant section of jenkins. yaml:

- Save the Docker client's X.509 certificates as a secret in the Jenkins credentials store:

```
credentials:
  system:
    domainCredentials:
    - credentials:
      - x509ClientCert:
          clientCertificate: |-
            -----BEGIN CERTIFICATE-----
            MIIFWzCCA0OgAwIBAgIUP3ixn0CHkruSk0EBH3SfcBIJ
            ...
            WgTeMtOfDU5FWQ247XxV2stSZksnq375Gl8TPlsTIw==
            -----END CERTIFICATE-----
          clientKeySecret: "{AQAAABAAAwO...5H0RBLkc5}"
          description: "docker-host-client"
          id: "docker-host-client"
          scope: SYSTEM
          serverCaCertificate: |-
            -----BEGIN CERTIFICATE-----
            MIIGKzCCBBOgAwIBAgIUUWfTDPdDwDs/3dr7KXVUqhZWN
            ...
            MmnH5/8Xh94Yn4YnCRPbr/pirxB8MwsVysv0LVyXTj+g=
            -----END CERTIFICATE-----
```

Notice that the public certificates, clientCertificate and serverCaCertificate, are stored in clear text, while the secret clientKeySecret is encrypted, even though the whole entry is a part of the credentials section. Jenkins does a good job of telling the actual secret apart from the non-secret configuration fields.

- Create and configure Docker cloud:

```
jenkins:
  clouds:
  - docker:
      dockerApi:
        connectTimeout: 60
        dockerHost:
          credentialsId: "docker-host-client"
          uri: "tcp://192.168.1.14:2376"
        readTimeout: 60
      name: "docker"
```

- Create a Docker agent template. This part goes under the cloud configuration on the same level as name:

```
      templates:
      - connector: "attach"
        dockerTemplateBase:
          cpuPeriod: 0
          cpuQuota: 0
          image: "jenkins/agent"
        labelString: "linux"
        pullStrategy: PULL_ALWAYS
        pullTimeout: 300
```

We spent a significant amount of effort in generating the certificates and configuring the Docker daemon service. However, that's all done outside of Jenkins, so JCasC can't be used to automate those tasks. I'm not blaming JCasC for this – Docker cloud host configurations and certificate generations are clearly outside of Jenkins' control. We just need to understand at this point that codifying Jenkins end to end requires more than just JCasC. To configure these specific items (the certificates and the Docker daemon), configuration management tools such as Chef, Puppet, Salt, or Ansible can fill the gap.

Merge the values into the same jcasc.yaml file and compare them against jcasc. controller.agent.dockercloud.yaml from the book's GitHub repo (https://github. com/PacktPublishing/Jenkins-Administrators-Guide/blob/main/ch6/jcasc.controller. agent.dockercloud.yaml). We have now codified the controller, agents, and Docker cloud configurations.

Let's continue to configure pipelines.

Converting the pipeline configurations to JCasC

Let's identify the changes we made for the CI/CD pipeline configurations, and see how to apply the same changes using JCasC by looking up the relevant section of jenkins.yaml:

- Save the GitHub personal access token, SSH key for GitHub, and Docker login information as credentials in the Jenkins credentials store:

```
credentials:
  system:
    domainCredentials:
    - credentials:
      - string:
          description: "github-calvinpark-pat"
          id: "github-calvinpark-pat"
          scope: GLOBAL
          secret: "{AQAAABAAAAAwjW... El/Jqf2SbXuq3ZSk}"
      - basicSSHUserPrivateKey:
          description: "github-calvinpark-priv"
          id: "github-calvinpark-priv"
          privateKeySource:
            directEntry:
              privateKey: "{AQAAABAAAAowlZeZqs9Uk...}"
          scope: GLOBAL
          username: "calvinpark"
      - usernamePassword:
          description: "dockerhub-calvinpark-userpass"
          id: "dockerhub-calvinpark-userpass"
          password: "{AQAAABAAAAAQk2...xfQKUbLhLVlY1dg=}"
          scope: GLOBAL
          username: "calvinpark"
```

- In System Configuration, set the GHPRB credentials and description.

  These can't be configured with JCasC natively because the GHPRB plugin doesn't fully support JCasC. There is the Configuration as Code – Groovy Scripting Extension plugin that allows us to configure the lower-level settings using Groovy, which could be useful if you're ready to dive into the plugin source code.

- In System Configuration, set Commit status Context:

```
unclassified:
  ghprbTrigger:
    extensions:
    - ghprbSimpleStatus:
        addTestResults: false
        commitStatusContext: "Jenkins CI"
        showMatrixStatus: false
```

This is the only GHPRB configuration that can be managed through JCasC.

- Configure the GHPRB plugin on the pipeline.

    JCasC doesn't handle pipeline configurations – that's the Job DSL plugin's responsibility. Therefore, these configurations must be managed outside of JCasC either manually or by creating a Job DSL project. This adds to the list of technologies required to fully configure Jenkins out of the box.

In addition, there are configuration changes on GitHub to send a push hook and require a successful build, but GitHub configurations are of course outside of Jenkins' responsibility, therefore JCasC doesn't manage those configurations.

Merge the values into the same jcasc.yaml file. We are done making a JCasC configuration file that represents our Jenkins! You can compare it with jcasc.controller.agent.dockercloud.pipeline.yaml on the book's GitHub repository (https://github.com/PacktPublishing/Jenkins-Administrators-Guide/blob/main/ch6/jcasc.controller.agent.dockercloud.pipeline.yaml). Let's use the configuration file to redeploy Jenkins.

Redeploying Jenkins using JCasC

We'll redeploy Jenkins using the new Dockerfile with plugins preinstalled and the JCasC configuration file:

1. SSH into the controller host.

2. Build a custom Docker image for the controller with plugins preinstalled and name it <Docker Hub ID>/jenkins:2.263.1-plugins:

```
controller:~$ docker build -t calvinpark/jenkins:2.263.1-plugins
-f jenkins.dockerfile .
```

During the image build, we can see that the plugins are installed.

3. Create a new directory named jcasc_home, then copy jcasc.yaml into the jcasc_home directory:

```
controller:~$ mkdir jcasc_home
controller:~$ cp jcasc.yaml jcasc_home/
```

4. Stop and delete the existing Jenkins controller to release the ports (don't worry, we'll get it back):

```
controller:~$ docker stop jenkins_controller
controller:~$ docker rm jenkins_controller
```

5. Finally, run the new Jenkins Docker image. Set jcasc_home as the data directory by using a new bind mount option, -v ~/jcasc_home:/var/jenkins_home, and name it jcasc_controller:

```
controller:~$ docker run \
    --detach \
    --restart on-failure \
    -u $(id -u):$(id -g) \
    -v ~/jcasc_home:/var/jenkins_home \
    -p 8080:8080 -p 50000:50000 \
    --name jcasc_controller \
    calvinpark/jenkins:2.263.1-plugins
```

6. In about a minute, the initial boot will complete. Look up the initial admin password from docker logs jcasc_controller, open a web browser and go to the Jenkins URL, then enter the admin password in the browser to unlock Jenkins.

7. On the Customize Jenkins page, click Install suggested plugins to move ahead. We've already preinstalled the additional plugins in the custom Docker image. We didn't, however, list all the default plugins in it, so we're installing them now.

8. On the Create First Admin User page, create an admin user as we did in the past. Make sure the username is admin because that's the username we've given Administrator permission in the JCasC configuration.

9. On the Instance Configuration page, Jenkins URL should already be correct (for example, https://jenkins-firealled.lvin.ca/). As an experiment, let's change it to http://changeme to see whether JCasC corrects it:

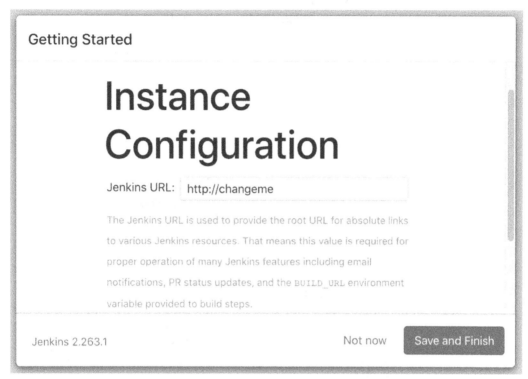

Figure 6.4 – Set Jenkins URL to an incorrect value to test whether JCasC corrects it

Jenkins is ready!

10. Let's see if the plugins are installed. Click Manage Jenkins, Manage Plugins, then Installed. Sure enough, the plugins are already installed:

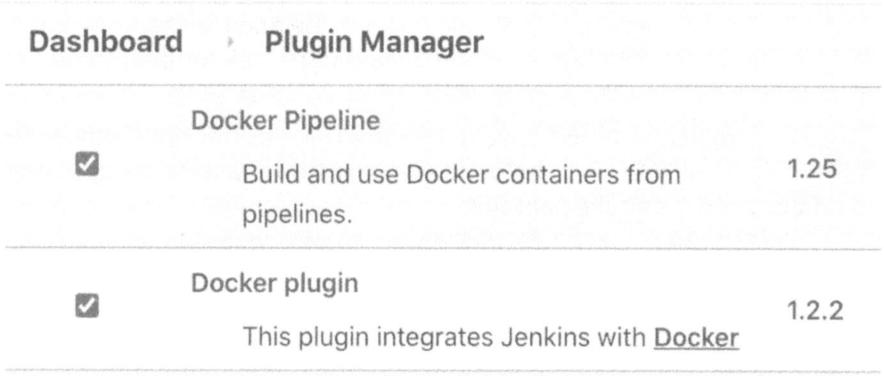

Dashboard › Plugin Manager

☑	**Docker Pipeline** Build and use Docker containers from pipelines.	1.25
☑	**Docker plugin** This plugin integrates Jenkins with <u>Docker</u>	1.2.2

Figure 6.5 – Plugin Manager showing preinstalled Docker plugins

11. You can tell that the other configurations aren't done yet because there are no agents attached and Build Executor Status is showing two executors for the controller. # of executors is one of the first things we changed:

Build Executor Status ⌃

1 Idle

2 Idle

Figure 6.6 – No agents and two executors configured for the controller

It's time to run JCasC. Click Manage Jenkins and Configuration as Code. On Path or URL, enter /var/jenkins_home/jcasc.yaml.

> **/var/jenkins_home/ not /var/jcasc_home/**
> Even though the new directory is ~/jcasc_home/, we have mounted it inside the container as /var/jenkins_home/ as we did before. Therefore, to the eyes of Jenkins, which lives inside the container, the JCasC configuration file is in /var/jenkins_home/jcasc.yaml.

12. Before clicking the Apply new configuration button, click anywhere outside of the textbox to have Jenkins validate the configuration file.

If the configuration file can't be validated, perhaps because the file doesn't exist, the file is incorrectly formatted, or the values are invalid, Jenkins either provides an error message or outright crashes showing the Fire Jenkins icon. In this case, look for a typo in the configuration file and try again. We can also see the Jenkins log (https://<Jenkins URL>/log/all) to look for information about the problem.

Once the problem is found, we can edit the YAML file directly without stopping the jcasc_controller container because the host directory ~/jcasc_home/ is bind-mounted inside the container:

Replace configuration source with:

Path or URL   `/var/jenkins_home/jcasc.yaml`

Oops!

Figure 6.7 – Jenkins couldn't parse the JCasC configuration file

If the configuration file is valid, Jenkins reports The configuration can be applied:

Dashboard   ›   Configuration as Code

Configuration as Code

Controller has no configuration as code file set.

Replace configuration source with:

Path or URL   `/var/jenkins_home/jcasc.yaml`

The configuration can be applied

Apply new configuration

Figure 6.8 – JCasC configuration file successfully validated

13. Click Apply new configuration. In a few seconds, the page reloads and reports that the configuration file is loaded:

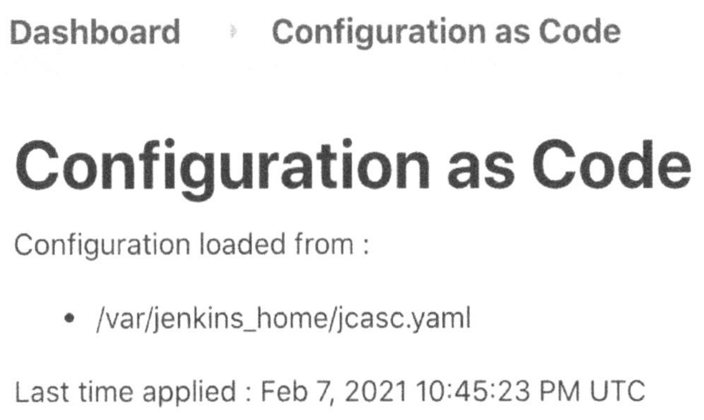

Dashboard   Configuration as Code

Configuration as Code

Configuration loaded from :

- /var/jenkins_home/jcasc.yaml

Last time applied : Feb 7, 2021 10:45:23 PM UTC

Figure 6.9 – JCasC configuration file successfully applied

Let's go through the configuration items and see how they worked out. Go to the Jenkins home page and you'll immediately see on the left that there are two agents. They are, however, disconnected – we'll examine why shortly.

First, go to System Configurations:

- # of executors is correctly set to 0.

- Jenkins URL is correctly changed from http://changeme to https://jenkins-firewalled.lvin.ca/.

- Pipeline Default Speed/Durability Level is correctly set to Performance-optimized.

- Global Build Discarders is correctly configured with Special Build Discarder set to keep 100 builds.

- GitHub Pull Request Builder has two sections. The first is setting the credentials, which wasn't supported by the GHPRB plugin, and the second is setting Commit Status Context to Jenkins CI. The second configuration is correctly set. The first configuration is a bit curious. Even though we didn't specify any configurations for setting the credentials, the plugin defaulted to the dockerhub-calvinpark-userpass credential, as shown in *Figure 6.10*. It's interesting that a default configuration is not *unconfigured* but instead *misconfigured*:

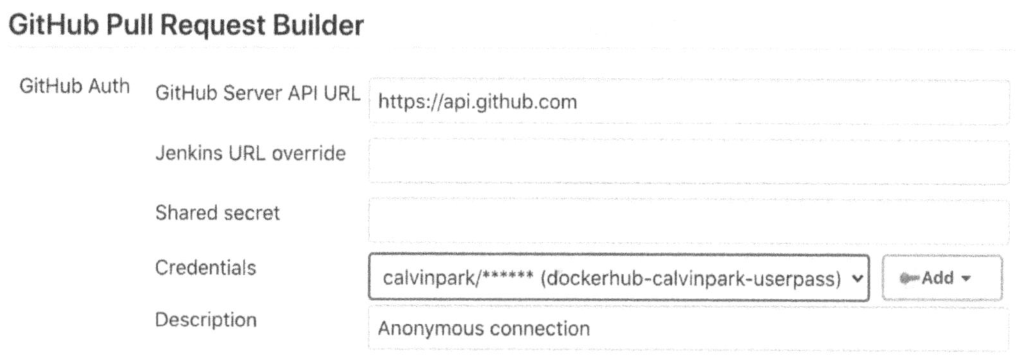

GitHub Pull Request Builder

GitHub Auth	GitHub Server API URL	https://api.github.com
	Jenkins URL override	
	Shared secret	
	Credentials	calvinpark/****** (dockerhub-calvinpark-userpass) ⌄ ⊕ Add ⌄
	Description	Anonymous connection

Figure 6.10 – GHPRB sets the Credentials field by itself, sometimes to a wrong value

Next, go to Global Security:

- Authorization Strategy **is correctly set to** Project-based Matrix Authorization Strategy **and the five permissions are correctly configured.**

- Markup Formatter **is correctly set to** Safe HTML.

- CSRF Protection/Crumb Issuer **has** Enable proxy compatibility **correctly checked.**

Next, go to Global Credentials:

- **The five credentials were created with the public fields** (ID, Description, Client Certificate, **and so on**) populated with correct values. However, we know from the previous demonstration that the private fields (Secret, Private Key, **and** Client Key) hold incorrect values. We'll have to update them.**

Next, go to Manage Jenkins **then** Manage Nodes and Clouds:

- **Both the SSH agent and the inbound agent are correctly created. However, the SSH agent connection log shows that the authentication failed. We will fix this issue after we look through the rest of the items:**

Dashboard › Nodes › firewalled-firewalled-agent

```
SSHLauncher{host='192.168.1.158', port=22, credentialsId='firewalled-agent-robot_acct-priv',
jvmOptions='', javaPath='', prefixStartSlaveCmd='', suffixStartSlaveCmd='',
launchTimeoutSeconds=60, maxNumRetries=10, retryWaitTime=15,
sshHostKeyVerificationStrategy=hudson.plugins.sshslaves.verifiers.ManuallyProvidedKeyVerifica
tionStrategy, tcpNoDelay=true, trackCredentials=true}
[02/08/21 00:53:19] [SSH] Opening SSH connection to 192.168.1.158:22.
[02/08/21 00:53:19] [SSH] SSH host key matched the key required for this connection.
Connection will be allowed.
ERROR: Server rejected the 1 private key(s) for robot_acct (credentialId:firewalled-agent-
robot_acct-priv/method:publickey)
ERROR: Failed to authenticate as robot_acct with credential=firewalled-agent-robot_acct-priv
java.io.IOException: Publickey authentication failed.
```

Figure 6.11 – SSH agent connection failed due to the corrupted private key

The inbound agent is offline as expected since it requires that a connection command is run from the agent host. It's not JCasC's fault that the inbound agent isn't fully configured.

Next, click Configure Clouds:

- Docker cloud is correctly configured, and the template is correctly created. Testing the connection fails because of the credential issue we saw previously:

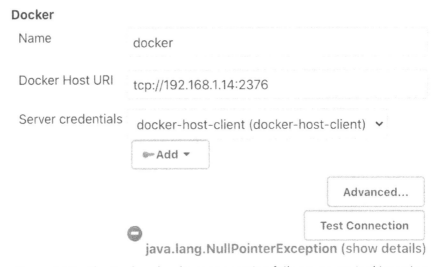

Figure 6.12 – The Docker cloud test connection fails on a new Jenkins using an imported secret

That's the entire list of settings we've configured in the JCasC configuration file. Let's fix the credentials so that agents come online and plugins begin working:

1. Go to Global Credentials and open all five credentials in separate tabs.

2. On each tab, click Update on the left and click Replace or Change Password to enter a new value.

   SSH keys are still available on the various hosts so you can look them up and add them again. The GitHub personal access token is a bit tricky since GitHub doesn't let us look up an existing key. The easiest way would be to simply create a new token. Once the secrets are updated, we can connect the agents and the cloud and validate the permissions on the GHPRB plugin.

And that, ladies and gentlemen, is how you use JCasC to configure a new Jenkins. Now let's return to the old Jenkins.

Reverting back to the original Jenkins

If you don't want to keep this Jenkins, it's easy to go back to the old one – simply delete the current controller and run the original again. Since everything in Jenkins is a flat file on the controller, all the configurations have persisted. You won't have to reconfigure anything, and you'll be back to right where you left it:

```
controller:~$ docker stop jcasc_controller
controller:~$ docker rm jcasc_controller
controller:~$ docker run \
    --detach \
    --restart on-failure \
    -u $(id -u):$(id -g) \
    -v ~/jenkins_home:/var/jenkins_home \
    -p 8080:8080 -p 50000:50000 \
    --name jenkins_controller \
    calvinpark/jenkins:2.263.1-lts
```

Open a browser and navigate to the Jenkins URL. You should be able to log in as adder-admin and start adding more pipelines to the adder folder. If you're familiar with Docker commands, you can run multiple Jenkins instances on the same machine using different ports to test new features.

Now that we're back in the comfort of the original Jenkins, let's look back and think about the state of JCasC.

Retrospective

JCasC is a powerful tool with active development still in progress. It gets a lot of the configurations codified, and while it's not perfect, it's *almost* there.

With the vast number of plugins in the Jenkins ecosystem, of course it's a challenge to wrangle every developer to make their plugins compatible with JCasC. It's disappointing when a plugin as popular as GHPRB still doesn't fully support JCasC, but GHPRB developers are aware of the missing feature and are working on addressing it[3]. For the people who really want to automate GHPRB configuration, the Configuration as Code Plugin - Groovy Scripting Extension plugin is available to fill the gap.

Also, Jenkins developers are working on the credentials import/export issue[4], so I think that in a couple more years JCasC will reach its full potential. As a CI/CD engineer who uses Ansible and dozens of Groovy scripts to manage Jenkins, I eagerly look forward to JCasC being completed.

Advanced: CasC Plugin – Groovy Scripting Extension

> **Required plugins**
>
> Configuration as Code Plugin - Groovy Scripting Extension

That's enough complaining! We're software engineers. Let's figure out how to use Configuration as Code Plugin – Groovy Scripting Extension (CasCP-GSE) to configure GHPRB.

> **Not required but still useful**
>
> As noted earlier, GHPRB has added full support for JCasC since the chapter was written, which means that CasCP-GSE is no longer required for configuring GHPRB. I have still kept this section because it demonstrates a way to configure a plugin using Groovy - this is a very useful skill to have because it can be applied to any plugins or any Jenkins configurations. In addition, it serves as a good introduction to learning about the internals of Jenkins for those who want to develop a custom plugin. The code and the development steps are still correct and would still configure GHPRB, so feel free to follow along to level up your knowledge.

---

3   https://github.com/jenkinsci/ghprb-plugin/pull/731#issuecomment-750466406

4   https://github.com/jenkinsci/configuration-as-code-plugin/issues/1141

The plugin home page (https://plugins.jenkins.io/configuration-as-code-groovy/) has an example configuration that allows us to run any Groovy script:

```
groovy:
  - script: >
      println("This is Groovy script!");
```

The home page also gives us a couple of pointers:

- It is recommended to use semicolons at the end of lines.

- There is no dry run implemented for the Groovy scripts feature.

If you are an experienced Jenkins administrator, this example and the pointers should remind you of an existing tool – this plugin is a gateway to the Script Console.

The Script Console allows us to modify the Jenkins object in the Java Virtual Machine. That means we can change anything we want on Jenkins, including installing plugins, configuring plugins, and creating pipelines. For the intermediate and advanced Jenkins administrators who have experience running Groovy scripts on the Script Console, it's easy to see that this plugin solves nearly every problem. The infrastructure and the credential management still require an external tool, but any configuration items inside Jenkins can be managed with Groovy scripts.

Let's go back and see the necessary GHPRB configuration. This is what we configured using the GUI

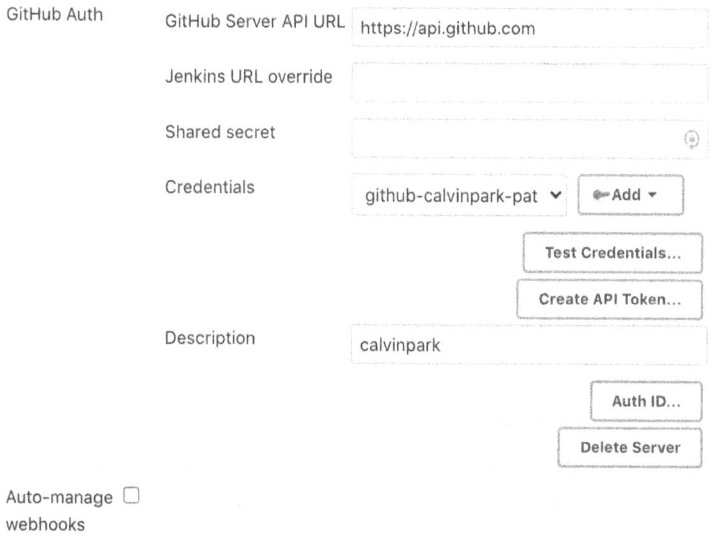

Figure 6.13 – GHPRB configurations in System Configuration

The GitHub page for the GHPRB plugin (https://github.com/jenkinsci/ghprb-plugin) has a Groovy script we can use to configure this:

```
import jenkins.model.*
import org.jenkinsci.plugins.ghprb.*

GhprbTrigger.DescriptorImpl descriptor = Jenkins.instance.
getDescriptorByType(org.jenkinsci.plugins.ghprb.GhprbTrigger.
DescriptorImpl.class)

List<GhprbGitHubAuth> githubAuths = descriptor.getGithubAuth()

String serverAPIUrl = 'https://api.github.com'
String jenkinsUrl = 'https://your.jenkins.url/'
String credentialsId = 'credentials-id'
String description = 'Anonymous connection'
String id = 'github-auth-id'
String secret = null

githubAuths.add(new GhprbGitHubAuth(serverAPIUrl, jenkinsUrl,
credentialsId, description, id, secret))

descriptor.save()
```

Reading the code, it's not difficult to match the String parameters to the configuration items seen in *Figure 6.13*. Here is the mapping from the variable to the input field name on the GUI, along with our desired value:

- serverAPIUrl: GitHub Sever API URL (https://api.github.com)
- jenkinsUrl: Jenkins URL override (empty)
- credentialsId: Credentials (github-calvinpark-pat)
- description: Description (calvinpark)
- id: Auth ID button (autogenerated UUID)
- secret: Shared secret (empty)

With the mapping information, we need to make four additional changes to the code:

- We can see in the GHPRB source (https://github.com/jenkinsci/ghprb-plugin/blob/master/src/main/java/org/jenkinsci/plugins/ghprb/GhprbGitHubAuth.java#L86-L94) that serverAPIUrl defaults to https://api.github.com if it's empty, therefore we can set it to null.
- The source also shows that id defaults to an autogenerated UUID if it's empty, so we should set that to null, too.
- We want to have only one GitHub Auth configured, so we need to modify the script to create a *new list* for githubAuths (as opposed to the current

behavior, which adds to the existing list) and set it as the `descriptor` object's `githubAuth` variable using the `setGitHubAuth()` method. This way, the list contains only one element and GitHub Auth is configured with just the new configuration.

- As the CasCP-GSE plugin home page advised, we should use semicolons at the end of lines.

Here is the final Groovy script for our configuration:

```
import jenkins.model.*;
import org.jenkinsci.plugins.ghprb.*;

GhprbTrigger.DescriptorImpl descriptor = Jenkins.instance.
getDescriptorByType(org.jenkinsci.plugins.ghprb.GhprbTrigger.
DescriptorImpl.class);

String serverAPIUrl = null;
String jenkinsUrl = null;
String credentialsId = 'github-calvinpark-pat';
String description = 'calvinpark';
String id = null;
String secret = null;

List<GhprbGitHubAuth> githubAuths = [
  new GhprbGitHubAuth(serverAPIUrl, jenkinsUrl, credentialsId,
                      description, id, secret)];

descriptor.setGithubAuth(githubAuths);

descriptor.save();
```

Since we know that CasCP-GSE runs the code on the Script Console, we can run it there ourselves to test the code. As an experiment, try running it with a wrong `credentialsId` (for example, dockerhub-calvinpark-userpass) and a different description so that you can see the changes in System Configuration. The script will run successfully with no error messages, and System Configuration will be updated with the changes. I won't bother with the screenshots since we're in an advanced section.

We have now validated the Groovy script. Let's create a YAML file with the code and run it through JCasC. Here are the first few lines – the rest of the file is the same as the preceding Groovy code, just indented four spaces to fit under the `script` node:

CasCP-GSE.yaml

```
groovy:
- script: >
    import jenkins.model.*;
    import org.jenkinsci.plugins.ghprb.*;
    ...
```

Copy the file into ~/jenkins_home/, validate it on JCasC, then apply it. The configuration is applied without an error, and we can see in System Configuration that GHPRB was properly configured.

There, we have configured GHPRB using JCasC.

Summary

In this chapter, we used JCasC to configure Jenkins to the equivalent state as the Jenkins that we set up from *Chapter 1, Jenkins Infrastructure with TLS/SSL and Reverse Proxy*, to *Chapter 4, GitOps-Driven CD Pipeline with Docker Hub and More Jenkinsfile Features*.

We first installed the JCasC plugin and downloaded the current configuration to understand how JCasC codifies Jenkins configuration. Based on the existing configuration, we extracted the relevant parts that define the controller, agents, Docker cloud, and pipeline configurations. We collected the configurations into a file to build a minimal configuration file that defines the Jenkins configuration we want.

We then stopped the existing controller, started a brand-new controller, then applied the configuration using the JCasC plugin. We saw that all the configurations were correctly applied, and also went through the items that couldn't be configured in order to understand the limitations of JCasC. We saw how we can revert to the old Jenkins, looked at the current state and the future of JCasC, then finally wrapped up by examining an additional plugin that allows us to use Groovy scripts in JCasC to fill the gaps.

With an understanding of JcasC, we're now embarking on a journey of administrating Jenkins – some would call it *Day 2 operations*. In the next chapter, we will continue learning administration skills by examining how to handle backup and restore and disaster recovery.

7
Backup and Restore and Disaster Recovery

Perhaps the most important administration skill in any technology is backup restoration. We'll look at the different ways of creating backups, decide which files should be backed up at which frequency, learn how to use the ThinBackup plugin, create a disaster by deleting a pipeline and restore it, and finally look at how to recover from an infrastructure failure.

One note I should make is that I will call out very specific timestamps for referring to backup archives. These obviously won't match with yours, but I didn't make them a generic number because backup and restore is closely tied to timestamps. Read through the chapter to understand the significance of each timestamp and apply the knowledge to your own timestamps.

In this chapter, we're going to cover the following main topics:

- A small change for testing backup and restore
- Backup strategies
- Deciding which files to back up and at what frequency
- Backing up and restoring with the ThinBackup plugin
- Disaster recovery from a user mistake
- Disaster recovery from an infrastructure failure

Technical requirements

You'll need the Jenkins instance that we made from *Chapter 1, Jenkins Infrastructure with TLS/SSL and Reverse Proxy*, to *Chapter 4, GitOps-Driven CD Pipeline with Docker Hub and More Jenkinsfile Features*.

The files in the chapter are available in the GitHub repository: https://github.com/PacktPublishing/Jenkins-Administrators-Guide/blob/main/ch7.

A small change for testing backup and restore

Before moving ahead with the chapter, let's make a small change that we will use later to test backup and restore. In System Configuration, add Hello Jenkins! to System Message, and click Save to exit, as seen in *Figure 7.1*:

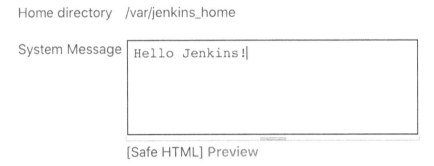

Figure 7.1 – Configuring System Message to "Hello Jenkins!"

On the Jenkins home page, we can see the new system message:

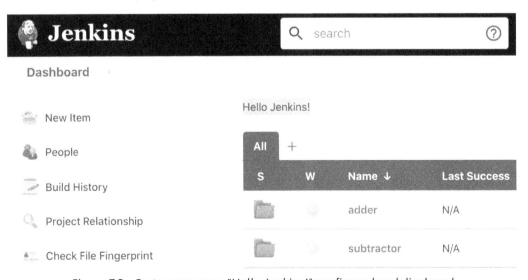

Figure 7.2 – System message "Hello Jenkins!" configured and displayed

This configuration is set in `jenkins_home/config.xml`, which is the main configuration file. Open the file to see the following change:

jenkins_home/config.xml

```
<?xml version='1.1' encoding='UTF-8'?>
<hudson>
[...]

  <buildsDir>${ITEM_ROOTDIR}/builds</buildsDir>
  <systemMessage>Hello Jenkins!</systemMessage>

[...]
</hudson>
```

Setting and removing the system message is just a matter of updating the string in the `config.xml` file. Subsequently, the system message configuration is backed up when `config.xml` is backed up.

With the system message configured, let's continue to discuss backup and restore.

Backup strategies

Everything in Jenkins is a flat file on the controller. This makes the backup very concise – just one directory, `jenkins_home`, needs to be backed up. We just need to determine the *frequency* and *methods* of backups.

Let's consider the tools that we have for a backup. Generally speaking, there are two methods: disk snapshots and file copies. We'll start with a disk snapshot, which is the simpler of the two.

Snapshotting the entire disk as an image

A disk snapshot creates an image that represents the controller's storage at a specific time. For example, on AWS Jenkins, the controller host's EBS volume would be snapshotted, and on firewalled Jenkins, the controller host's disk would be snapshotted.

The biggest advantage of a disk snapshot is that the backup and restore responsibilities can be delegated to a different group. The snapshot mechanism is uniform across all OSes and applications, so corporate IT-provisioned VMs almost always come with a daily backup as a service. If IT manages EC2 instances, they will typically also configure the daily EBS snapshots for instances used in production environments.

It also helps that disk snapshotting is a mature and established process, so there is no shortage of professional tools for the job. Amazon EBS has a very robust infrastructure for creating an EBS snapshot while it's in use, and other vendors such as VMware also provide similar tools. Online forums also help in finding information about these popular tools.

On a more technical level, snapshotting doesn't depend on the content of the disk, therefore it allows for a very reliable backup system that isn't interrupted upon unexpected OS updates or network outages. Even when the backup script, application, or entire OS is misconfigured or broken, the disk can still be snapshotted because the mechanism is external to what the disk contains.

Another big advantage is that a snapshot contains the entire disk as a working environment. During the critical minutes of a disaster recovery, an operator doesn't need to figure out where the files should go, and instead can simply rely on the fact that the whole disk is a working environment.

It's not without fault, of course. Most notably, a snapshot is an opaque image.

When an application misconfiguration is found, identifying the first snapshot with the misconfiguration often requires restoring several snapshots because it's difficult to list the changes in a snapshot since the previous snapshot.

It is also difficult to do a partial restoration where only certain directories or files are restored. Sometimes there are advanced tools that allow for this, but in most cases, we need to start a whole new VM in parallel and then copy the necessary files over.

The final disadvantage is that the snapshots are big because all the files on the disk are saved. It's usually not possible to snapshot specific directories, so even the unnecessary files and directories in the disk are backed up. Recall that we only need to back up jenkins_home, but a snapshot also contains everything else on the disk, increasing the cost of storage. This is the key disadvantage that disallows creating a snapshot more than once or twice per day – when there are hundreds of VMs, saving the terabytes of snapshots can get expensive.

Due to these characteristics, disk snapshots are good for infrequent low-maintenance backups if we're using the popular EXT filesystem.

The good news is that some of the more advanced filesystems, such as ZFS, have come a long way in addressing these disadvantages. ZFS supports taking diffs for the snapshots and even allows us to list and search files from the snapshots. Admins who rely only on the disk snapshots should explore ZFS and other advanced filesystems as a way of improving the quality of the disk snapshot backups. More on this in a bit.

Next, let's look at a more precise backup method of saving specific files.

Saving the directory content as files

A file-level backup is almost exactly the opposite of a snapshot.

It allows us to backup and restore just the files we need. Because we're picking up only a small number of files, the process is faster than a snapshot and results in a smaller output. The backup script can run every hour to yield 24 backups per day, resulting in 10 to 20 times more granular backups. In an extreme case, the most important files, such as logs, can be backed up to an external monitoring system in real time.

Because we're operating at a file level, we can even compare with a previous version and backup only the files that have changed. Combined with the fact that most backed-up files are text files, which can be heavily compressed, the storage cost is much lower than a snapshot system even though the backups are done a lot more frequently.

It's not without faults, of course. Most notably, *you* need to do the work.

As a Jenkins administrator, only you know which files are important. The entire jenkins_home directory of a decent-sized Jenkins is often impractical to back up at a file level because of the hundreds of thousands of small files generated by the build history. Also, there are plenty of files, such as those in the plugin cache, that simply don't need to be backed up because Jenkins recreates them automatically if they're missing.

Even if we have narrowed down the list of files to back up, IT often doesn't assist in a file-level backup (AWS most certainly does not). This means that you need to find the backup tool and configure it correctly. You need to validate the backed-up files and run quarterly disaster recovery drills. You need to set the retention period and delete the old files. Most importantly, *you are responsible* for the restoration in the case of a disaster. You need to remember which file goes where, and you need to write and maintain the operation manual for a disaster recovery plan so that your peers can handle outages while you're away.

Despite its vastly superior frequency and resolution, file-level backup should be avoided as much as possible due to its vastly higher operational costs. File-level backup should be reserved for very specific use cases, which we will discuss soon.

Backing up a large Jenkins instance

In a large Jenkins instance with over 1,000 pipelines, the disk bottleneck on the controller becomes too overwhelming to use file-level backups. A small number of configuration files can be backed up at a file level, but attempting to read and copy the hundreds of thousands of build directories will grind Jenkins to a halt. To back up the build directories of a large-scale Jenkins, a disk snapshot becomes the only remaining choice.

Thankfully, due to the advances made in disk snapshotting technologies, we don't lose much for not having a file-level backup. As we discussed earlier, a more advanced filesystem such as ZFS allows for more precise and transparent snapshots, remedying the main disadvantages of a disk snapshot. In addition, using a popular networked storage solution such as NetApp as the controller's storage allows us to take daily snapshots of the entire jenkins_home directory easily.

As we discuss in *Chapter 8*, *Upgrading the Jenkins Controllers, Agents, and Plugins*, scaling Jenkins to have over 1,000 pipelines requires very careful planning and maintenance. I would advise you to look for the best backup strategies for the storage backend that you're using, but in reality you wouldn't have a large Jenkins instance unless you're already doing that. But for those of you aspiring to scale your Jenkins instance beyond 1,000 pipelines, be advised that each storage backend has a unique set of tools for disk snapshot backups. The feature set of the backup tools as well as the backup policy you'll have should be a factor in deciding the controller's storage medium.

Deciding which files to back up and at what frequency

The Jenkins User Handbook does an excellent job of describing the content of jenkins_home (https://www.jenkins.io/doc/book/scaling/architecting-for-scale/#anatomy-of-a-jenkins_home). Here is a table that is derived from the guide that identifies the files and directories that need to be backed up and how often they should be backed up:

Path from $JENKINS_HOME	Need backup	Can't be recreated	Large # of files	Backup Frequency
Files in $JENKINS_HOME	Y	Y		High
.cache/				
.groovy/	Y			
.java/	Y			
caches/				
config-history/	Y			
fingerprints/	Y		Y/N	
jobs/<job>/builds/	Y	Y	Y	Medium
jobs/<job>/workspace/			Y	

Path from $JENKINS_HOME	Need backup	Can't be recreated	Large # of files	Backup Frequency
jobs/<job>/*	Y			
labels/	Y			
logs/	Y	Y		High
nodes/	Y			
plugins/	Y		Y	
secrets/	Y			
updates/	Y			
userContent/	Y			
users/	Y/N			
war/			Y	
workflow-libs/	Y			
workspaces/			Y	

Let's look at what each column means:

- Y in the Need backup column means that the data should be backed up. Some directories, such as cache, don't need to be backed up, but they're still backed up by daily disk snapshots. The users directory does not need to be backed up if AD or LDAP is used for authentication.

- The Can't be recreated column means that the data is generated by Jenkins rather than by the users. For example, if a secret is created but the secret is somehow lost, the same secret can be created again by an admin. On the other hand, logs can't be created again once they're lost.

- Large # of files means that the directory contains too many files to effectively handle at a file level and should be backed up by a daily disk snapshot.

Data in the three highlighted rows identified by Y in the Can't be recreated column must be backed up more frequently at a file level because they are *time-sensitive* data. If the VM running the Jenkins controller suddenly crashes, you would want the logs right up to the moment of the crash. Let's examine how to handle the three directories.

Directories for live backup

XML and other files in the root of the `jenkins_home` directory are critical for disaster recovery. Also, the `logs` directory contains important clues to understanding the root cause of the failure. These should be configured with an external logging system such as the ELK stack (Elasticsearch, Logstash, Kibana, which you can learn more about at https://www.elastic.co/what-is/elk-stack) to stream the changes live. If a live monitoring service is not available, they should be backed up every minute.

Directories for hourly backup

`jobs/<job>/builds/` is the build history for a pipeline. Depending on the pipeline, its logs may be important to preserve to the last minute. However, there are just too many files in these directories and it's impractical to copy them frequently. Copying alone can take over an hour on a medium-sized Jenkins with 100-1,000 pipelines, and attempting to copy these on a large-scale Jenkins with over 1,000 pipelines will saturate the disk access and pretty much break Jenkins – it cannot serve pages to the users on the UI, and it can't even communicate to the agents, so the connections drop off.

If there are critical pipelines whose build logs are important to the business, use the live backup strategy and stream them into an external logging system.

For other pipelines, save the log as a file and archive it in an artifact store such as JFrog Artifactory or Sonatype Nexus. This way, all logs for the completed builds are archived externally and only the logs for builds that were running at the time of the crash are at risk. This has an added benefit of reducing the load on the controller (we'll learn more about logs management in *Chapter 9, Reducing Bottlenecks*).

If an artifact store is not available, pick just a few important pipelines and back up their logs every hour.

All directories other than the three listed here can be backed up by daily disk snapshots.

Backing up and restoring with the ThinBackup plugin

Required plugins

ThinBackup

There are a handful of backup plugins, and ThinBackup is one of the most effective plugins to meet our backup requirements:

- Back up specific files and directories rather than the entire `jenkins_home` directory.
- Back up only the changed files.
- Back up at a custom time interval.
- Compress the backed-up files.
- Back up to external file storage.

The last point is an important one. Unlike a disk snapshot, which is stored outside of the disk by design, file-level backup initially generates the backup archive on the same disk that it is backing up. Unless the archives are moved out of the disk, the backup data is at risk of being destroyed, in the case of disk corruption. ThinBackup doesn't provide this natively, but it does allow us to specify the backup directory so that we can easily expose it out of the controller's Docker container. Let's explore how we can use this to our advantage.

Moving the backup archives out of the disk

There are many ways to move files out of a disk. We can use `scp`, `rsync`, or even `git` on an interval using `cron`, to name a few. Another good method is using Network File System (NFS) like this:

1. Create a ~/backup directory on the controller host.
2. Mount an NFS file share to the directory.
3. Bind-mount the ~/backup directory on the controller host to /backup in the controller Docker container.
4. Configure ThinBackup to save the backup archives in /backup.

ThinBackup creates backup archives on /backup inside the container, which is bind-mounted to the host ~/backup directory, where the NFS share is mounted. The end result is that the backup archives are saved on an NFS server outside of the disk automatically without having to push the files. This is what it looks like:

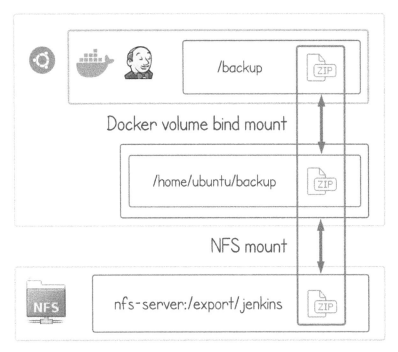

Figure 7.3 – Jenkins backup archives automatically saved on an external NFS server

---

Mounting an S3 bucket instead of NFS

An alternative to using an NFS server is mounting an S3 bucket using FUSE. This saves us the maintenance cost of running an NFS server, but beware that the performance will be much lower and there may be issues with the S3 bucket's eventual-consistency policies. For a small Jenkins with fewer than 100 pipelines, it may work just fine.

Let's see how we can set this up:

1. Start by creating the ~/backup directory. If we skip this step, the docker run command will create it as the root user and cause permission errors:

```
controller:~$ mkdir backup
```

2. Setting up an NFS server is out of scope for this book so I'll leave that to you. Once an NFS share is mounted onto the ~/backup directory, the controller Dockerfile must be updated to add the /backup directory.

3. Update jenkins.dockerfile to create the /backup directory and change the owner to the VM host user. If we don't make the directory and change the

ownership during image creation, the docker run command will create it as root and cause permission issues. AWS Jenkins should change the UID 123 and GID 30 to 1000 and 1000:

jenkins.dockerfile

```
FROM jenkins/jenkins:2.263.1-lts
ARG j_uid=123
ARG j_gid=30
USER root
RUN  usermod -u ${j_uid} -g ${j_gid} jenkins
RUN  mkdir /backup && chown ${j_uid}:${j_gid} /backup
USER jenkins
```

4. Build the image and create a container again with the ~/backup host directory bind-mounted to the /backup container directory:

```
controller:~$ docker build -t calvinpark/jenkins:2.263.1-lts -f
jenkins.dockerfile .
controller:~$ docker stop jenkins_controller
controller:~$ docker rm jenkins_controller
controller:~$ docker run \
    --detach \
    --restart on-failure \
    -u $(id -u):$(id -g) \
    -v ~/jenkins_home:/var/jenkins_home \
    -v ~/backup:/backup \
    -p 8080:8080 -p 50000:50000 \
    --name jenkins_controller \
    calvinpark/jenkins:2.263.1-lts
```

The backup directories are configured. When ThinBackup creates backup archives in the /backup directory in the container, the archives will be written into the NFS share that is mounted on the ~/backup directory on the host.

With the directories set up, let's move on to configuring ThinBackup.

Experimenting with ThinBackup

Click Manage Jenkins, scroll down to the bottom to the Uncategorized section, click ThinBackup, and then click Settings. Let's back up the `logs` directory every minute to learn how the ThinBackup plugin works. Configure as follows:

- Backup directory: `/backup/`
- Backup schedule for full backups: `*/5 * * * *` (every 5 minutes)
- Backup schedule for differential backups: `* * * * *` (every minute)
- Max number of backup sets: `10` (older archives will be deleted automatically)
- Wait until Jenkins/Hudson is idle to perform a backup: **Uncheck**
- Backup build results: **Check (checked by default)**
- Backup additional files: `logs|tasks|slaves|.*\-agent|.*\.log.*` (explained in a bit)
- Clean up differential backups: **Check**
- Move old backups to ZIP files: **Check**

Here is what it should look like:

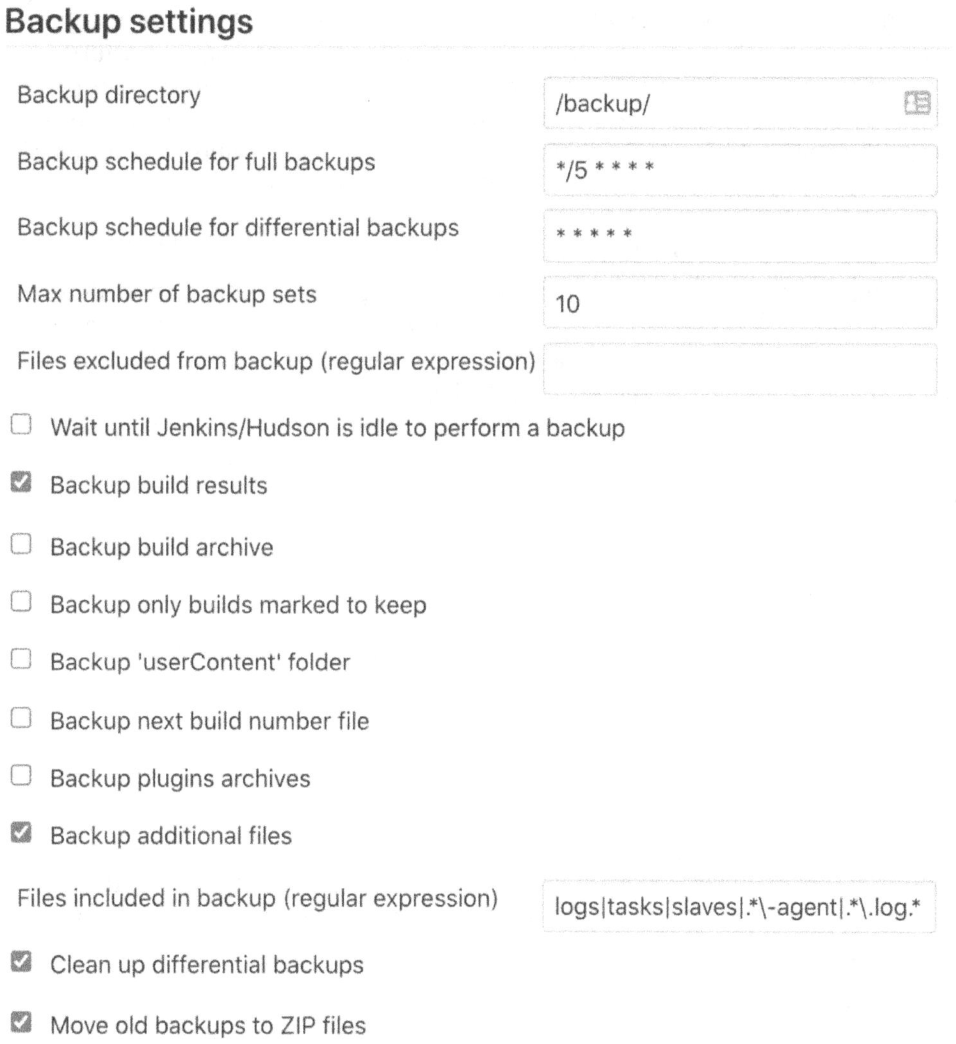

Backup settings

Backup directory	/backup/
Backup schedule for full backups	*/5 * * * *
Backup schedule for differential backups	* * * * *
Max number of backup sets	10
Files excluded from backup (regular expression)	

☐ Wait until Jenkins/Hudson is idle to perform a backup

☑ Backup build results

☐ Backup build archive

☐ Backup only builds marked to keep

☐ Backup 'userContent' folder

☐ Backup next build number file

☐ Backup plugins archives

☑ Backup additional files

Files included in backup (regular expression)	logs|tasks|slaves|.*\-agent|.*\.log.*

☑ Clean up differential backups

☑ Move old backups to ZIP files

Figure 7.4 – ThinBackup configuration for a backup every minute

A differential backup is for backing up just the files that have changed since the last full backup. This is a good way to save disk space, but more importantly, it's a good way to *identify the changes since the last backup*. Just by looking at the content of the backup, we can tell whether a file was changed since the last backup.

Backup additional files has a pretty strange regex because of the strange way the plugin applies the regex (https://issues.jenkins.io/browse/JENKINS-64490). Rather than comparing the regex against the file path, it compares it against each directory and filename. For example, `logs/tasks/task.log` doesn't match the `logs.*` pattern in ThinBackup because each of `logs`, `tasks`, and `task.log` is compared against the pattern. Specifically, `tasks` and `task.log` do not match the `logs.*` pattern, so the file is not backed up. In order to match and back up `logs/tasks/task.log`, the regex has to be `logs|task(s|\.log)`. This means that every subdirectory name has to be explicitly listed, which sort of defeats the purpose of using regex. This is one of the few flaws in the otherwise perfect ThinBackup plugin.

Other configurations should be self-explanatory. Click Save, then click Backup Now.

Go to the controller host and look at the ~/backup directory. Within a minute, we will see the first full backup. Examine the directory content to find that there are a lot more files than just the `logs` directory content. This is because the plugin always backs up a certain set of files and directories, and there's no way to exclude them. Thankfully, the number and the size of the mandatory files is small enough that we can simply ignore them. In this case, the 3.4M-sized directory will be compressed to less than 700K when it's rotated out:

```
controller:~$ ls backup/
FULL-2021-02-19_05-05
controller:~$ find backup/FULL-2021-02-19_05-05 | wc -l
401
controller:~$ du backup/FULL-2021-02-19_05-05 --max-depth 0 -h
3.4M   backup/FULL-2021-02-19_05-05/
```

After a few more minutes we'll start seeing differential backups. Examine the differential backup directories to see that there are a lot fewer files in the differential backups, because a differential backup saves only the files that have changed since the last backup:

```
controller:~$ ls backup/
DIFF-2021-02-19_05-06    FULL-2021-02-19_05-05
controller:~$ find backup/DIFF-2021-02-19_05-06 | wc -l
31
controller:~$ du backup/DIFF-2021-02-19_05-06 --max-depth 0 -h
104K   backup/DIFF-2021-02-19_05-06/
```

After 5 minutes, a new full backup is made and all previous differential backups are compressed into a ZIP file. In the following example, I have waited an hour to get a complete list of zipped backup sets, diff backups, and a full backup. Since we've configured to keep 10 sets of backups, we'll at most have 9 ZIP files and a set of full and differential backup directories. Once the 10th ZIP file is created, the oldest ZIP file gets automatically deleted. It's pretty cool to see it work so well:

```
controller:~$ ls backup/

BACKUPSET_2021-02-19_05-10_.zip
BACKUPSET_2021-02-19_05-15_.zip
BACKUPSET_2021-02-19_05-20_.zip
BACKUPSET_2021-02-19_05-25_.zip
BACKUPSET_2021-02-19_05-30_.zip
BACKUPSET_2021-02-19_05-35_.zip
BACKUPSET_2021-02-19_05-40_.zip
BACKUPSET_2021-02-19_05-45_.zip
BACKUPSET_2021-02-19_05-50_.zip
DIFF-2021-02-19_05-56
DIFF-2021-02-19_05-57
DIFF-2021-02-19_05-58
DIFF-2021-02-19_05-59
FULL-2021-02-19_05-55
```

Since we've mounted an NFS share to the ~/backup directory, all these files are automatically being saved outside of the controller's disk.

Now that we understand how to create backups using the ThinBackup plugin, let's try restoring one of the backups.

Restoring a backup using ThinBackup

At the beginning of the chapter, we set the system message as "Hello Jenkins!". Let's delete that message and recover it using ThinBackup.

Go to System Configuration, delete the message, and click Save to exit. On the Jenkins home page, the "Hello Jenkins!" message is no longer there. Look at the clock to remember the time of the disaster (in our case it's 9:43 p.m. PST), then go take a coffee break for 10 minutes.

Now restore the backup to get the agent back. Click Manage Jenkins, ThinBackup, and Restore. From the drop-down menu, pick the time slot before the change was made. The timestamps are in UTC, which means that the disaster occurred at 5:43 a.m. Choose 05:40 since that's the last backup before the disaster, then click Restore as shown in *Figure 7.5*.

 # Restore Configuration

Restore options

restore backup fron ✓ 2021-02-19 05:51
              2021-02-19 05:50
☐ Restore next bui   2021-02-19 05:45   backup)
              2021-02-19 05:40
☐ Restore plugins   2021-02-19 05:35
              2021-02-19 05:30
              2021-02-19 05:25   **Restore**
              2021-02-19 05:20
              2021-02-19 05:15
              2021-02-19 05:10
              2021-02-19 05:05

Jenkins 2.263.1

Figure 7.5 – Restore a version before deleting the "Hello Jenkins!" system message

To our disappointment, the "Hello Jenkins!" system message did not return. This is because even though the ThinBackup plugin restored the configuration file, the Jenkins controller is still running with the old configuration in memory. Click Manage Jenkins, Reload Configuration from Disk, and OK as seen in *Figure 7.6:*

Figure 7.6 – Reload the configuration from disk to apply the restored configuration

Once Jenkins is reloaded, the system message "Hello Jenkins!" is restored.

Now that we know how to use ThinBackup for both creating and restoring backups, let's configure it correctly.

Configuring ThinBackup

Configure the ThinBackup settings by clicking Manage Jenkins, ThinBackup, then Settings.

Earlier in the chapter, we discussed that the logs directory should be backed up every minute. However, the current configuration is problematic because the jobs/builds directories are also getting backed up every minute. The jobs/builds directories grow very large, so backing them up at this rate strains the controller's disk too much. There are two options for mitigating this.

The first option is to stream the logs directory content to an external logs management system so that the logs are saved in real time outside of Jenkins, then slow down the frequency of backups to every hour for the jobs/builds directories. This is the best option if your jobs/builds directories don't grow too large, but it requires an external logs management system. If you choose this option, configure as follows:

- Set Backup schedule for full backups to H H * * * so that a full backup is taken once a day. H in this context stands for hash – more on that is in the Jenkins User Handbook (https://www.jenkins.io/doc/book/pipeline/syntax/#cron-syntax).

- Set Backup schedule for differential backups to H * * * * so that a differential backup is taken every hour.

- Set Max number of backup sets to 5 to save 5 days' worth of backup. Remember that there are nightly disk snapshots, therefore we don't need any more than a day's worth. However, it's much easier to restore from ThinBackup than from a disk snapshot, so it's not a bad idea to keep a few more around – older backups get deleted automatically anyway. Also, this value is for *backup sets*, which means five full backups, not five differential backups.

- Finally, uncheck Backup additional files to ignore the logs directory.

This is what it should look like:

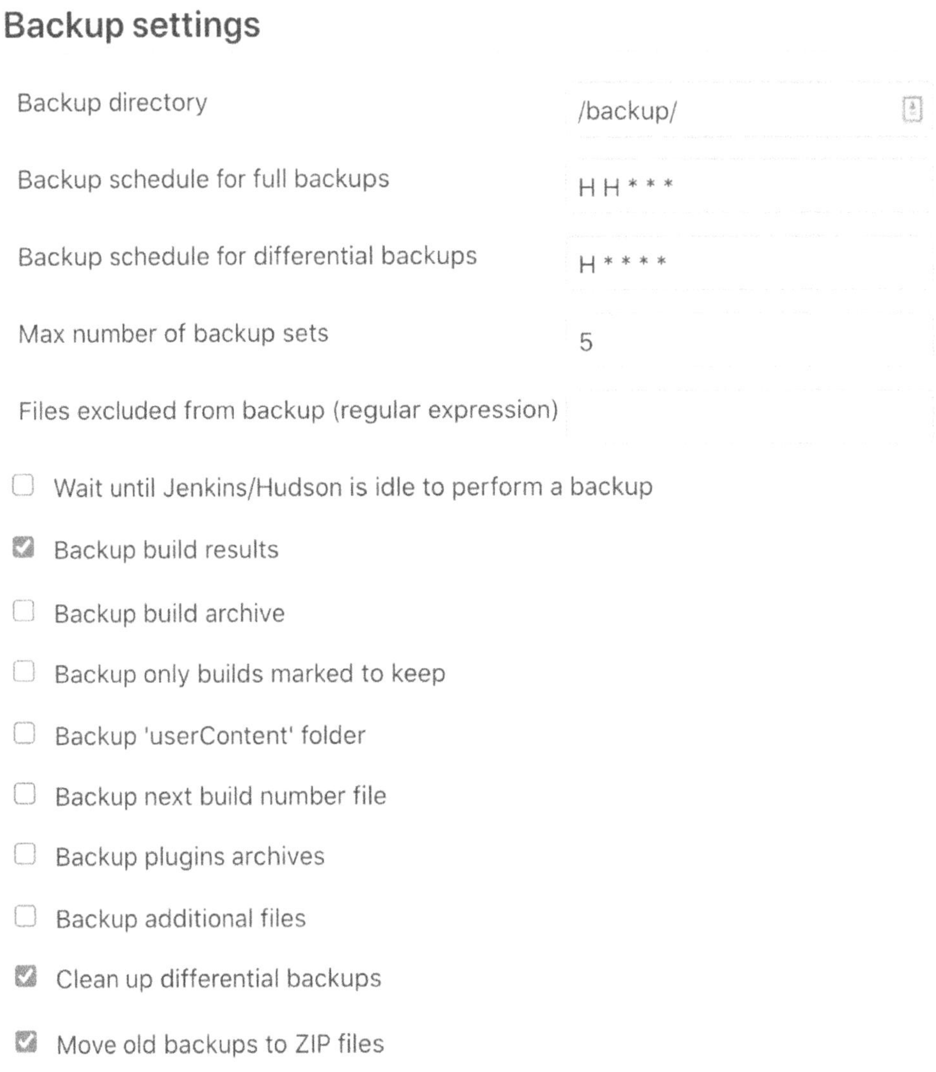

Backup settings

Backup directory	/backup/
Backup schedule for full backups	H H * * *
Backup schedule for differential backups	H * * * *
Max number of backup sets	5
Files excluded from backup (regular expression)	

☐ Wait until Jenkins/Hudson is idle to perform a backup

☑ Backup build results

☐ Backup build archive

☐ Backup only builds marked to keep

☐ Backup 'userContent' folder

☐ Backup next build number file

☐ Backup plugins archives

☐ Backup additional files

☑ Clean up differential backups

☑ Move old backups to ZIP files

Figure 7.7 – Configuration for an hourly differential backup and a daily full backup

The second option is to keep the frequency at every minute and not back up jobs/
builds using ThinBackup. Instead, the jobs/builds directories will be backed
up during the nightly disk snapshots. This has a risk of data loss in the jobs/
builds directories but the system logs are still saved outside of Jenkins frequently,
which can be more important than build logs. If you choose this option, configure
as follows:

- Set Backup schedule for full backups to H * * * * so that a full backup is taken every hour.

- Set Backup schedule for differential backups to * * * * * so that a differential backup is taken every minute.

- Set Max number of backup sets to 120 to save 5 days' worth of logs. We can decrease this further since there are nightly disk snapshots.

- Finally, uncheck Backup build results to ignore the jobs/builds directories.

This is what it should look like:

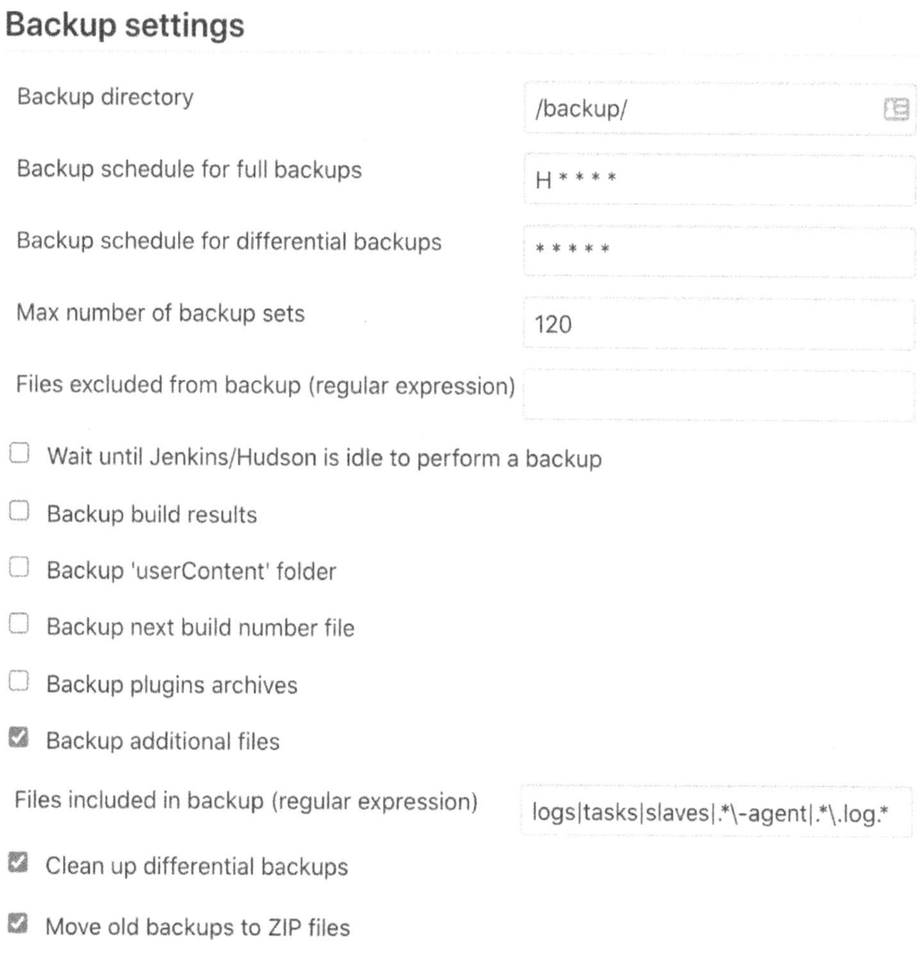

Backup settings

Backup directory	/backup/
Backup schedule for full backups	H * * * *
Backup schedule for differential backups	* * * * *
Max number of backup sets	120
Files excluded from backup (regular expression)	

☐ Wait until Jenkins/Hudson is idle to perform a backup

☐ Backup build results

☐ Backup 'userContent' folder

☐ Backup next build number file

☐ Backup plugins archives

☑ Backup additional files

Files included in backup (regular expression)   logs|tasks|slaves|.*\-agent|.*\.log.*

☑ Clean up differential backups

☑ Move old backups to ZIP files

Figure 7.8 – Configuration for a differential backup every minute and an hourly full backup

Click Save to exit.

Our backup configuration is complete. Next, we'll look into a few common disaster recovery scenarios and use the backups created by ThinBackup to recover the lost data.

Disaster recovery from a user mistake

One of the most common mistakes that a project admin can make is accidentally deleting a pipeline and its build history. We'll create a simulated disaster with ambiguous information to see how we can recover from a deleted pipeline.

Disaster

Let's go over the details of our disaster scenario:

- Sometime in the last 48 hours, the adder/postmerge pipeline was mistakenly deleted. The exact time is unknown.
- ThinBackup is configured to back up the jobs/builds directories every hour, which is the first backup configuration option from the previous section.
- Within the last 48 hours, the adder/premerge (a different pipeline, not postmerge) pipeline had run three times, and its last build number is now 35.
- Within the last 48 hours, the system message was changed from "Hello Jenkins!" to "Hello ThinBackup!".

Here is the timeline of events:

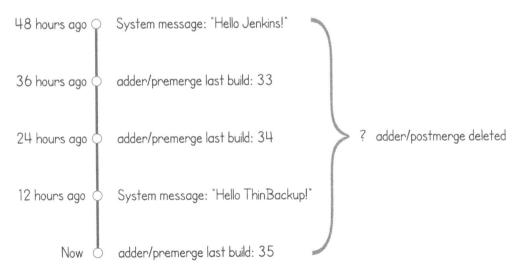

Figure 7.9 – Timeline of events for the disaster recovery scenario

The success criteria for a recovery are as follows:

1. It's okay to lose the history of the adder/postmerge pipeline for builds that ran during the last hour before a backup. Suppose that the last backup ran at 3:00 p.m. and the last build number was 10. At 3:05 p.m, another build ran, thus the last build number became 11. The next backup is scheduled at 4:00 p.m, therefore build number 11 hadn't been backed up yet. If the pipeline was deleted at 3:10 p.m, the build number 11 was never backed up, so it can't be recovered. This loss is acceptable.

2. Other pipelines or configurations must not be modified in any way. Rewinding time and removing the build history from other pipelines is not okay. In our specific case, the build history for adder/premerge must have 35 builds, not have 32, 33, or 34 builds. In addition, the system message should remain as "Hello ThinBackup!" and not revert to "Hello Jenkins!".

Success criterion 1 is about risk management. In a perfect world, we would create backups every few milliseconds, which means everything is recoverable and nothing is lost. That of course comes at a very high cost. In our Jenkins, reducing disk bottleneck on the controller is critical to Jenkins' performance, therefore we've set the backup frequency at 1 hour. If you have pipelines that need a higher backup frequency, instead of increasing the backup frequency, consider other safety measures to reduce the risk of the disaster itself. For example, Shelve Project is a plugin that allows you to delete a project into a recycle bin that later you can recover by simply *unshelving* it. By removing the delete project permission from project admins and instead using Shelve Project, the pipelines can be easily recovered without a costly increase of backup frequency. In a large-scale Jenkins instance, removing the delete permission from everyone might be a good policy that leads to eliminating the maintenance cost of recovering a deleted pipeline entirely. In a car analogy, this is akin to investing in better brakes rather than better airbags – focus on preventing disasters in the first place.

Success criterion 2 is perhaps obvious but still important – if the entire Jenkins failed, then we can restore the last snapshot as is. In this scenario, only one pipeline was deleted, therefore only one pipeline must be recovered. In other words, we need a *partial recovery*. Let's continue to see how that's done.

Recovery

Go into the ~/backup directory and examine its content. ZIP files and directories show that the last full backup was made on 2021-02-24 12:00 a.m. UTC and a differential backup was made every hour since. Older backups are in ZIP files:

```
controller:~/backup$ ls
BACKUPSET_2021-02-21_20-00_.zip    DIFF-2021-02-24_03-00
BACKUPSET_2021-02-23_00-00_.zip    DIFF-2021-02-24_04-00
DIFF-2021-02-24_01-00              FULL-2021-02-24_00-00
DIFF-2021-02-24_02-00
```

The first thing we can do is to search the *directories* for the postmerge pipeline. Since a full backup is created once a day and converted to a *ZIP file*, searching the *directories* means that we're searching for the pipeline in the backups made *today*. If the pipeline was deleted sometime today, the following find command would show that the backup directories contained the postmerge pipeline. Unfortunately for us, the pipeline was deleted before today, so the find command didn't turn up anything. Try changing the command to look for premerge instead to see that the backup directories contain the build history for the premerge pipeline:

```
controller:~/backup$ find . -name postmerge -type d
<empty result>
controller:~/backup$ find . -name premerge -type d
./DIFF-2021-02-24_03-00/jobs/adder/jobs/premerge
./DIFF-2021-02-24_01-00/jobs/adder/jobs/premerge
./FULL-2021-02-24_00-00/jobs/adder/jobs/premerge
./DIFF-2021-02-24_02-00/jobs/adder/jobs/premerge
./DIFF-2021-02-24_04-00/jobs/adder/jobs/premerge
```

Next, look through the content of all *ZIP files* and look for the postmerge pipeline. Since a full backup is created once a day and converted to a *ZIP file*, searching the content of the *ZIP files* means that we're searching for the pipeline in the backups made *yesterday or older*. We have two ZIP files, and the search shows that one of the two ZIP files contains the postmerge pipeline. The shell command is pretty ugly, but in essence, it looks through the content of all ZIP files and reports the ZIP filenames that contain postmerge:

```
controller:~/backup$ for f in *.zip; do unzip -l $f | if grep -q
postmerge; then echo $f; fi ; done
```
BACKUPSET_2021-02-21_20-00_.zip

Let's try restoring this copy. Go to Jenkins, Manage Jenkins, ThinBackup, and Restore. Under Restore Options, we can see the list of backups that we can restore. This list should match the directory listing of the ~/backup directory. Select 2021-02-21 20:00, which is the ZIP file that contains the postmerge pipeline.

Before clicking Restore, let's think about what's about to happen. At the time of the backup that we're about to restore, the adder/premerge pipeline's last build was 34, which is one before the current last build 35. Also, the system message was "Hello Jenkins!", which is different from the current message "Hello ThinBackup!". Will clicking Restore rewind the last build to 34? Will it change the system message back to "Hello Jenkins!"? Let's find out. Click Restore, Manage Jenkins, Reload Configuration from Disk, then OK.

The first thing we notice is that the system message has indeed been reverted back to "Hello Jenkins!". We'll get back to what this means in a moment. Go into the adder folder and voilà, the postmerge pipeline is back! Also, to our relief, the premerge pipeline's last build was not reverted to 34 and is still 35. That's two out of three things restored correctly – not bad for a first try.

Here is how the ThinBackup restore works. Clicking the Restore button is a simple file copy and overwrite process – the plugin takes the content of the backup ZIP file and writes it on jenkins_home. If there is already a file of the same name in jenkins_home, such as config.xml, the existing file is overwritten with the file from the backup. At the beginning of the chapter, we saw that the system message is configured in config.xml, so restoring an old copy of config.xml reverted the system message to the old message "Hello Jenkins!". Directories and files for the build history of the premerge pipeline also already exist in jenkins_home, and they were overwritten by the directories and files from the backup. It just happens that the current set of directories and files is identical to the ones from the backup, so there are no visible changes. In addition, existing files in jenkins_home are never deleted by ThinBackup's restore function – again, it's merely a file copy that is sometimes an overwrite but never a delete. As a result, the files for build 35 for the premerge pipeline remained as is, so the last build number for the premerge pipeline remained as 35. Here is what it looks like:

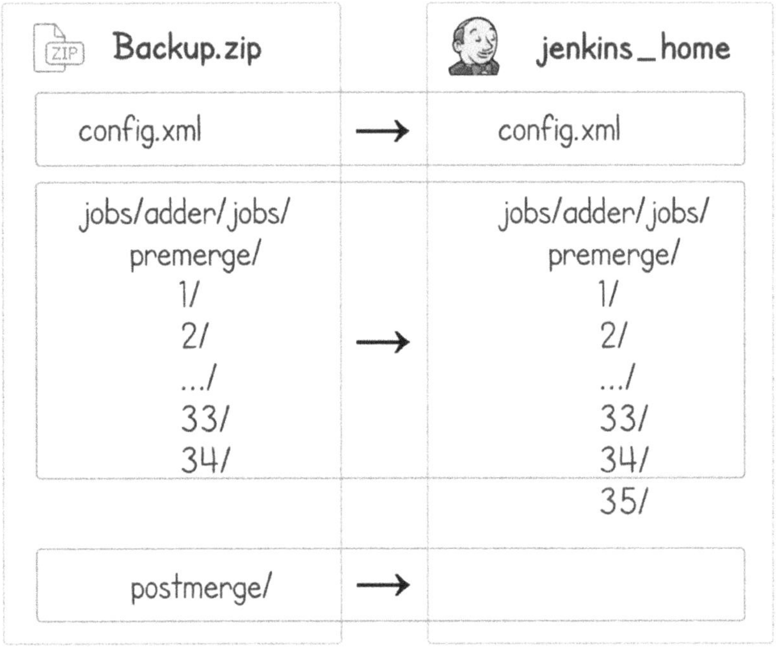

Figure 7.10 – ThinBackup's restore function writes the directories and files from the backup zip/directory onto jenkins_home

Let's discuss how we should handle the unwanted changes to config.xml. In fact, unwanted changes were also made in the premerge build history from 1 to 34, as well as in all other files that were in the ZIP file – it's just that they didn't have any visible effects.

A quick way to undo all unwanted changes is to restore again from the latest backup. Since the postmerge pipeline is already restored, restoring the latest backup will re-apply the changes in config.xml while leaving postmerge restored. Since a disk reload is required after a restore anyway, we could even plan to restore first the backup with the recovery material to get the postmerge pipeline back, followed by a restore of the latest backup to get config.xml back, then finally reload Jenkins. An advantage of this approach is that the whole operation can be done through the UI if you prefer the UI over managing files in the jenkins_home directory.

Another approach is restoring by hand using the backup ZIP files from ThinBackup. As we've discussed, ThinBackup's restore function is a simple file copy – why not do this on our own? First, identify the ZIP file we want to restore, unzip it outside of the ~/backup directory, and finally copy the directories and files we want to restore into jenkins_home. In our case, the jobs/adder/jobs/postmerge directory contains everything we need. Copy that into jenkins_home, reload Jenkins, and the recovery is complete without touching any other files. This technique can also be used to restore from a disk snapshot backup. Mount a snapshot somewhere to get access to the files, pick the files we want to restore, and copy them over.

Let's try a manual recovery to see how it works. In this exercise, you will "restore" a pipeline that you didn't even have before – I will provide you a backup of *my* pipeline, Hello-Backup-Restore, and guide you to restore that in *your* Jenkins. SSH into the controller, download the backup tarball from the book's GitHub repo, then extract it to see its content:

```
controller:~$ wget -q https://github.com/PacktPublishing/Jenkins-
Administrators-Guide/raw/main/ch7/Hello-Backup-Restore.tar.gz

controller:~$ tar -xf Hello-Backup-Restore.tar.gz

controller:~$ tree Hello-Backup-Restore
Hello-Backup-Restore
├── builds
│   ├── 1
│   │   ├── build.xml
│   │   ├── log
│   │   ├── log-index
│   │   └── workflow
│   │       └── flowNodeStore.xml
│   ├── 2
│   │   ├── build.xml
│   │   ├── log
│   │   ├── log-index
│   │   └── workflow
│   │       └── flowNodeStore.xml
│   ├── legacyIds
│   └── permalinks
├── config.xml
└── nextBuildNumber

5 directories, 12 files
```

The tarball contains the pipeline definition config.xml for a pipeline named Hello-Backup-Restore (identified by the directory name), along with its build history. The directory structure resembles what we saw for the postmerge pipeline. Let's restore this pipeline inside the subtractor folder. Simply move the directory under the subtractor jobs folder:

```
~$ mv Hello-Backup-Restore jenkins_home/jobs/subtractor/jobs/
```

Go to Jenkins and load the backup by clicking Manage Jenkins, Reload Configuration from Disk, and OK. Once Jenkins reloads, go into the subtractor folder to see that you have the Hello-Backup-Restore pipeline and its history *restored*:

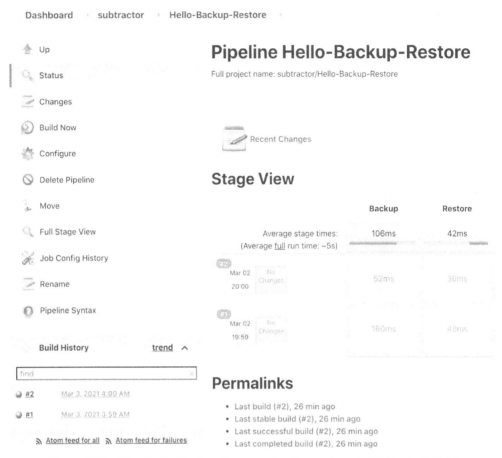

Figure 7.11 – The Hello-Backup-Restore pipeline restored with the build history under the subtractor folder

We can see that the build has already been run twice by yours truly. Click build #2 and look at its logs to see that the build was started by the user Calvin Park (Admin), even though you don't have the Calvin Park (Admin) user in your Jenkins (if you're another Calvin Park reading this, please contact me for a free copy of the book):

Dashboard › subtractor › Hello-Backup-Restore › #2

 # Console Output

```
Started by user Calvin Park (Admin)
Running in Durability level: PERFORMANCE_OPTIMIZED
[Pipeline] Start of Pipeline
[Pipeline] node
Running on firewalled-firewalled-agent in
/home/robot_acct/firewalled-firewalled-agent/workspace/Hello-
Backup-Restore
[Pipeline] {
[Pipeline] stage
[Pipeline] { (Backup)
[Pipeline] echo
Backup!
[Pipeline] }
[Pipeline] // stage
[Pipeline] stage
[Pipeline] { (Restore)
[Pipeline] echo
Restore!
[Pipeline] }
[Pipeline] // stage
[Pipeline] }
[Pipeline] // node
[Pipeline] End of Pipeline
Finished: SUCCESS
```

*Figure 7.12 – Console output for build #2 shows that it was run
by a user that's not in this Jenkins*

This demonstrates that a disaster recovery from a configuration issue ultimately boils down to simple file copies because everything in Jenkins is a flat file on the controller. Next, let's take a look at a lower-level infrastructure failure.

Disaster recovery from an infrastructure failure

Imagine a scenario where Jenkins doesn't come up. https://jenkins-firewalled.lvin. ca/ simply fails to load. What are the recovery steps? Let's recall the architecture and understand that there are three major components: the controller host, containers, and the DNS.

The first thing we should try is to SSH into the controller host. If the SSH fails for a host in AWS, make sure that your IP hasn't changed since we created the EC2 routing rules with the My IP option. The IP would change if it were renewed on the router or if we were using a phone's hotspot instead of a home internet connection. In this case, go back to the EC2 dashboard on AWS and update the IP range for My IP. If the SSH fails for reasons other than network issues, restart the host. If SSH continues to fail, we can try extending the size of the disk – SSH fails when the disk is 100% full. If the SSH still fails, then it's time to create a new VM from a backup disk image. More on this later.

Once we're on the host, check to make sure that the containers are running using the `docker ps -a` command. On an AWS controller there should be just one container for Jenkins, and on a firewalled controller there should be two: NGINX and Jenkins. If they are not running, look through the logs using the `docker logs jenkins_controller` command. Read the error message and fix the issue.

If the containers are running and Jenkins is still down, try accessing it from the controller host itself. `curl localhost:8080/login` should connect and receive the HTML output for the login page as follows. If it doesn't, that means the controller is either not listening to port 8080 or the controller process is unavailable. We can restart the container with the `docker restart jenkins_controller` command:

```
controller:~/backup$ curl localhost:8080/login

    <!DOCTYPE html><html lang="en">...<h1>Welcome to Jenkins!</
h1>...</html>
```

If the container is responding on port 8080 but Jenkins is still down, try accessing it on the https port. For an AWS controller, we need to find the URL for the ELB and use it like this: `curl -k https://<ELB URL>/login`. For a firewalled controller, we can continue to use `localhost` like this: `curl -k https://localhost/login`. The -k option is required for ignoring the TLS certificate issues because the TLS certificate we're using is for the Jenkins domain, not the ELB URL or `localhost`. The curl command should connect and receive the HTML output for the login page just as it did for port 8080. We can't really restart an ELB, but we can restart the NGINX container with the docker restart nginx command.

If the container is responding on https but Jenkins is still down, make sure that the domain is pointing to the VM. If your VM uses a dynamic IP using DHCP, it's possible that the IP of the VM changed. On AWS, we can prevent this by getting a static IP using Elastic IP. On a corporate network, talk to your network admin about assigning a static IP to the VM.

On AWS, one other thing we can check is the routing rule. VPC is a complex system with its public and private subnets, network ACLs, and security groups – it's easy to have a misconfiguration that disables the necessary access.

If all else fails, we can create a new Jenkins controller using a backup disk image. A backup disk image already has all the necessary files in the right locations and everything correctly configured, so we can simply turn it on and adjust the DNS to point your domain to the new controller. Since the disk image was created while the VM is running, it probably even has Jenkins and NGINX containers created. SSH into the host and simply start the two containers with the docker start jenkins_ controller nginx command. Once Jenkins is running, use ThinBackup to restore the latest backup. Since a disk snapshot is taken only once a day, an additional restore from ThinBackup will give us a more current restoration point.

Once Jenkins is operational, look through the logs to examine what happened. There are logs for Jenkins, NGINX, the host, and DHCP.

Summary

In this chapter, we've covered all aspects of backup and restore and disaster recovery.

First, we added a system message to learn that most configurations are saved in the `config.xml` file.

We've reviewed the benefits and shortfalls of two backup strategies: disk image snapshot and file copy. We've learned that both strategies should be used since they complement each other.

We went through the list of files in `jenkins_home` to learn which directories should be backed up more frequently than others.

We've learned how to use ThinBackup and we've learned its strengths and weaknesses. We've learned two different ways of configuring ThinBackup for high-frequency `logs` directory backup or medium-frequency `jobs/builds` directory backup. We've also learned that we can use an NFS mount to automatically save backup files offsite without having to push the files manually.

We created an actual disaster by deleting the `postmerge` pipeline and even added additional conditions to simulate a real disaster by running `premerge` a few more times and changing the system message. We've learned how to recover from this disaster by using ThinBackup and we observed that it restores a backup by overwriting existing files. We've also discussed how to restore just the directories we want without affecting other files, and we ran through an exercise where you recovered a pipeline that you didn't even have before by restoring a backup that I created.

Finally, we've gone through the steps of debugging an application failure from the bottom of the stack to the top. We've also discussed the steps for recreating a new Jenkins controller using a disk backup image.

In the coming chapter, we will use the recovery skills we've learned to upgrade Jenkins.

8
Upgrading the Jenkins Controller, Agents, and Plugins

Combining the knowledge we've gained from all previous chapters, now it's time to handle upgrades. The upgrade process for Jenkins is unfortunately one of Jenkins' weak points, and a successful upgrade requires a good understanding of Jenkins and plugins, coupled with good planning and execution.

In this chapter, we're going to cover the following main topics:

- Understanding the challenges of plugin version management
- Upgrade strategies
- Upgrading plugins using Plugin Manager
- Upgrading the controller

Technical requirements

We need the Jenkins instance that we built in *Chapter 2, Jenkins with Docker on HTTPS on AWS and inside a Corporate Firewall*. Also, we'll be modifying the Dockerfile that we've developed throughout the earlier chapters.

The files for this chapter are available in the GitHub repository at https://github.com/PacktPublishing/Jenkins-Administrators-Guide/blob/main/ch8.

Understanding the challenges of plugin version management

The difficult part of upgrading Jenkins is managing the plugin versions. Let's go through a few scenarios to understand why.

Upgrading to the next immediate LTS version of Jenkins

Upgrading Jenkins can be trivially easy – just run a new container with an upgraded version. Container image versions are well defined after all (https://hub.docker.com/r/jenkins/jenkins/tags). This works well when the delta between the old Jenkins version and the new Jenkins version is small. Nearly all plugins installed on one long-term support (LTS) version of Jenkins will be compatible with the next immediate LTS version of Jenkins. This means that starting a new container with an upgraded version of Jenkins using the existing plugins will most likely work:

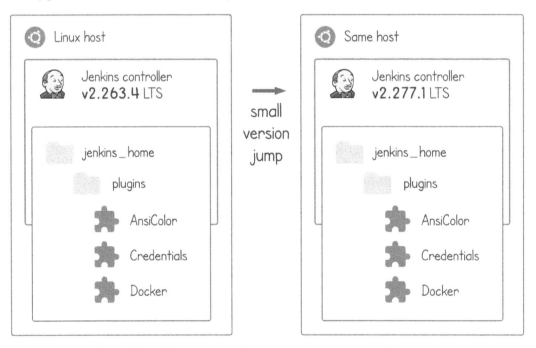

Figure 8.1 – A simple upgrade by using a newer controller image

Once the new version of Jenkins is running, we can go to Plugin Manager to update the plugins to the latest version that is compatible with the new Jenkins.

Upgrading while skipping many versions of LTS releases

Difficulties arise when we're jumping a large number of releases for a Jenkins upgrade. The existing old plugins in our `jenkins_home` directory won't be compatible with the new Jenkins version, and the incompatible plugins will fail to load. If something as fundamental as `credentials` fails to load, the entire Jenkins will fail to load:

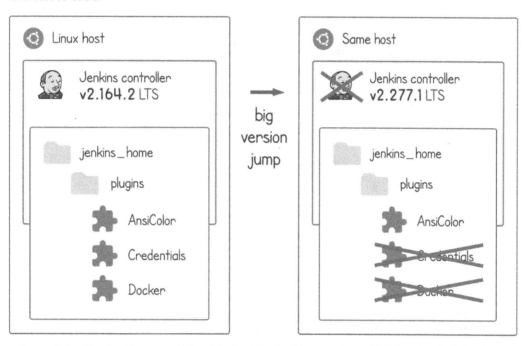

Figure 8.2 – Plugins incompatible with the new Jenkins version will fail to load. If the failed plugin is a critical component, the entire Jenkins will fail to load

Since Jenkins fails to load, we can't go to Plugin Manager to upgrade the old incompatible plugins.

Pitfalls of preinstalling failed plugins

One way to overcome the plugin version incompatibility issue is by preinstalling an upgraded version of the plugins that are failing to load. Using the technique we learned in *Chapter 6, Jenkins Configuration as Code*, we can modify the Dockerfile to preinstall the necessary plugins. The following is an example of an updated Dockerfile that uses the latest LTS version of Jenkins and installs the current latest credentials plugin version 2.3.15:

Jenkins.small-medium.dockerfile

```
FROM jenkins/jenkins:2.277.1-lts

USER root

RUN jenkins-plugin-cli --plugins \
    active-directory \
    ansicolor \
    cobertura \
    configuration-as-code \
    credentials:2.3.15 \
    docker-plugin \
    docker-workflow \
    ghprb \
    junit

RUN  usermod -u 123 -g 30 jenkins

RUN  mkdir /backup && chown 123:30 /backup

USER jenkins
```

Unfortunately, this technique has many critical pitfalls, and the problems get worse as more versions are jumped during an upgrade.

The biggest problem is compatibility between the old and new versions of the same plugin. If a plugin is upgraded while skipping a few versions, there's a chance that the old and the new versions of the same plugin manage their configuration data (such as the configs for ThinBackup) in different formats. The specific plugin version that changed the data format would have the code to convert the old format to the new, but a much later version may no longer have that code. As a result, the configuration could be lost, or worse, the plugin could fail to load and bring Jenkins down with it. The worst-case scenario is that the plugin could start to convert the data and then crash in the middle, resulting in corrupted plugin data that can't be processed by both the old and the new versions of the plugin. If the crashing plugin is one that the Jenkins controller requires to load, neither the old nor the new Jenkins can start. Effectively, we have a corrupted jenkins_home and the paths for both an upgrade and a rollback are blocked. At this point, we are forced to restore from a backup and try a different method.

Another problem with this approach is that we don't know which version of each plugin we're installing in the Dockerfile. We can specify the version number of each plugin in the Dockerfile as we did previously for the credentials plugin, but does that mean we need to go to the home page for all nine plugins to look up the version numbers? Sure, we can do that, but also remember that there are many more plugins in our Jenkins such as Folders and Git plugins that are not listed in the Dockerfile.

There are over 100 plugins installed in our Jenkins because even the core features such as Pipelines are technically plugins. What versions should those be? How do we specify them? How do we get a list of all plugins anyway?

The final problem with this approach is that we won't know what other plugins will break until we try again. Under the Jenkins umbrella, we are effectively running over 100 small programs that could break Jenkins. We'll end up repeating the loop of updating the Dockerfile, building a new image, and running it to see what else breaks. Each time we go through the loop, we face the risk of corrupting plugin data.

As we can see, the biggest challenge associated with upgrading Jenkins is *plugin versions management*, not Jenkins version management. A number of strategies are available for a successful upgrade.

Upgrade strategies

The amount of risk we can take during an upgrade in exchange for an easier operation depends on the cost of a possible outage. On a small- to medium-scale Jenkins, a whole day of outage on a Saturday may be acceptable. On a large-scale Jenkins, not only is it important that we minimize the chance and the duration of an outage, but the upgrade must also be rehearsed in a staging environment before it is attempted in a production environment. Let's look at the upgrade strategy for each case, starting with the smaller of the two.

Upgrade strategy for a small- to medium-scale Jenkins instance

The key points of the first strategy are as follows:

- Upgrade the plugins using Plugin Manager.
- Upgrade all plugins and Jenkins each quarter (four times per year).
- Upgrade Jenkins only to the LTS releases, as opposed to the very latest releases.
- Upgrade the plugins before and after the Jenkins upgrades to minimize the chances of version incompatibility issues.

The biggest difference between the upgrade strategies for a small- to medium-scale versus a large-scale Jenkins is that the former uses Plugin Manager to upgrade the plugins. Using a graphical user interface (GUI) to manage a system state can be risky because we can't easily code-review the actions, have a full record of events, and have a way to roll back. Despite these disadvantages, the GUI is still recommended because using the code to manage the state for Jenkins is very tedious as we will see shortly. Since we should have good backups in place that we can restore easily, the risks can be worth the amount of reduced work, especially on a small- to medium-scale Jenkins where downtime on a Saturday is acceptable.

The second point of upgrading regularly is important. As we will see soon in the coming *Upgrading plugins using Plugin Manager* section, Plugin Manager allows us to upgrade plugins only to the latest version. If we defer the upgrade too long, the latest version becomes incompatible with the current Jenkins and the upgrade is blocked. In other words, there's a small window of opportunity where we can upgrade plugins – missing the window will increase the chance of a problem during an upgrade.

Using an LTS release rather than the very latest release is a best practice for running anything in a production environment. There are a large number of users for the LTS releases, so the plugin issues will likely be ironed out by the time we upgrade. Some organizations choose to stay back one version, but this isn't recommended on Jenkins because of the upgrade window problem. If the latest LTS was released only a few days before the quarterly upgrade, we can defer the Jenkins upgrade for a month or maybe even a quarter. Do not wait until the next LTS is released.

The last point is about reducing the chance of a version incompatibility issue. Before upgrading Jenkins to the next LTS, upgrade the plugins to the latest version so that the plugins have a higher chance of being compatible with the new LTS. After the Jenkins upgrade, upgrade the plugins again so that we have the latest versions supported by the new LTS.

We'll go through an actual upgrade scenario in a bit. Before that, let's see how the upgrade strategy differs for a large-scale Jenkins.

Upgrade strategy for a large-scale Jenkins instance

A large-scale Jenkins with over 1,000 pipelines should also be upgraded quarterly to the next LTS version, but the plugin upgrades require a more precise strategy. Breaking a large Jenkins can be very costly to the business, so each step must be carefully planned, reviewed, and executed. Unlike the previous strategy, we should not blindly take the latest plugins, but instead pin them to specific versions and test them before applying them to the production Jenkins.

This means that we need to write down the version numbers of all 100+ plugins used in our Jenkins to our Dockerfile. Once we have a Dockerfile with all plugins and their versions listed, a Docker image created from the Dockerfile can be used to create an identical *staging* Jenkins where we can validate the upgrade plans. While this is no small task, it gives us the freedom to experiment and make mistakes without incurring downtime.

Here are the broad strokes of the upgrade steps:

1. Capture the list of currently installed plugins and their versions by running the following command in the Script Console:

```
Jenkins.instance.pluginManager.plugins.collect {
 plugin -> "    ${plugin.shortName}:${plugin.version} "
}.sort().join('\\\n')
```

The output is a list with four leading spaces that fits perfectly into our Dockerfile:

 Script Console

```
1  Jenkins.instance.pluginManager.plugins.collect {
2    plugin -> "    ${plugin.shortName}:${plugin.version}   "
3  }.sort().join('\\\n')
4
```

Run

Result

```
Result:      ace-editor:1.1 \
    active-directory:2.24 \
    ansible:1.1 \
    ansicolor:0.7.5 \
```

Figure 8.3 – Script Console command to list all installed plugins and versions

2. Copy and paste the list into our Dockerfile for Jenkins under the `jenkins-plugin-cli` line as follows. Also, update the tag in the FROM line to the latest LTS version:

jenkins.large.dockerfile

```
FROM jenkins/jenkins:2.277.1-lts

USER root

RUN jenkins-plugin-cli --plugins \
    ace-editor:1.1 \
    active-directory:2.24 \
    ansible:1.1 \
    ansicolor:0.7.5 \
    [...]
    workflow-step-api:2.23 \
    workflow-support:3.8 \
    ws-cleanup:0.39

RUN   usermod -u 123 -g 30 jenkins

RUN   mkdir /backup && chown 123:30 /backup

USER jenkins
```

This Dockerfile now contains the new LTS Jenkins version along with the plugin versions that are installed in the existing Jenkins.

3. On a new host, start a staging Jenkins using an image from the updated Dockerfile. We should be on a new host different from the controller so that we don't end up with port conflicts. It's also not a good practice to use the production server as a testbed. I won't repeat the meaning of each flag because we're now in the big league, handling the large-scale Jenkins:

```
newhost:~$ docker build -f jenkins.large.dockerfile . -t
calvinpark/jenkins:2.277.1--lts

newhost:~$ docker run \
    --detach \
    --restart on-failure \
    -u $(id -u):$(id -g) \
    -v ~/jenkins_home:/var/jenkins_home \
    -p 8080:8080 -p 50000:50000 \
    --name jenkins_controller \
    calvinpark/jenkins:2.771.1-lts
```

Since this is a test instance mainly focusing on the version compatibility issues, the reverse proxy and the TLS can be skipped.

4. Once Jenkins is up and running, go to Plugin Manager and upgrade all plugins. Even in a code-driven upgrade, we still need to use the Plugin Manager GUI because there are no other easy ways to find out the latest version of each plugin.

5. Once all the plugins are upgraded, run the Script Console command again to capture the updated plugin versions. This gets us the latest plugin versions that are compatible with the new LTS Jenkins version.

6. Update the Dockerfile again with the updated plugin versions. This Dockerfile now has both the latest Jenkins version as well as the latest plugin versions. This is our release candidate for the upgrade. Get this change reviewed by your peer through a PR.

7. Schedule a maintenance window and communicate it to the users (more on that soon).

8. Upgrade Jenkins using the Jenkins image from the latest Dockerfile.

The steps for upgrading a large-scale Jenkins are admittedly hacky, tedious, and error-prone. During the research for writing this book, I talked to several large-scale Jenkins admins, and everyone shared the same sentiment that this is a problem yet to be solved.

Scaling Jenkins is a difficult task for various reasons such as performance bottlenecks, permission models, and credential management. If you have succeeded in scaling Jenkins to have over 1,000 pipelines, I'm willing to bet that you have developed your own way of organizing the pipelines and managing the controller state, ending up with a unique Jenkins management solution that you have developed. At that point, you alone would know how best to upgrade your unique Jenkins.

The list of steps shared previously is just one example of a custom solution that obviously won't fit your unique Jenkins. Rather than taking it as a ready-made list of steps, understand the key points such as the challenges of managing the version compatibility issues, and incorporate them into your own solution.

Regardless of whether our Jenkins is big or small, it's important that we *upgrade Jenkins and the plugins quarterly* at a minimum. An upgrade becomes more likely to break the more we delay.

Let's now move on to upgrade the plugins.

Upgrading plugins using Plugin Manager

We'll upgrade our Jenkins using the first strategy. We are upgrading from one newish LTS version to the next LTS version, therefore we're not expecting any plugin version incompatibility issues. Even so, to minimize the possibility, let's upgrade all the plugins before upgrading the controller.

Log in to Jenkins as the admin user, click Manage Jenkins, and then Manage Plugins. On the Updates tab, check all the boxes that are available for an upgrade. In *Figure 8.4*, notice that LDAP cannot be upgraded because version 2.4 is not compatible with the current version of Jenkins. But wait a minute. The current version of the LDAP plugin is 1.26, which is far back from 2.4. Would version 2.0, 2.1, 2.2, or 2.3 be compatible with the current Jenkins? How do we look up the version compatibility chart for older versions of a plugin?

Updates	Available	Installed	Advanced			
Install	Name ↓			Version	Released	Installed
☑	Bootstrap 4 API `api-plugin` Provides Bootstrap 4 for Jenkins plugins.			4.6.0-2	19 days ago	4.6.0-1
☑	bouncycastle API `api-plugin` `Library plugins (for use by other plugins)` This plugin provides an stable API to Bouncy Castle related tasks.			2.20	13 days ago	2.18
	LDAP `UNAVAILABLE` `Authentication and User Management` Adds LDAP authentication to Jenkins This version of the plugin exists but it is not being offered as an update. This is typically the case when plugin requirements, e.g. a recent version of Jenkins, are not satisfied. See the plugin documentation for information about its requirements.			2.4		1.26

Figure 8.4 – Not all plugins can be upgraded due to version incompatibility issues

The answer is that we can't. And herein lies the reason why we must upgrade frequently. The Updates tab of Plugin Manager only allows us to upgrade to the latest available version of a plugin. If we miss the window to upgrade while the latest plugin is compatible with our Jenkins, we can no longer use the Updates tab to upgrade the plugin, even if there is an older version that's compatible with our Jenkins. It's technically possible to go to each plugin's Git repository, look up the release history, guess the compatible version for our Jenkins, and then update the Dockerfile for Jenkins to upgrade, but in reality, that's not a scalable solution. If we delay the upgrade long enough, we'll eventually get to a point where nearly all plugins can't be upgraded. Let *Figure 8.5* be a cautionary tale and upgrade quarterly:

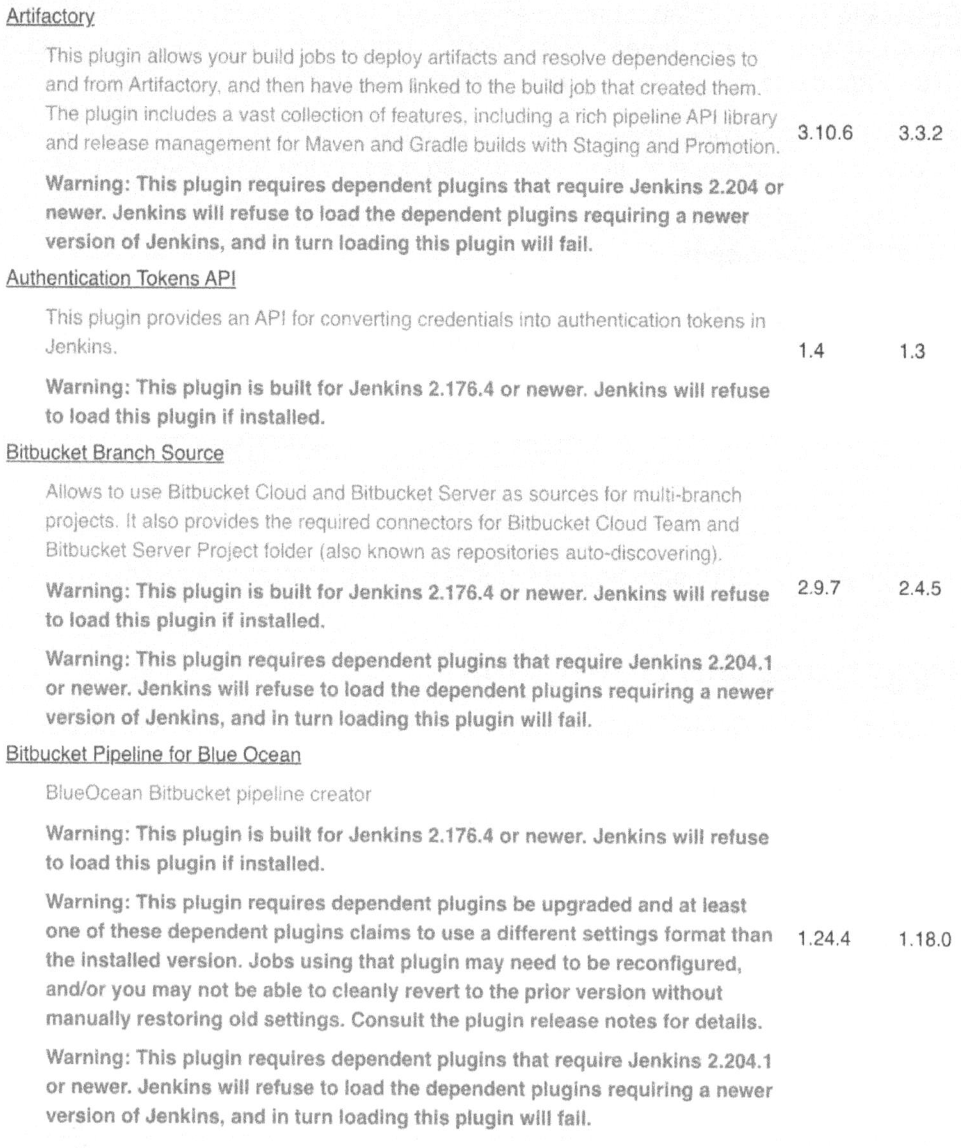

Artifactory

This plugin allows your build jobs to deploy artifacts and resolve dependencies to and from Artifactory, and then have them linked to the build job that created them. The plugin includes a vast collection of features, including a rich pipeline API library and release management for Maven and Gradle builds with Staging and Promotion. 3.10.6   3.3.2

Warning: This plugin requires dependent plugins that require Jenkins 2.204 or newer. Jenkins will refuse to load the dependent plugins requiring a newer version of Jenkins, and in turn loading this plugin will fail.

Authentication Tokens API

This plugin provides an API for converting credentials into authentication tokens in Jenkins.   1.4   1.3

Warning: This plugin is built for Jenkins 2.176.4 or newer. Jenkins will refuse to load this plugin if installed.

Bitbucket Branch Source

Allows to use Bitbucket Cloud and Bitbucket Server as sources for multi-branch projects. It also provides the required connectors for Bitbucket Cloud Team and Bitbucket Server Project folder (also known as repositories auto-discovering).

Warning: This plugin is built for Jenkins 2.176.4 or newer. Jenkins will refuse to load this plugin if installed.   2.9.7   2.4.5

Warning: This plugin requires dependent plugins that require Jenkins 2.204.1 or newer. Jenkins will refuse to load the dependent plugins requiring a newer version of Jenkins, and in turn loading this plugin will fail.

Bitbucket Pipeline for Blue Ocean

BlueOcean Bitbucket pipeline creator

Warning: This plugin is built for Jenkins 2.176.4 or newer. Jenkins will refuse to load this plugin if installed.

Warning: This plugin requires dependent plugins be upgraded and at least one of these dependent plugins claims to use a different settings format than 1.24.4   1.18.0
the installed version. Jobs using that plugin may need to be reconfigured, and/or you may not be able to cleanly revert to the prior version without manually restoring old settings. Consult the plugin release notes for details.

Warning: This plugin requires dependent plugins that require Jenkins 2.204.1 or newer. Jenkins will refuse to load the dependent plugins requiring a newer version of Jenkins, and in turn loading this plugin will fail.

Figure 8.5 – If we don't upgrade for long enough, we get to a point where we can't upgrade at all

Once all the boxes are checked, click Download now and install after restart. That will take us to the Installing Plugins/Upgrades page where we can see the plugins being prepared for an upgrade. Check the Restart Jenkins when installation is complete and no jobs are running box to restart Jenkins and complete the plugin upgrade. The page will seem like it's hung – wait for 10 seconds and refresh the page manually.

Go back to the Updates tab. Sometimes, we'll find that more plugins became available for an upgrade. Plugins depend on each other, so upgrading one could allow another to be upgradable. Repeat the upgrade loop until all plugins are upgraded.

Also, a nice feature of Plugin Manager is that it allows a downgrade. If we upgraded a dozen plugins and one of them fails to load, we can downgrade just that plugin:

Figure 8.6 – Plugin Manager allows a downgrade

With the plugins fully upgraded, let's continue to the controller upgrade.

Upgrading the controller

The mechanics of upgrading the controller are rather simple – just update the Jenkins image tag, build a new image, and restart. However, the upgrade can't be completed without stopping and starting a new container, which means that there will be an unavoidable downtime. Let's see how to handle this effectively.

Announcing the upgrade plans to the users

There are three main ways to communicate with users: emails, messengers, and Jenkins system messages. Let's first look at emails.

Sending emails to all users

When AD or LDAP is used for authentication, the best practice is to create an AD group and have all users join the group as part of onboarding to Jenkins. With an AD group, we can easily contact all users through the group's email address, and also have an accurate count of the number of users. There can be multiple AD groups for a more specific user/group permission as we saw for adder or subtractor projects, but even with the specific AD groups, it's still useful to have a catch-all AD group for all Jenkins users.

For those of us using Jenkins' own user database, it's a bit trickier. We can run the following code in the Script Console to collect the email addresses of all users:

```
import hudson.tasks.Mailer
def allUsers = Jenkins.instance.securityRealm.allUsers
for (User u : allUsers) {
    email = u.getProperty(Mailer.UserProperty.class).address
    print(email + " ; ")
}
```

Here's what it looks like on the Jenkins that we've made in *Chapter 1, Jenkins Infrastructure with TLS/SSL and Reverse Proxy*, to *Chapter 4, GitOps-Driven CD Pipeline with Docker Hub and More Jenkinsfile Features*:

```
1  import hudson.tasks.Mailer
2
3  def allUsers = Jenkins.instance.securityRealm.allUsers
4  for (User u : allUsers) {
5      email = u.getProperty(Mailer.UserProperty.class).address
6      print(email + " ; ")
7  }
```

`Run`

Result

```
adder-admin@lvin.ca ; adder-user@lvin.ca ; calvinspark@gmail.com ;
subtractor-admin@lvin.ca ; subtractor-user@lvin.ca ;
```

Figure 8.7 – List of email addresses from all users

We can communicate with users by copying the email addresses and putting them in BCC (don't use To or CC unless you want an annoying email thread that everyone reply-alls to).

Next, let's look at messengers.

Group messengers

Similar to having an AD group for all users, it's a good idea to have a dedicated channel for Jenkins users in Slack, Teams, or other group messengers. Administrators can use the channel to make announcements, and users can post any support questions. Since messengers often support threaded discussions, it's a good place to announce the more incremental status updates during an upgrade as compared to sending emails.

Finally, let's look at the Jenkins system message.

Jenkins system message

Go to System Configuration and use HTML code in System Message to make the announcements more visible. Click Preview to preview the message and then click Save to apply it on the home page:

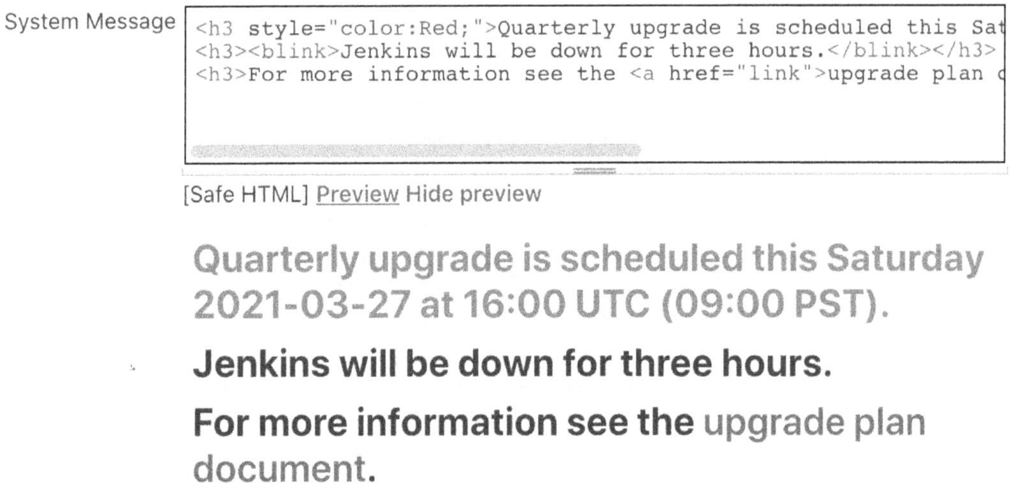

System Message

```
<h3 style="color:Red;">Quarterly upgrade is scheduled this Sat
<h3><blink>Jenkins will be down for three hours.</blink></h3>
<h3>For more information see the <a href="link">upgrade plan o
```

[Safe HTML] <u>Preview</u> Hide preview

Quarterly upgrade is scheduled this Saturday 2021-03-27 at 16:00 UTC (09:00 PST).

Jenkins will be down for three hours.

For more information see the upgrade plan document.

Figure 8.8 – Previewing the HTML code in the system message

It's a good idea to announce the upgrade plans using the system message about a week before the upgrade. We can even start with all black text and then change the text to red and increase the font size 2 days before the actual upgrade to remind users.

Make the appropriate upgrade announcements using the three communication tools. Next, let's discuss the upgrade plan document.

Upgrade plan document

Even a simple upgrade ends up having many small steps, and it's easy to miss a step, especially under stress if an upgrade fails. Developing an upgrade plan document and following the list one item at a time helps to reduce operator mistakes.

An upgrade plan document should aim to contain the high-level description of the activity, schedule, list of actions, and the business continuity plan. The format and the amount of detail can vary depending on your team's dynamic. Continue reading the chapter to see an example you can follow, or search online for "SRE runbook" to find additional examples.

Let's continue prepping for the upgrade.

Building a new controller image

Our controller is defined in a Dockerfile, and this file should be kept in a Git repository to be version-controlled, just like a source code file:

1. Clone the Git repository that contains the Dockerfile and make a branch.

2. Modify the Dockerfile and change the FROM image tag to 2.277.1-lts, which is the latest LTS version at this time:

jenkins.dockerfile

```
FROM jenkins/jenkins:2.277.1-lts
RUN jenkins-plugin-cli --plugins \
    active-directory \
    ansicolor \
    cobertura \
    configuration-as-code \
    docker-plugin \
    docker-workflow \
    ghprb \
    junit
USER root
RUN  usermod -u 123 -g 30 jenkins
RUN  mkdir /backup && chown 123:30 /backup
USER jenkins
```

3. Save the change, make a commit, and push it upstream to create a PR.

4. In the PR description, add a link to the upgrade plan document. During the review for this PR, the following items should be discussed:

   - New Jenkins version
   - Upgrade plan document
   - The two upgrade strategies for small- and large-scale Jenkins

Approving this PR should mean that our peers agree with us on the new Jenkins version, the information in the upgrade plan document, and the upgrade strategy we've chosen. The merged PR can serve as a tracker for the upgrade history.

5. Once the Dockerfile changes have been approved and merged, build a new image on the controller host. We want to minimize the downtime during an upgrade, which means we should do everything we can in preparation for the upgrade before the day of the upgrade:

```
controller:~$ docker build -f jenkins.dockerfile . -t
calvinpark/jenkins:2.277.1-lts
```

The new Docker image is ready. Next, let's go through the pre-upgrade checklist.

Pre-upgrade checklist

Go through this list to make sure we're ready to upgrade:

- Upgrade announcement email sent a week before the upgrade, and again the day before the upgrade date.
- Jenkins system message updated with the upgrade announcement a week before the upgrade.
- All plugins upgraded to the latest version.
- New Docker image for the controller reviewed, approved, and built.
- Back up `jenkins_home`. Have the backup ready in case it needs to be restored.

We are now ready to upgrade. Let's dive in.

Finally, the actual upgrade

In this section, I have listed not only the technical aspects of an upgrade, but also the communication requirements, timeline, and the business continuity plan. Let's go through a mock upgrade scenario together.

Checkpoint 1 – drain (08:00 PST)

1. Announce the following to the #devops Slack channel: *The quarterly maintenance begins now. Jenkins is put in maintenance mode to drain the builds running right now.*

2. Put Jenkins in maintenance mode to stop new builds from running. Builds that are already running will continue to run until completion. On Jenkins, click Manage Jenkins, **followed by** Prepare for Shutdown. A red banner with the message *Jenkins is going to shut down* will appear:

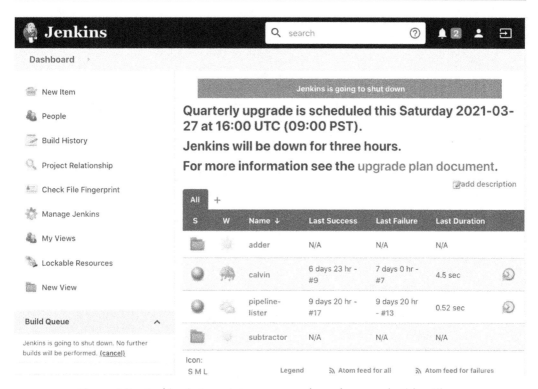

Figure 8.9 – Jenkins is in maintenance mode and no new builds will run

As we're waiting for the builds to drain, watch out for the deadlocks. If a build for pipeline A is waiting for a new build on pipeline B, A will never finish because B cannot start a new build in maintenance mode. Contact the owners of the deadlocked builds and ask whether the builds can be canceled. If they must finish, take Jenkins out of maintenance mode briefly to let the queued builds start, and then put it back in maintenance mode. We can move on to the actual upgrade once there are no running builds.

Checkpoint 2 – upgrade (09:00 PST or sooner if all the builds are drained)

3. Announce the following to the #devops Slack channel: *All running builds are drained. Bringing down Jenkins.*

4. SSH to the controller's host: $ ssh controller.

5. Create the following file in jenkins_home:

jenkins_home/init.groovy

```
import jenkins.model.Jenkins;

Jenkins.instance.doQuietDown();
```

This ensures that the upgraded Jenkins starts in maintenance mode so that the queued builds do not start until we're done with the upgrade.

6. Stop and delete the existing Jenkins container:

```
controller:~$ docker stop jenkins_controller

controller:~$ docker rm jenkins_controller
```

7. Start a new Jenkins container using the new image:

```
controller:~$ docker run \
    --detach \
    --restart on-failure \
    -u $(id -u):$(id -g) \
    -v ~/jenkins_home:/var/jenkins_home \
    -v ~/backup:/backup \
    -p 8080:8080 -p 50000:50000 \
    --name jenkins_controller \
    calvinpark/jenkins:2.277.1-lts
```

8. Monitor the logs to make sure that Jenkins has started successfully.

```
controller:~$ docker logs -f jenkins_controller
```

Look for the message Jenkins is fully up and running. The timestamp will be different, but the message should look like this:

```
                robot_acct@firewalled-controller: – – ssh robot_acct@jenkins-firewalled.lvin.ca
2021-03-19 05:32:10.722+0000 [id=27]    INFO    o.s.c.s.AbstractApplicationContext#obtainFreshBeanFactor
y: Bean factory for application context [org.springframework.web.context.support.StaticWebApplicationCon
text@34a9a39]: org.springframework.beans.factory.support.DefaultListableBeanFactory@55975d15
2021-03-19 05:32:10.723+0000 [id=27]    INFO    o.s.b.f.s.DefaultListableBeanFactory#preInstantiateSingl
etons: Pre-instantiating singletons in org.springframework.beans.factory.support.DefaultListableBeanFact
ory@55975d15: defining beans [filter,legacy]; root of factory hierarchy
2021-03-19 05:32:10.833+0000 [id=61]    WARNING h.p.sshslaves.JavaVersionChecker#resolveJava: Java is no
t in the PATH nor configured with the javaPath setting, Jenkins will try to guess where is Java,this gue
ss will be remove in the future.
2021-03-19 05:32:10.878+0000 [id=62]    WARNING h.p.sshslaves.JavaVersionChecker#resolveJava: Java is no
t in the PATH nor configured with the javaPath setting, Jenkins will try to guess where is Java,this gue
ss will be remove in the future.
2021-03-19 05:32:10.922+0000 [id=26]    INFO    jenkins.InitReactorRunner$1#onAttained: Completed initia
lization
2021-03-19 05:32:11.018+0000 [id=20]    INFO    hudson.WebAppMain$3#run: Jenkins is fully up and running
[03/19/21 05:32:19] SSH Launch of calvin on 192.168.1.158 completed in 10,546 ms
[03/19/21 05:32:19] SSH Launch of firewalled-firewalled-agent on 192.168.1.158 completed in 10,549 ms
2021-03-19 05:32:19.560+0000 [id=137]   INFO    hudson.model.AsyncAperiodicWork#lambda$doAperiodicRun$0:
 Started Update IdP Metadata from URL PeriodicWork
```

Figure 8.10 – Jenkins has loaded successfully

9. Open a web browser to verify that Jenkins is up.

10. If Jenkins is up, announce the following to the #devops Slack channel: *New Jenkins came up successfully. Beginning validation.* Depending on your team's preference, this can be announced on a thread to reduce the noise on the Slack channel. In general, the announcements for a service disruption like we saw in *steps 1* and *3* should go to the main channel, and the details can go on a thread. Continue to *Checkpoint 4*.

11. If Jenkins is not up, announce the following to the #devops Slack channel: *New Jenkins failed to start. Debugging now.* Continue to *Checkpoint 3*.

Checkpoint 3 – fix or roll back (09:15 PST or sooner)

12. Examine the log and identify the cause of the failure. Use the disaster recovery guide (either developed in-house or as specified in *Chapter 7, Backup and Restore and Disaster Recovery*, of this book) to further debug the issue.

13. If the issue is fixed by 10:30 PST, announce the following to the #devops Slack channel: *Issues were resolved, and new Jenkins came up successfully. Beginning validation.* Continue to *Checkpoint 4*.

14. If the issue isn't fixed by 10:30 PST, consider the upgrade a failure and begin a rollback.

15. Announce the following to the #devops Slack channel: *Jenkins upgrade has failed. Beginning rollback.*

16. Stop and delete the existing Jenkins container:

```
controller:~$ docker stop jenkins_controller
controller:~$ docker rm jenkins_controller
```

17. Start a new Jenkins container using the old image:

```
controller:~$ docker run \
    --detach \
    --restart on-failure \
    -u $(id -u):$(id -g) \
    -v ~/jenkins_home:/var/jenkins_home \
    -v ~/backup:/backup \
    -p 8080:8080 -p 50000:50000 \
    --name jenkins_controller \
    calvinpark/jenkins:2.263.1-lts
```

18. Monitor the logs to make sure that Jenkins has started successfully:

```
controller:~$ docker logs -f jenkins_controller
```

Look for the message Jenkins is fully up and running.

19. Open a web browser to verify that Jenkins is up. If Jenkins is up, continue to *Checkpoint 4*.

20. If the old image is also failing to load, jenkins_home has likely been corrupted during the failed upgrade.

21. Stop and delete the container and then restore the backup copy of jenkins_home.

22. Start Jenkins using the old image and the restored backup copy of jenkins_home.

23. Announce the following to the #devops Slack channel: *Jenkins is rolled back. Beginning validation.*

Checkpoint 4 – validate (11:00 PST or sooner)

24. Go through the configurations in the System Configuration, Global Security, and plugin pages to validate that the settings remained unchanged.

25. Take Jenkins out of maintenance mode so that we can run the validation pipelines. This means that the queued builds will start, which is unfortunate because we're not fully done with the validations. More on this in a bit.

26. Run several pipelines to validate the following: premerge and postmerge hooks, agent availability, Docker cloud availability, pipeline build, and the connections to external services such as Slack or Artifactory.

> **Create a dedicated set of validation pipelines and Script Console validation scripts**
>
> Jenkins doesn't provide a way to run the validation pipelines when it's in maintenance mode. This means that we need to take it out of maintenance mode and allow the queued builds to start building before we have a chance to fully validate it.
>
> One way to alleviate this issue is by developing validation scripts to be run on the Script Console because the Script Console is available even in maintenance mode. This still doesn't fully fix the issue because not all features can be validated through the Script Console (for example, a premerge hook), not to mention that developing such scripts requires advanced knowledge of the Jenkins internals as well as Groovy programming skills.
>
> At a minimum, we should create a set of pipelines dedicated to validation. This way, we can run just a few validation pipelines to test all aspects of Jenkins quickly, as opposed to running the production pipelines.

27. Announce the following to the #devops Slack channel: *Jenkins is validated successfully.*

Checkpoint 5 – finish (12:00 PST or sooner)

28. Remove the Jenkins system message about the upgrade.

29. Delete `jenkins_home/init.groovy` so that a restart doesn't bring Jenkins to maintenance mode.

30. Announce the following to the #devops Slack channel: *Jenkins upgrade has completed (un)successfully.*

31. If the upgrade was unsuccessful, develop a Root Cause and Corrective Action (RCCA) document describing the incident, identifying the root cause, and planning corrective action. The messages and the timestamps on the Slack channel and the threads will be useful in developing the RCCA document.

32. Announce to the email distribution list that the upgrade has been completed (un)successfully. If the upgrade was unsuccessful, add the link to the RCCA and the plan for the next upgrade.

Whew. Even a *simple upgrade* turned out to be a 32-step process (or more if you count the pre-upgrade steps). It wouldn't be a bad idea to capture these steps directly in the upgrade plan document.

> **What about the agent upgrades?**
> The agents are automatically upgraded when the controller is upgraded. This is one of Jenkins' best features!

Jenkins is fully upgraded. Don't forget to do this every quarter.

Summary

In this chapter, we learned about the challenges of upgrading Jenkins and then went through a mock-upgrade scenario.

We learned that the biggest challenge is managing the plugin versions because of both the compatibility issues with a new Jenkins version as well as the compatibility issues with an older version of the same plugin.

With an understanding of the challenges, we learned two ways of handling upgrades. The first method, using Plugin Manager to upgrade the plugins on the production Jenkins instance, is more suitable for a small- to medium-scale Jenkins due to the risks we're taking by installing the latest plugins. The second method of modifying the Dockerfile and testing the changes by creating a staging Jenkins is more suitable for a large-scale Jenkins where every change must be carefully controlled, even though it takes a lot more work than the first method.

We then followed the first upgrade method and learned how to use Plugin Manager to upgrade the plugins. We learned that Plugin Manager offers an upgrade only to the latest version even if the latest version is incompatible with current Jenkins, even if there are older versions that are compatible.

Then we learned three ways of communicating to users about the upgrade plans: email, group messengers, and Jenkins system messages. We learned that an upgrade plan document should be prepared, reviewed, and approved. We built a controller image with the latest LTS release of Jenkins. Then, we finally upgraded Jenkins.

Instead of covering just the mechanics of an upgrade, we went through a mock-upgrade scenario where we learned the end-to-end process of communicating with users and handling failures.

This concludes *Part 2, Jenkins Administration,* of the book. With the knowledge from Parts 1 and 2, we should be able to successfully manage a small- to medium-scale Jenkins instance. In *Part 3, Advanced Topics,* we will dive deeper into the skills beyond simple Jenkins maintenance, which will help us grow a medium-scale Jenkins into a large-scale Jenkins. The following *Chapter 9, Reducing Bottlenecks,* is full of useful tips and tricks for readers of all technical levels. For those of you with advanced Jenkins administration skills, this will be the first time in a long time that you have learned something new about Jenkins. We are very excited to include advanced material for veterans – we hope you enjoy reading it as much as we enjoyed writing it.

Part 3

Advanced Topics

In this section, you will learn various optimization techniques to scale your Jenkins. You will learn the various ways to provide and consume shared libraries. You will also learn about the Jenkins security model, and the ways to protect Jenkins and pipelines in a shared environment.

This part of the book comprises the following chapters:

- *Chapter 9, Reducing Bottlenecks*

- *Chapter 10, Shared Libraries*

- *Chapter 11, Script Security*

⑨
Reducing Bottlenecks

Any long-time server administrator managing Jenkins will have their Jenkins battle stories where wide-reaching critical problems occurred within infrastructure and they swooped in to save the day with the perseverance to meet problems head-on and fix them. Sometimes, fixes can be temporary workarounds, and other times permanent fixes are available. Whatever the case you'll encounter, this chapter hopes to battle-harden you for your journey ahead by taking lessons learned from other Jenkins administrators and sharing these experiences.

> **A word of caution**
>
> Before diving too deep into this chapter on Jenkins performance and bottlenecks, please bear in mind to not prematurely optimize. Some optimizations can be delayed until you actually start experiencing performance issues, so it's important not to over-engineer your setup if you don't need to.

It is also worth noting that every year since 2011, Jenkins has several global community events where presentations are recorded and published online for free for users to view. Over the years, the event has been branded Jenkins User Conference, Jenkins World, and, more recently, DevOps World. Each year, users and contributors share their stories on performance issues they've experienced and learned from while operating Jenkins infrastructure within organizations across the globe. I encourage you to check these out every year, especially where Jenkins performance, bottlenecks, and workarounds are presented. This chapter will recommend some relevant videos from past Jenkins conferences that you can watch as supplemental material.

A small note is that this chapter is written by Sam Gleske, so the writing style may be slightly different from the other chapters.

In this chapter, we're going to cover the following main topics:

- Recommendations for hosting Jenkins to avoid bottlenecks
- Quick performance improvements in an existing Jenkins instance
- Improving Jenkins uptime and long-term health
- Pipeline as Code practices
- Controller bottlenecks created by an agent
- Storing controller and agent logs in CloudWatch
- Other ways to reduce agent log output

Technical requirements

As the chapter progresses, it becomes more technical. The introduction at the beginning and the summary at the end of the chapter is intended for all readers. Readers are recommended to have a GitHub account, AWS account, a running Jenkins instance, and knowledge of Java/Groovy is recommended.

Files in the chapter are available in GitHub: https://github.com/PacktPublishing/ Jenkins-Administrators-Guide/tree/main/ch9.

Recommendations for hosting Jenkins to avoid bottlenecks

Jenkins has many benefits in terms of its versatility to deliver automation solutions. As a Jenkins server grows, it is more likely a first-time setup will start having growing pains in terms of performance. This section is going to cover some recommendations for Jenkins hosting, application server configuration, and even recommendations for tracking operational costs if you're in the cloud.

General server recommendations

Several bottlenecks can be avoided by having good hosting for Jenkins from the beginning.

Jenkins controller should have fast storage, but can be light on CPU and memory. In general, at least 2 vCPUs and 16 GB of system memory is recommended if you plan to configure thousands of jobs. The Jenkins controller uses the /tmp directory a lot, so it could be mounted on a tmpfs if you're using more than 16 GB of memory. Agents tend to have higher computing and memory requirements due to them being the workhorse of Jenkins infrastructure, which needs to compile software.

AWS recommendations for Jenkins controllers (assuming distributed architecture with no builds on the controller) are as follows:

- Instance: c5.2xlarge (8 vCPUs, 16 GB of memory)
- OS storage: 50 GB GP2 (general purpose SSD with 150 IOPS)
- $JENKINS_HOME storage: 256 GB GP2 (general purpose SSD with 768 IOPS)

AWS recommendations for Jenkins agents are as follows (may vary depending on application requirements). Use the EC2 plugin for autoscaling:

- Instance: c5.2xlarge (8 vCPUs, 16 GB of memory)
- OS storage: 100 GB GP2.

  - To provision storage within the Amazon EC2 plugin, configure the advanced setting, Block device mapping, to the following value: /dev/sda1=:100:true:gp2
  - Generic Block device mapping is <block-device>=[<snapshot-id>]:[<size>]:[<delete-on-termination>]:[<type>]:[<iops>]:[encrypted]

Beyond the general server recommendations, there are some optimizations that can be made via Jenkins settings and plugins, such as the GitHub Pull Request Builder plugin.

How to keep Jenkins memory footprint light

The culmination of all practices introduced in this chapter will help keep your Jenkins controller heap size manageable:

- The regular maintenance of jobs and artifacts helps to significantly reduce used memory. Configure build cleanup globally so as not to rely on individual job cleanup policies. If using multibranch pipeline jobs, then refer to *Log cleanup for multibranch pipeline job types* for ways to properly manage log cleanup.

- Use a custom landing page such as the one provided by the Dashboard View Jenkins plugin and organize Jenkins jobs into Jenkins folders. This helps limit the number of objects required to render the web page when users browse the Jenkins web UI.

- Disable Jenkins features and services that are not used. For example, disable the Jenkins CLI by adding a $JENKINS_HOME/init.groovy.d/disable-jenkins-cli.groovy script (https://github.com/samrocketman/jenkins-script-console-scripts/blob/jenkins-book/disable-jenkins-cli.groovy). Other features to potentially disable include the following:

  - Disable SSH to the controller if it's unused.

  - Disable JNLP agents in global security settings if it's unused.

- Regularly terminate long-running pipelines that may be stuck. This is covered later in this chapter.

- Make use of @NonCPS pipeline code when developing complex groovy string processing and rely on agents for heavy-load processing through the sh (shell) step. @NonCPS is covered later in this chapter.

Memory and garbage collection tuning

Specifying JVM memory and garbage collection tuning settings is a must for a lean-running and well-performing Jenkins controllers. Settings discussed in this section must be applied differently depending on how Jenkins was installed:

- RPM Jenkins package: /etc/sysconfig/jenkins in the JENKINS_JAVA_OPTIONS variable.

- DEB Jenkins package: /etc/default/jenkins in the JAVA_ARGS variable.

- Apache Tomcat: $CATALINA_HOME/bin/setenv.sh in the CATALINA_OPTS variable.

- Official Docker image: The --env JAVA_OPTS environment variable[1] must be set with tuning settings.

Memory and garbage collection tuning is a well-covered topic in the Jenkins community[2]. We'll cover JVM tuning for settings for all heap sizes, settings for under 4 GB of heap, and settings for over 4 GB of heap. The following table lists JVM settings applicable to Jenkins infrastructure when hosted with Oracle JDK 8, OpenJDK 8, or OpenJDK 11. Earlier or later versions of Java may not be stable for

---

1   https://github.com/jenkinsci/docker

2   https://www.jenkins.io/blog/2016/11/21/gc-tuning/

these settings. JVM settings vary depending on how large you configure the Java heap. In the following table, based on the heap size of your Jenkins controller, use the recommended settings:

Heap size	Recommended settings
All heap sizes	-Djava.awt.headless=true -server -XX:+AlwaysPreTouch
CMS algorithm 2GB–4GB heap -Xms4G -Xmx4G	-XX:+UseConcMarkSweepGC -XX:+ExplicitGCInvokesConcurrentAndUnloadsClasses -XX:+CMSParallelRemarkEnabled -XX:+ParallelRefProcEnabled -XX:+CMSClassUnloadingEnabled -XX:+ScavengeBeforeFullGC -XX:+CMSScavengeBeforeRemark -XX:NewSize=512m -XX:MaxNewSize=2g -XX:NewRatio=2
G1 algorithm Over 4GB of heap -Xms10G -Xmx10G	-XX:+UseG1GC -XX:+ExplicitGCInvokesConcurrent -XX:+ParallelRefProcEnabled -XX:+UseStringDeduplication -XX:+UnlockExperimentalVMOptions -XX:G1NewSizePercent=20 -XX:+UnlockDiagnosticVMOptions -XX:G1SummarizeRSetStatsPeriod=1

Based on the preceding table, the following are some high-level points for operating Jenkins on dedicated infrastructure:

- Provision a host dedicated to only running Jenkins for both the controller and the agents separately. Do not mix multiple purposes on the same hosting server.

- Set your minimum and maximum heap to be equal. For example, I have managed a Jenkins controller with over 7,000 jobs and its heap is set to 10 GB (-Xms10G -Xmx10G) with 16 GB RAM for the host. This recommendation comes from general Java application server management because there's a performance cost for the JVM to allocate more memory when it is a range.

- Configure the JVM to be headless with the -server option and always pre-touch the heap. The -server option is designed for headless servers (servers with no GUI) and pre-touching the heap forces the JVM to allocate all of the memory it could use initially with no additional memory allocation later.

- The CMS garbage collection algorithm is well suited for heap sizes of 4 GB and smaller. This is recommended if your system memory is 8 GB.

- The garbage first (G1) garbage collection algorithm has the most performance for heap sizes larger than 4 GB. The recommended RAM size is 16 GB for the host. If allocating more RAM to the host, then it may be worthwhile to go through the bottlenecks section again and address some additional cleanup.

- After optimizing Jenkins, if you still require more than 16 GB of RAM for the host, then consider getting deeper performance metrics using application performance monitoring (APM). jmxtrans[3] is a good APM solution, along with the Jenkins metrics plugin being installed. APM is not fully covered in this book since it is a more advanced topic.

- If you recall *Chapter 2, Jenkins with Docker on HTTPS on AWS and inside a Corporate Firewall*, the official Jenkins controller Docker image was called with docker run. The following example illustrates the same command, but with the extra Java options discussed in the table for 16 GB of host memory, where the JVM will use a 10 GB heap:

---

3 https://github.com/jmxtrans/jmxtrans-agent

```
controller:~$ docker run \

    --env JAVA_OPTS="-Djava.awt.headless=true -server
-XX:+AlwaysPreTouch -Xms10G -Xmx10G -XX:+UseG1GC
-XX:+ExplicitGCInvokesConcurrent -XX:+ParallelRefProcEnabled
-XX:+UseStringDeduplication -XX:+UnlockExperimentalVMOptions
-XX:G1NewSizePercent=20 -XX:+UnlockDiagnosticVMOptions
-XX:G1SummarizeRSetStatsPeriod=1" \

    --detach \
    --restart on-failure \
    -u $(id -u):$(id -g) \
    -v ~/jenkins_home:/var/jenkins_home \
    -p 8080:8080 -p 50000:50000 \
    --name jenkins_controller \
    calvinpark/jenkins:2.263.1-lts
```

You'll specify the preceding example Java options differently depending on how you installed Jenkins on your server. The preceding Java option recommendations have worked well for me in several large-scale Jenkins deployments.

Periodic triggers versus webhook triggers

Before we can talk about optimizing controller load for jobs being built with periodic triggers, it may be helpful to define different ways in which a build can be triggered within Jenkins. Builds can be started in one of several ways:

- Periodic trigger: Also called a timer trigger or cron trigger. This scans SCM periodically and if there are changes pushed since the last build, then it will start a new build. It could also be configured to start a new build every time, regardless of the SCM state.

- Webhook trigger: This will initiate a build on push to Git. There are multiple ways to configure Jenkins to automatically build a job when developers push to a Git repository. BitBucket, Gitea, GitHub, GitLab, and other SCM hosting platforms have plugins for webhooks. Jenkins also supports generic Git triggering provided by the Git plugin. This requires no server load when idle. Jenkins will wait for an external service to make a REST API call to initiate a build and otherwise does nothing.

- Custom trigger: Some Jenkins plugins provide custom methods for triggering builds. For example, the GitHub plugin allows a build to be triggered by a pull request comment. This is a feature of GitHub webhooks.

- User trigger: When a user logs in and starts a build via the Build Now button.

A common bottleneck as a Jenkins controller grows is the fact that periodic triggering takes up a significant CPU load on the Jenkins scheduler. Periodic triggers suffer from the following issues:

- Not all jobs are active forever. Sometimes, development moves on and jobs associated with older projects are no longer used. Inactive projects using periodic triggering produce an unnecessary load.

- You might vertically scale the Jenkins controller by adding more memory and vCPUs to compensate for the increased load. This may significantly increase operational costs. Vertical scaling can be avoided if regular cleanup occurs or a different trigger type is used.

- Another common trick for solving periodic trigger CPU bottlenecks is for admins to increase the period during which periodic checking occurs. This increases delays in development because developers may have to wait an extended period before their build launches. If the timer period is 15 minutes, then a developer opening a change for peer review, merging to the main branch, and launching a tag build may be delayed by as much as 45 minutes. This artificial delay increases development costs as opposed to triggering builds immediately as soon as developers are ready.

Webhook triggers is a solution created specifically to *solve the periodic triggering bottleneck*. Webhook-triggered builds have the following benefits:

- Stale jobs associated with inactive repositories do not produce load on the Jenkins controller. There are no push events to trigger jobs, so they take minimal resources to exist. This also makes it easy for an admin to clean up by finding jobs that haven't been built in a long time.

- The Jenkins controller CPU load is significantly reduced because there is no CPU load associated webhooks; unlike periodic checking. No vertical scaling is required because it doesn't matter how many jobs exist; the jobs will be idle until a webhook is received. A webhook is sent when a developer pushes to a Git repository.

- Webhooks are nearly instant where a change pushed will cause a build to trigger a few seconds later. Developers get more immediate feedback and less time is wasted waiting for a build to start. Due to the event-based design of the webhook architecture, there's little to no artificial delay in users' development process.

Vertically scaling the Jenkins controller to compensate for poorly planned job triggering has a real cost associated with it. Vertical scaling usually involves either buying a new server in the data center or increasing resources in the cloud. Increased operational costs eat into profits and may even increase the risks associated with poor uptime due to bad Jenkins controller configurations.

Tracking operational costs in the cloud

Often, an afterthought is operational costs for CI/CD at scale in the cloud. In this section, we'll cover how to track cost upfront as part of infrastructure design. I'll also discuss strategy for reducing costs. It is important to understand how money is spent in the cloud, because it will help you to determine where your time and effort should be invested; so as to cut costs to maximize your gains for cost optimization.

AWS has a feature called resource tags. Resource tags are metadata attached to resources that can help give admins more insight into how their money is spent across different projects hosted in AWS. Adding tags to resources does not add any extra costs to the resource and, in general, you cannot have too many tags on a resource in AWS. To start, here are some recommended tag names that you could use to identify various parts of your Jenkins infrastructure, which would be hosted fully in AWS:

- Name, team, project, environment, half, infra_type, git_repo, git_hash, and git_tag

The following table covers the potential values of these tags. AWS uses Name as a standard for human-readable names across most resources. Keep in mind that it is up to the user to determine which of these tags are useful for billing purposes, so any that don't make sense in relation to your setup can be omitted:

Tag name	Tag value
Name	Can be a human-readable name describing the resource
business_unit	Or bu, this tag can track multiple teams under the same business unit
team	Your team name; consistent for all services managed by your team
project	jenkins or another name unique to your project
environment	dev, staging, qa, or prod
half	blue or green, typically used in a blue/green deployment pattern
infra_type	controller, agent, storage, backup, or artifact
git_repo	The URL of the Git repository deploying your Jenkins infrastructure
git_tag	The Git tag name that deployed your infrastructure (for production)
git_hash	The Git full hash of the commit that deployed the Jenkins infrastructure

The team tag should be your team name and be consistent across all services your team manages. team is mostly useful if you're in a multi-tenant account where infrastructure within the account is shared across multiple teams. business_unit may also be useful in a multi-tenant account.

The project tag should be declared on all resources related to one particular project hosted in AWS such as Jenkins. In addition to Jenkins, it is typical for large-scale setups to include other services such as Sonatype Nexus for artifact hosting and caching, SonarQube code quality and security metrics, and DependencyTrack for continuous component analysis for dependency vulnerabilities. There are many quality open source projects that can be used to shift security left and improve deployment reliability.

environment and half are related to the usage intent of the environment. For example, common environments would be dev, staging, qa, or prod (for production). A prod environment means it is meant for users to use and is likely a critical piece of software.

half refers to blue or green where blue/green deployments are a kind of deployment pattern in operations. The blue/green deployment pattern is where you create a complete duplicate of your infrastructure (in this case, Jenkins controller and agents) and cut over to the inactive half. A blue/green deployment pattern enables easy rollback if issues are found with the controller maintenance.

Jenkins hosted in AWS will utilize several different AWS services and features as part of its hosting and operation. So it's good to define different infrastructure types within the project to help with a cost breakdown. The infra_type tag recommendation breaks the Jenkins-type infrastructure down into five categories:

- controller: The Jenkins controller that schedules jobs to perform work on agents. It is also the web UI where users interact with their projects in Jenkins.
- agent: An agent that performs work for jobs that are building or deploying software.
- storage: $JENKINS_HOME persistent data that is attached to the Jenkins controller.
- backup: EBS snapshots of the $JENKINS_HOME persistent data serving as snapshot backups.
- artifact: When practicing immutable infrastructure, the Amazon Machine Image (AMI) of the controller and agent would have its own artifact type to differentiate it from other kinds of data storage costs.

Finally, if you're deploying Jenkins infrastructure using Jenkins as code, then it's a good idea to add SCM tags to your deployed resources. SCM tags include the Git repository (git_repo), the Git tag used for deployment (git_tag), and the Git long hash (git_hash). The long hash is recommended as an unchanging reference because Git tags can be overwritten but Git hashes are immutable. Including the Git references in your metadata helps with traceability and troubleshooting when controller or agent issues arise. It also aids in recovery if you are aware of the last stable deployment that is hosting Jenkins.

A useful tool in AWS is the AWS Cost Explorer. Other clouds may have a similar tool. Cost Explorer helps you break down your usage and spending in AWS so that you can identify how you can improve and reduce your overall AWS hosting cost. Take *Figure 9.1*, for example:

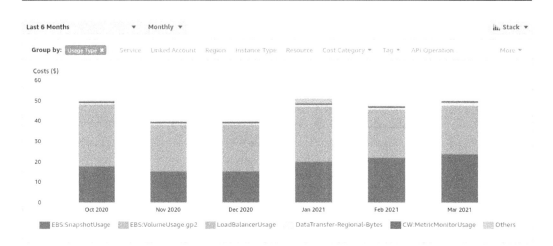

Figure 9.1 – AWS Cost Explorer broken down by usage type

Based on your actual costs, you can determine how to improve your hosting costs. It looks like 30-40% of costs in the example graph are dedicated to EBS snapshot storage. This could be improved by refining snapshot backups, and perhaps doing a more aggressive snapshot cleanup.

Take the following real-world example by one of the authors, Sam Gleske: *Because of resource tagging and taking the time to explore costs in AWS Cost Explorer, I have been able to reduce a $3,000 monthly AWS bill to $500 per month for Jenkins hosting. I determined that static agents were a huge cost and configured the* Amazon EC2 plugin *to use autoscaling agents based on Jenkins labels. That constitutes an 84% cost reduction in hosting. When calculated annually that represents $30,000 in savings.*

Tagging resources is an important aspect of being able to inspect and improve your cloud hosting operational costs.

Quick performance improvements in an existing Jenkins instance

Some Jenkins configuration out of the box is not necessarily the best setting for scaling up Jenkins usage in the long term. Some settings provided by plugins can even be hidden behind application start up options. This section will recommend top-level settings and other tips for improving the load time of the Jenkins controller landing page.

GitHub Pull Request Builder plugin boot optimization

If you are using the GitHub Pull Request Builder (GHPRB) plugin, then you may notice it has a steep start up bottleneck when booting Jenkins controllers. Once you've reached over 100 jobs, you'll start to realize that the Jenkins start up time becomes very delayed when booting or restarting your controller. On Jenkins controllers as large as 1,000+ jobs, I have seen Jenkins startup delayed by up to 20 minutes and more.

The root cause behind the start up delay is the fact that the GHPRB plugin forces all jobs to synchronously reregister its webhooks on boot with GitHub repositories. It takes around 2-3 seconds per job to register its webhook on a good internet connection. To mitigate the GHPRB boot issue, add the following Java property to the Jenkins boot options:

```
-Dorg.jenkinsci.plugins.ghprb.GhprbTrigger.
disableRegisterOnStartup=true
```

If you're configuring a new Jenkins controller, then the Jenkins community is moving toward multibranch pipelines using the GitHub Branch Source plugin instead of configuring standalone jobs with GitHub Pull Request Builder.

Another issue with a new Jenkins controller is that its default view includes the *weather* health display on jobs and folders. This will cause severe performance issues as your usage of Jenkins increases.

Frontpage load delay due to the "weather" health display

A common bottleneck is the *weather* column for Jenkins jobs and folders; the sun icon in the following image. The purpose of the weather icon is to give users a general sense of build health for jobs run previously. However, it is extremely CPU-intensive and disk read-intensive for a Jenkins controller. The larger a Jenkins instance, the greater the performance hit the *weather* calculation takes when the front page is loaded. Every time a user visits or reloads the Jenkins front page, a significant delay occurs while *weather* build health is calculated recursively for all jobs and folders throughout Jenkins. If you disable the weather calculation on the front page, then users will experience a significant improvement in terms of page load time. General performance will improve for Jenkins due to reduced load and disk I/O usage, which is important if there are a lot of users and jobs:

Icon: S M L

Figure 9.2 – Screenshot of the "weather" column (W) as a sun

By default, the Jenkins All view is what users see when first logging in to the Jenkins web interface. The All view includes the *weather* column. To resolve this issue, follow these steps:

1. Create a new List View that will display all jobs. Customize it so that the weather column is not included.
2. Set the new List view as the global default when users log in.
3. Optional: Delete the old All view.

Let's look at these steps in detail in the following sections.

Creating a new List view

Let's create a new view without the weather column:

Jenkins

Dashboard

- New Item
- People
- Build History
- Manage Jenkins
- My Views
- Open Blue Ocean
- Lockable Resources
- New View

View name

All Jobs

○ List View
Shows items in a simple list format. You can choose which jobs are to be displayed in which view.

○ My View
This view automatically displays all the jobs that the current user has an access to.

OK

Figure 9.3 – Creating a new dashboard view

On the Jenkins instance home page, click New View in the menu on the left. You can name the view anything you want. In the preceding screenshot, it is named All Jobs so it doesn't conflict with the All view in terms of name. Choose List View as the type and then press OK:

Figure 9.4 – The All Jobs filter in the new list view settings

While in the view configuration page, you'll want to click Add Job Filter and choose All Jobs as the filter. If you skip this step, you'll end up with an empty view containing no jobs. You can reach this page by clicking Edit View in the Jenkins menu to the left in case you miss this step and need to reconfigure:

Columns

Figure 9.5 – New list view settings; deleting the Weather column

While in the settings, be sure to *Delete* the weather column so that it's not included in the page view. *Figure 9.5* shows a view with the weather column before it is deleted. Once you've finished creating your newly optimized view for jobs, it is time to set it as the landing page when users log in.

Setting the new List view as the default view

Go to System Configuration and scroll down the settings page until you reach Default view:

Figure 9.6 – Changing the Default view in the global configuration

Change Default view to be All Jobs, which is the List View created in the previous section. Save the settings and visit the main home page of your Jenkins instance:

Figure 9.7 – Deleting the former All view as an optional cleanup step

With a new default view selected, you now have the option to delete the old All view. There are several Jenkins plugins that provide more custom job filters and dashboard views, such as Dashboard View and View Job Filters. Dashboard View plugin allows us to use HTML and CSS to create a customized dashboard. View Job Filters plugin provides an admin with extended job filters that can be applied to the item name or type, such as for jobs or folders.

Pipeline speed/durability settings

Jenkins pipelines make a lot of writes to disk in order to regularly persist state as a pipeline runs. Persisting pipeline state enables Jenkins to pause and resume pipelines. By often writing to disk, this can create a bottleneck in Jenkins home storage due to disk I/O. Changing the pipeline durability settings for optimized performance is most recommended. Go to System Configuration and update speed/ durability settings to Performance-optimized:

Pipeline Speed/Durability Settings

Pipeline Default Speed/Durability Level

Performance-optimized: much faster (requires clean shutdown to save running pipelines)

Figure 9.8 – Screenshot of speed durability set to Performance-optimized

Speed/durability settings will positively impact the performance of all steps, including the echo step, which is often used by first-time Pipeline as Code writers. If Jenkins is not cleanly shut down or it crashes, then it will not automatically resume pipelines that were running. There are other durability settings, such as max durability, that can account for an unclean shut down for a performance cost. I strongly recommend evaluating your own needs to see whether the trade-off on performance is worth it.

Improving Jenkins uptime and long-term health

As Jenkins usage goes up in an organization, it becomes mission-critical for orchestrating deployments and improving developer efficiency. As a result, Jenkins tends to run 24x7 with as much uptime as humanly possible. One of the ways in which you can help improve the overall health and reliability of Jenkins is by automating periodic maintenance.

What is a periodic maintenance job and how do you create one?

> **Required plugins**
> Groovy

A periodic maintenance job is any job that runs on a schedule that performs some kind of server maintenance. The maintenance tasks can vary, such as creating a backup of configuration, terminating jobs, restarting agents, or any other task that a Jenkins administrator determines must be done on occasion. I run a lot of system Groovy scripts in Jenkins to help with keeping controller and agent operations smooth. Ensure that you have the *Groovy Plugin for Jenkins* installed[4] and follow these steps:

1. Create a freestyle job.
2. Insert a system Groovy script shell step. This is equivalent to running a script console script, which is what we want to run in order to terminate long-running jobs.
3. In the system Groovy script, enter the contents of a Groovy script of your choice.
4. At the top of the job configuration, under build triggers, configure the job to build periodically.

The following are some examples of periodic maintenance jobs I like to run, organized with a view, to keep Jenkins healthy.

Terminating long-running pipelines

Jenkins pipelines have a thread in the controller runtime known as a flyweight executor, as opposed to heavyweight executors that run workloads on agents. Flyweight executors can run indefinitely if users use an input step without a timeout. Occasionally, there are Jenkins plugin bugs encountered that exhibit similar behavior even with timeouts configured. From an end user and admin perspective, the pipelines just hang without aborting. While it's generally a good practice to encourage users to use timeouts, an admin shouldn't rely on it. Especially in large-scale Jenkins instances with over 500 users, there will inevitably be users of varying skill levels. Therefore, it makes sense to have a maintenance job dedicated to running daily to terminate all user pipelines that have been running for a long period of time.

---

4   https://plugins.jenkins.io/groovy/

I run the `kill-all-day-old-builds.groovy` (https://github.com/samrocketman/jenkins-script-console-scripts/blob/jenkins-book/kill-all-day-old-builds.groovy) Groovy script in production. I have run this script in production every 24 hours for years and maintained Jenkins controller stability. Please note, the script will hard kill Jenkins builds and, because of this, you might run into occasional bugs with other plugins that do not handle hard killing well. The lockable resources plugin does not handle hard killing builds well, so I run another maintenance job to release stale locked resources.

Releasing stale locks in lockable resources from force killing builds

If you're using the Lockable Resources plugin, you'll also want to periodically reset stale locks that can occur when a user or script hard kills a build. Refer to `reset-stale-lockable-resources.groovy` (https://github.com/samrocketman/jenkins-script-console-scripts/blob/jenkins-book/reset-stale-lockable-resources.groovy).

When a build is force killed, it gets terminated in such a way that it does not allow for any kind of post-build actions or resource cleanup. If the job was holding on to a lockable resource, then the lock will never be released. I configure this job to run periodically and I allow any user in Jenkins to run it because it is a safe script to run. This means that if users get stuck by this issue, then they can be given a workaround, which is self-service.

Log cleanup for beginners

A common bottleneck for long-running Jenkins controllers that have 24x7 operation is running out of disk space on the $JENKINS_HOME directory. Jenkins logs can be very verbose and generate hundreds of megabytes of data per build. In order for a Jenkins controller to continue healthy operation, regular log cleanup should be configured. How you clean up logs depends on your needs, but Jenkins has built-in log cleanup available as a feature.

You can apply log cleanup for all jobs. Go to System Configuration and scroll down to the Global Build Discarders section. Add Specific Build Discarder, which can limit jobs by the number of builds, the age of builds, or both for cleanup. Log cleanup can also be configured within individual jobs. Log cleanup is especially important for jobs that contain periodic triggers to build because they'll automatically grow with builds according to the periodic schedule of the job. Growing build logs are not the only performance issue with periodically triggered jobs. There's also a controller load associated with periodic building that can be improved by avoiding periodic triggers, with exception for maintenance jobs.

Log cleanup for multibranch pipeline job types

Due to the design of multibranch pipelines, discarding old build configurations does not work well. In multibranch pipelines, there are three different kinds of Jenkins pipelines when using Git SCM as the repository source:

- A pull request or merge request (also called an automated peer review) pipeline: These are pipelines that run before a merge to a shared branch. After a pull request is merged, the pipeline continues to exist but is disabled.

- Branch pipeline: Typically run from the main branch to create releases.

- Tag pipeline: Typically used to publish compiled artifacts and deploy applications.

The cleanup strategy is not necessarily straightforward depending on your needs. Expiring builds is one solution, but it's not really necessary to hold on to pull request builds for a long time. You can't configure the build discarder to limit to a specific number of builds because there's one pipeline per pull request. So pipelines will quickly grow in number and cause performance issues at scale without cleanup. You could ask users to use different build discarders in their pipeline as code; however, it doesn't scale very well because every user would have to include custom setup code for the build discarder. Builds and pipelines left over from old jobs take up unnecessary disk space and memory on the Jenkins controller. Having a cleanup policy tailored for multibranch pipelines keeps Jenkins running lean with good performance. I configure two periodic cleanup jobs that run the following Groovy scripts:

- `delete-builds-pr-older-than-30days.groovy` (https://github.com/ samrocketman/jenkins-script-console-scripts/blob/jenkins-book/delete-builds- pr-older-than-30days.groovy).

- `delete-builds-branch-and-tag-older-than-120days.groovy` (https:// github.com/samrocketman/jenkins-script-console-scripts/blob/jenkins-book/ delete-builds-branch-and-tag-older-than-120days.groovy) will delete builds associated with branches and tag release pipelines older than 4 months.

This is my alternative mechanism to discarding old builds. I've decided that I don't want pull request builds to last longer than 30 days. Pull requests can be tuned even further. I've also decided that I don't need tagged builds or branch builds to last longer than 4 months. This is specific to my environment and your needs may be different. However, these serve as a great example, which you can tweak for your own needs. Defining your source of truth for build and deployment may be important if you have compliance considerations to make. In this case, hosting evidence in an external system may be better than using Jenkins as the source of truth due to its performance constraints.

Pipeline as Code practices

Writing pipeline as code can be tricky because seemingly innocent code can have huge implications on performance.

Avoiding and reducing the use of echo step

The echo pipeline step is often used by pipeline writers to print some form of debug code or enhance logging messages with more rich text. However, this is not the same thing as echo in shell code. Every pipeline step has some persistence around its speed/durability and the echo step is no exception. According to the Jenkins World 2017 presentation, *Pipelines At Scale: How Big, How Fast, How Many?* [5] all pipeline steps perform 4–5 writes to disk in the Jenkins home directory. This also means an echo step will perform 4–5 *writes to disk* per echo operation on the Jenkins controller $JENKINS_HOME. This creates a disk I/O bottleneck on the Jenkins controller when pipelines are scaled up. Let's discuss an author strategy for writing pipelines. The following echo step code can be improved. Please note that in a Jenkins pipeline, parentheses are optional, so one of the steps includes parentheses as an example:

```
echo 'step 1'
echo 'step 2'
echo 'step 3'
echo('step 4')
```

Instead of performing four separate echo statements, it would be better to combine them as a single *String* to be printed with a single echo step:

```
echo(['step 1', 'step 2', 'step 3', 'step 4'].join('\n'))
```

In the first code example, four echo steps will create *16 read/write operations* on the Jenkins controller $JENKINS_HOME. However, in the second code sample, we've combined all statements into a single *String* and used one echo step. One echo step will print the same information, but reduces the load on $JENKINS_HOME down to 4 *read/write operations*. Being conscious of disk speed and usage is part of maintaining a healthy Jenkins controller. Jenkins logs can be verbose and fill up disk, which also impacts disk performance and, by extension, will impact the end user experience when browsing Jenkins on the web.

Another way to avoid using the echo step is to write bash shell scripts, which provide better output for users to debug when building software. For example, the bash

---

5   JW2017: *Pipelines at Scale: How Big, How Fast, How Many?* https://youtu.be/p0qX409wwPw?t=1770

shell command, set -exo pipefail, will cause commands to exit on the first error (set -e), output to stderr commands before they're executed (set -x), and to fail commands in a bash pipeline (set -o pipefail). You'll want to run this as the first line in your script. It creates a lot of verbosity in a build log, but it also means that there's less need to write echo pipeline steps as a mark between build commands to try to track down build and deployment issues.

CPU bottleneck: NonCPS versus CPS pipeline as code

> **Required plugins**
> Pipeline: Groovy

This is important for shared pipeline libraries. Jenkins scripted pipelines and Jenkins declarative pipelines are interpreted by a modified Groovy runtime known as continuation-passing style (CPS) provided by the Pipeline: Groovy plugin (plugin ID: workflow-cps). Here is another description of CPS: It behaves like a REPL (read, evaluate, print, and loop) where each pipeline statement is read line by line. It is interpreted (in other words, evaluated), the result of the step is printed, and it continues to the next step. This interpretation is done per step and is extremely slow when compared to compiled Java bytecode.

In scripted pipeline and shared pipeline libraries, you can use the @NonCPS annotation on functions. The function will then be just in time (JIT) compiled into Java bytecode before it is executed. @NonCPS Groovy code can be more than 100x faster than the same Groovy code as a CPS pipeline. The following is an example of a @NonCPS annotated function:

```
@NonCPS
String addTwoStrings(String a, String b) {
    return a + b
}
```

Here are a few simple rules you can follow that make writing NonCPS functions easy:

- @NonCPS functions should be treated like functional programming and not object-oriented. It only knows about its arguments and only returns its results. Don't try to reference pipeline steps or other pipeline features, and instead only rely on the function arguments for processing. It is possible to reference some pipeline features such as env variables, but this should be avoided because you're introducing a possibility for inconsistent pause and resume behavior.

- The same inputs (arguments) should always result in the same output (return value). This is important for pipelines that pause because @NonCPS methods are not persisted to disk when paused. Instead, @NonCPS functions are called again on resumption and are expected to have the same return values. Otherwise, you'll get inconsistent pause and resume behavior.

The script security plugin executes the user Jenkinsfile within a pipeline sandbox that has security restrictions. Global shared libraries containing @NonCPS functions (no sandbox) run up to 5–10x faster than the same functions defined in user Jenkinsfiles that run in a sandboxed interpreter.

A great YouTube video to watch on the subject of pipeline performance is a Jenkins World 2017 presentation: *Pipelines At Scale: How Big, How Fast, How Many?* where the presenter, Sam Van Oort, covers his findings on testing the performance limits of executing pipeline code.

Here are a few simple questions you can ask to determine whether you should put the pipeline code you're writing into a NonCPS function or let the function be CPS:

- Are you writing Jenkins pipeline steps in the function? NonCPS functions can't execute pipeline steps, so the function should be CPS.

- Are you writing pure Groovy logic to do some kind of extra processing, such as on a string? Then, the Groovy logic should be refactored into its own NonCPS annotated function and referenced.

Using NonCPS reduces server load by executing bytecode faster than CPS steps because it bypasses any disk I/O of a normal serialized pipeline step. However, you may find the extensive use of NonCPS to be cumbersome and difficult to write test code. As your NonCPS code base grows, it may make more sense to compile it into a Jar.

Pre-compiled code is more performant than even NonCPS code because the JIT compiling of code has a CPU cost associated with it in the Jenkins controller. In a shared pipeline library, you can import remote Jar files into the shared pipeline classpath by using Apache Grape annotations to pull in jars depended on by your shared library; specifically, the @Grab annotation. This means that you can pre-compile your code and dynamically include it into the classpath runtime via @Grab. The following section will cover this in detail.

Pre-compiling all NonCPS code as an external jar

Once you start building out extensive NonCPS code in your shared pipeline library, you may find testing a challenge. I recommend moving all of your NonCPS code into a dedicated jar to be compiled and tested separately from shared pipeline libraries once your code base becomes more of a maintenance challenge. While we're describing your code as NonCPS, it is worth noting that we're just talking about plain old Java objects (POJOs) compiled into a JAR file. You would use a build tool such as Maven or Gradle to package your code into a jar and publish it to a hosted Maven repository. Sonatype Nexus 3 is an example of a self-hosted Maven repository. You can write in any language, which results in a JAR file containing classes, with Java bytecode being generated. Languages which create Java bytecode when compiled include: Java, Groovy, Kotlin, or Scala. In this example, we'll cover generating a Groovy-based library that compiles into a jar using Gradle as its build tool. We're relying on Groovy because Jenkins already has its own Groovy runtime and dependencies embedded within its core controller.

Before you create your project, determine the following:

- Determine the version of Java and Groovy used by your Jenkins controller. The version of Java used by the controller should be the same version run by Jenkins' agent.jar for stability.

- You can run the following script console script in Jenkins to get the versions:

```
println "Java ${System.getProperty('java.version')}"
println "Groovy ${GroovySystem.version}"
```

While there are ways for Gradle to target specific Java versions at compile time, it is simplest to just match the compile runtime of your library with the same one hosting your Jenkins controller. If you desire maximum compatibility, then, as of this writing, targeting Java 8 will give you support for all versions of Java supported by Jenkins (Java 8 and Java 11).

You can refer to the gradle init documentation[6] for other language support. The following code example assumes you've installed OpenJDK 11 and Gradle:

---

6 https://docs.gradle.org/current/userguide/build_init_plugin.html

```
$ mkdir jenkins-pipeline-jar
$ cd jenkins-pipeline-jar
$ git init
$ touch README.md
$ git add README.md
$ git commit -m 'initial commit'
# setup your gitignore file for gradle
$ echo -e '.gradle\nbuild\n*.class\n*.jar\n!gradle/wrapper/gradle-
wrapper.jar' > .gitignore
# create your first groovy library
$ gradle init --type groovy-library
$ git add -A
$ git commit -m 'initial project code'
```

Don't forget to update the `build.gradle` file so that the version of groovy used for compiling Groovy matches the same version used by Jenkins. This simplifies dependency loading in Jenkins. You can compile your JAR file with the following code:

```
# build your code using gradle wrapper
$ ./gradlew clean Jar
```

If you would like to customize the name of your project later, update the `settings.gradle` file in the root of your repository. When writing code, be sure to put all code under a common package for your library, for example, package `com.example`. Developing Groovy is outside the scope of this book.

You'll want to publish your library to a Maven repository with its own Maven coordinates. Using Maven coordinates (group, artifact, and version) will allow you to include this library as a dependency. Later in this chapter, when I discuss packaging your library into a Jenkins plugin, it will be referenced as a dependency. It is also necessary for referencing your jar library from a Jenkins shared pipeline library[7].

---

7   https://www.jenkins.io/doc/book/pipeline/shared-libraries/

Referencing your external jar in shared pipeline libraries

Jenkins pulls in dependencies in shared pipeline libraries using Apache Grape annotations. Here are annotations supported for importing a shared pipeline library:

```
@GrabResolver(name='nexus', root='https://nexus.example.com/
repository/maven-public/')

@Grab(group='com.example', module='jenkins-pipeline-jar',
version='0.1', transitive=false)
```

To avoid memory leaks in shared pipelines, you should limit your library to no external dependencies and perform all API communication using pure Groovy. If this is not possible for whatever reason, then it is recommended to create a library-only plugin that should include all of your dependencies pre-packaged as a Jenkins plugin. This is covered later in this section.

Concurrency concerns

In large-scale Jenkins instances, as well as multibranch pipelines, it's not uncommon to encounter a concurrency error while grabbing grapes when using the @Grab annotation in shared pipeline libraries. The error is a long-standing issue unlikely to be resolved any time soon, tracked by JENKINS-48974[8].

Pitfalls of Grab

Grab can be buggy. Not only does it suffer from concurrency issues at scale, but its option for transitive dependencies does not reliably stop transitive dependencies from being loaded. The most reliable method for avoiding transitive dependencies is by relying on the @GrabExclude annotation. The following is an example of its usage. Let's exclude SnakeYAML, which is already pre-packaged as a Jenkins library plugin:

```
@GrabResolver(name='nexus', root='https://nexus.example.com/
repository/maven-public/')

@GrabExclude(group='org.yaml', module='snakeyaml')

@Grab(group='com.example', module='jenkins-pipeline-jar',
version='0.1', transitive=false)
```

---

8 https://issues.jenkins.io/browse/JENKINS-48974

Including a NonCPS library as a plugin

For the best performance for NonCPS code referenced in your shared pipeline libraries, you can package your library as a Jenkins plugin. By it being loaded in Jenkins as a plugin, the libraries become available for import in shared pipeline libraries. It's also useful if your library will require several dependencies. By including your own library as a Jenkins plugin, you can also avoid Apache Grape entirely due to its outstanding bugs. A drawback to including your library as a plugin is the fact that it introduces more complexity into your own Jenkins infrastructure since you're using a custom plugin.

Before relying on other libraries as dependencies, check to see whether Jenkins already provides a library as a plugin. If so, then you'll reference other Jenkins plugins as a plugin dependency of your own. Jenkins has provided several libraries as plugins, such as SnakeYAML. You can find all library-only plugins by searching for plugins with the label library[9].

Creating your first library plugin with Maven and Java

As of the time of writing, Jenkins has good documentation on how to create a plugin[10]. Hence, information included here will follow the Jenkins documentation, with the intended purpose of only including your own NonCPS jar as a library plugin. The plugin itself should not contain any Java or Groovy code since its purpose is to include the library you wrote in earlier sections. Because library-only Jenkins plugins' name end with -lib, I recommend following this convention for your own plugin. From the terminal, run the following Maven command. This assumes you have Java 1.8 and Maven 3+ already installed:

```
$ mvn -U archetype:generate -Dfilter="io.jenkins.archetypes:"
```

The preceding command will give you an interactive dialog where options are to be filled in by the user. Choose the archetype io.jenkins.archetypes:empty-plugin. When prompted, select the highest version available of the empty-plugin archetype. Fill in your plugin ID, for example, jenkins-pipeline-jar-lib. Type N to change details. Update the desired details and finish by choosing Y. I also recommend creating a Git repository for your Jenkins plugin:

---

9   https://plugins.jenkins.io/ui/search/?labels=library
10 https://www.jenkins.io/doc/developer/tutorial/create/

```
$ cd jenkins-pipeline-jar-lib/
$ git init
$ touch README.md
$ git add README.md
$ git commit -m 'initial commit'
$ echo 'target' > .gitignore
$ git add -A
$ git commit 'initial Jenkins plugin'
```

Your Jenkins plugin can be hosted in a Maven repository such as Sonatype Nexus. Compiling your plugin is easy; just run a Maven command and upload its HPI file to Jenkins:

```
$ mvn clean package
$ ls target/*.hpi
target/jenkins-pipeline-jar-lib.hpi
```

The resulting jenkins-pipeline-jar-lib.hpi file is the compiled Jenkins plugin. It should be uploaded to your Jenkins instance under the Advanced page of the Jenkins plugin manager.

Managing secrets

This is more of a best practice for managing secrets when you organize your NonCPS pipeline code into a Jar or Jenkins plugin. At times, when developing shared library code, you will inevitably need to authenticate with internal APIs. Authentication typically requires managing some sort of secret in your code; such as a username and passphrase or some kind of token. Don't store your secret as a String. If you store your secret as a string, then your secrets will be written as plain text to disk on the Jenkins controller when the pipeline step state is persisted for pause/resume behavior. Instead of using the java.lang.String class for secret text (for example, passwords or API tokens), Jenkins provides the core API hudson.util.Secret class. This class should be used because it automatically ciphers secrets when they're serialized to disk, which secures the data at rest when a pipeline is paused. This is especially important if you write a class that implements Serializable, which is meant to be called from the CPS pipeline. It is worth noting that java.io.Serializable is implicitly imported within Java similar to java.lang.String. Classes intended to be used within Jenkins pipeline code must implement the Serializable interface. Without it, an exception will be thrown when the class is instantiated in a running pipeline. Jenkins uses classes implemented with the Serializable interface to serialize and deserialize Java objects from $JENKINS_HOME.

Controller bottlenecks created by an agent

While Jenkins infrastructure best practices offer a distributed builds architecture[11], Jenkins has only one controller to schedule work across agents. Because of this, as more agents and processes are connected to a controller, it can create CPU bottlenecks, disk I/O bottlenecks, and network I/O bottlenecks by agents provisioning and streaming data from builds. This section will explain the bottlenecks and share proposals for solving them in terms of the relationship between the controller and the agent. Topics in this section are also pretty well covered in the Jenkins World 2017 presentation by Jesse Glick: *How to use Jenkins less*[12].

Defining agent and controller interaction bottlenecks

Agents can both download and upload a lot of data as part of normal operation:

- The start up process of an agent booting for the first time has an I/O cost that gets worse as more Jenkins plugins are installed on the controller.

- Stashing, unstashing, and archiving artifacts use the Jenkins controller as an intermediary by default.

- Agents stream annotated log output back to the controller. This can be several hundred megabytes and even gigabytes of uncompressed plain text data.

In the following sections, we'll address each of these issues and provide a recommendation to avoid the problem.

Agent booting start up bottleneck

> **Required plugins**
> Amazon EC2

Solving this bottleneck is a more advanced topic and is not fully covered in this book. This short snippet will describe the issue and solution, but the solution (deploying immutable infrastructure) has high development costs for Jenkins management, so we've decided to limit the scope of this topic.

The agent booting start up bottleneck mostly affects ephemeral agents. For long-running Jenkins agents, there will be a one-time start up cost, but then it

---

11   Best practice: distributed builds architecture: https://www.jenkins.io/doc/book/scaling/architecting-for-scale/

12   JW2017: *How to use Jenkins Less*, https://youtu.be/Zeqc6--0eQw

will not occur again because the agent is long-running. Provisioning agents on Jenkins can have trade-offs no matter which strategy you choose. Sometimes, admins will provision long-running agents that can degrade over time and become unmaintainable opaque boxes. Long-running agents are hard to recover if they're ever deleted (or more likely crash). Another strategy is to use ephemeral agents, which tend to be short-lived, automatically provisioned, and automatically deleted agents on a regular basis. However, ephemeral agents have their own hidden costs associated with them.

When Jenkins launches its agent JAR file, it first downloads all Jenkins plugins installed on the controller. The same plugin runtime set must be available locally in its agent cache as the controller. This is how Jenkins distributes work via extensions provided by plugins. However, a modest Jenkins controller will have over 100 Jenkins plugins installed totaling over 100 MB of data. So, when an agent boots, it will be synchronizing hundreds of files and hundreds of megabytes of data as a part of its agent start up procedure. This procedure occurs before the agent is ready to take on work.

One strategy to improve agent start up and bootstrapping is to practice immutable infrastructure and use the *Amazon EC2* Jenkins plugin to provision dynamic agents that have been pre-baked with tools and agent plugins.

Stashing and archiving artifacts

The Jenkins pipeline feature for stashing artifacts, unstashing artifacts, and archiving artifacts by default uses the controller as the intermediary. Stashed and unstashed artifacts persist only as long as a pipeline runs and archived artifacts persist long after a pipeline finishes until a build is deleted. Artifacts can be stored per job and sometimes even per build should the user developing the pipeline choose to do so. User pipelines that stash artifacts can be widely varying at scale. Some stashes could be small and contain millions of files, which creates a significant CPU bottleneck. Other stashes can contain large amounts of data (gigabytes) with very few files introducing network I/O and disk I/O bottlenecks. This means that the more stashing and archiving is used, the more likely the Jenkins controller is loaded with CPU and I/O. The effect experienced by users will be performance degradation when browsing the web UI and development delays as processing new builds becomes delayed.

This isn't to say you shouldn't use the steps to archive or stash. It should be limited to temporary debugging. Jenkins works best as an event-based task runner. Its design is not meant to handle millions of small and large artifacts, so it tends to perform poorly in this scenario. The following two sections will detail solutions to avoiding storing artifacts in the Jenkins home configuration directory.

Artifact Manager on S3 plugin

> **Required plugins**
> Artifact Manager on S3

I have covered how stashing, unstashing, and archiving artifacts affects the Jenkins controller. The Jenkins community has published the *Artifact Manager on S3* Jenkins plugin[13]. This replaces the stash, unstash, archiveArtifacts, and unarchive pipeline steps. Artifacts will be uploaded and downloaded directly from Amazon S3 and will not involve transferring artifacts to or from the Jenkins controller. A pipeline code developer uses the same steps in Jenkins pipeline code for stashing and archiving artifacts, with the added benefit that S3 will be used transparently. It does not require changes to the user pipeline code. Users won't even know it is stored in S3 when they download artifacts from the controller web UI.

Later in this chapter, the *Storing controller and agent logs in CloudWatch* section covers IAM policies for CloudWatch, which also grant S3 access for the *Artifact Manager on S3 Plugin*. This policy is combined with the CloudWatch policy because only one policy can be associated with an EC2 instance profile.

Using third-party artifact storage

Because Jenkins is not designed to host artifacts, it makes the most sense to rely on third-party solutions designed for the task. The benefits of relying on third-party artifact storage include the following:

- Jenkins becomes less critical for artifact retrieval, which helps to decentralize critical process points of failure.
- Jenkins load and storage requirements are reduced because Jenkins does not need to host artifacts and users download artifacts from external object storage instead of from the Jenkins UI.

Multiple third-party solutions offer artifact storage with far more performance than Jenkins as an artifact store. Examples of artifact storage include the following:

- GitHub Packages or GitHub Releases
- Sonatype Nexus 3
- JFrog Artifactory
- Amazon S3 (discussed in a later section)
- Any other generic object store designed to host files

---

13   https://plugins.jenkins.io/artifact-manager-s3

Choosing your external artifact storage solution is highly dependent on the needs of your organization, budget, and internal best practices. I recommend Sonatype Nexus 3 because you get the most out of the box in terms of artifact hosting features for free in the OSS version when compared with other listed solutions for the same price.

Storing controller and agent logs in CloudWatch

One way to completely reduce the load on a Jenkins controller for agent logs is to store the logs in a completely different backend than the controller $JENKINS_HOME on disk. This can be done with Jenkins plugins and Docker integrations. For the controller logs, admins would visit CloudWatch to see any controller application errors. For build logs, users would interact with Jenkins the same way they always have.

In general, doing all of your work on an agent helps to prevent controller bottleneck issues because agents are dedicated to performing the most CPU- and I/O-intensive work. As you solve one bottleneck, you tend to discover another. There are limits and diminishing returns to solving bottlenecks, but log output is another category that has some decent community solutions for it.

Agent logs can be several hundred megabytes of data and, in some cases, even multiple gigabytes of data. A mature development process will have features such as unit testing, integration testing, smoke testing, user acceptance testing, test report publishing, artifact publishing, and deployments (with or without approvals). All of these stages and steps generate logs. More verbose processes help debug software problems but generate a lot of log data in order to be useful.

Jenkins, by default, does not compress its own logs and stores logs associated with builds on the controller. So the gigabytes of output generated on the agent will be uploaded to the controller and the controller writes it to disk in $JENKINS_HOME. Additionally, when users browse the Jenkins web UI with their web browser, the data is read from $JENKINS_HOME. Jenkins tries to manage this by showing the user part of the log data, but users could opt to view the full log output.

If you're hosting the Jenkins controller in AWS, there are a few seamless solutions for exporting logs to CloudWatch.

Pipeline Logging over CloudWatch plugin

Required plugins

Pipeline Logging over CloudWatch

Agent logging creates two issues on the Jenkins controller:

- It increases storage requirements to store large amounts of logs.

- It creates a network I/O bottleneck as logs are streamed to the controller.

The *Pipeline Logging over CloudWatch*[14] Jenkins plugin bypasses the Jenkins controller entirely for logs produced on agents to use AWS CloudWatch as the backend storage. When you browse the Jenkins Web UI, the logs displayed come from CloudWatch.

Benefits include the following:

- CloudWatch is used for log storage so there's no storage used in $JENKINS_ HOME for build logs.

- Because logs are streamed directly to CloudWatch from the agent host, the Jenkins controller does not consume network I/O since it does not take part in collecting agent logs.

From an end user perspective, they won't even know CloudWatch is being used to store their build logs.

Controller logging over CloudWatch

If you're running the Jenkins controller from Docker, it's worth mentioning that Docker natively supports streaming logs to CloudWatch via the awslogs log driver[15]. This will bypass writing controller logs to disk and write them directly to AWS CloudWatch. For Docker to write logs, the Docker daemon will need to be configured by editing /etc/docker/daemon.json with the following contents:

---

14   https://plugins.jenkins.io/pipeline-cloudwatch-logs/

15   https://docs.docker.com/config/containers/logging/awslogs/

```
{
    "log-driver": "awslogs",
    "log-opts": {
        "awslogs-region": "us-east-1",
        "awslogs-group": "JenkinsCloudwatchLogs",
        "awslogs-stream": "ControllerLogs-Prod"
    }
}
```

Users don't normally look at controller logs. With the Docker logging over CloudWatch integration, an admin can view logs either in the Jenkins controller or directly in CloudWatch. This will avoid filling up logs on the host where the Jenkins controller Docker container is running.

In order for logging to work seamlessly without having to configure credentials, an instance profile with an IAM role can be used to grant AWS permissions to the EC2 host of a controller or agent.

AWS IAM roles for controller and agent CloudWatch logging

To simplify authentication and authorization for AWS integration, the Jenkins controller should have an IAM profile associated with the EC2 instance running the controller. The instance policy grants it read and write access to S3 and CloudWatch for plugins discussed in earlier sections. Before creating the IAM profile with its IAM role, there are some prerequisites to keep in mind:

- The instance should be dedicated to only running the Jenkins controller.
- The S3 bucket for storing archived artifacts must already exist.
- The JenkinsCloudwatchLogs log group will automatically be created by Docker when the daemon is restarted with its awslogs log driver.

Because AWS policy documents can be quite large, I've uploaded both the controller and agent instance policies to a GitHub repository with additional instructions on how to use them. Visit the following link:

https://github.com/PacktPublishing/Jenkins-Administrators-Guide/tree/main/ch9/aws-instance-policies

To use the IAM policies, you must do the following:

1. Create an IAM role with the contents being the policies, one for the controller and one for the agent.

2. Create an instance profile for the controller and agent that will reference their respective IAM role.

3. When launching a controller or agent instance, choose the instance profile. This profile will grant the EC2 instance AWS permissions via the IAM role.

Refer to the Amazon website, *AWS Identity and Access Management User Guide*, for more detailed information under the *Using instance profiles*[16] section.

Other ways to reduce agent log output

Part of solving agent bottlenecks related to logs for Jenkins can come from training users on how to reduce log output within their builds. While this might not scale uniformly across all user pipelines in a very large environment, every little bit of optimization helps.

Strategy – Writing logs to the agent disk

One strategy could be to write more verbose logs to disk on the agent, compress them on the agent, and then archive them for later viewing by the user to debug issues. This helps to drastically reduce the amount of data streamed to the controller as logs because plain text data compresses really well.

Benefits include the following:

- A faster build time because processes write logs to files on disk faster than the agent can stream logs to the Jenkins controller.

- Less data streamed to the controller means less load on the network and less storage used by $JENKINS_HOME when compressed logs are uploaded as archived artifacts.

Take, for example, the following shell script. It is written in such a way that it writes all build log output to a file on the agent, but it still allows the Jenkins pipeline code developer to output select logs to the Jenkins UI:

---

16   https://docs.aws.amazon.com/IAM/latest/UserGuide/id_roles_use_switch-role-ec2_instance-profiles.html

```
#!/bin/bash
function agentout() { echo "$*" >&3; }
function agenterr() { echo "$*" >&4; }
set -exo pipefail
exec 3>&1 4>&2 1> debug_output.log 2>&1
trap 'gzip debug_output.log' EXIT
agentout 'Logging to agent stdout: Building software'
make build
agenterr 'Logging to agent stderr: Success'
```

Let's define each line:

1. The shell interpreter for the script, commonly called the shebang (#!). This is a script for bash (/bin/bash).

2. This is followed by two bash functions, agentout and agenterr, which will be discussed in other points.

3. set -exo pipefail modifies the behavior of the bash shell. -e and -o pipefail will exit the script when an error occurs for a command or any error within a series of piped commands. The -x option will cause bash to print out a command before executing it, which is useful when debugging longer shell scripts.

4. exec, when called without arguments, will redirect log output from commands run within the script. Its following arguments are redirecting file descriptors, which is where the output for commands gets logged. Please note the order in which output redirection occurs changes how it gets redirected based on prior arguments. Let's define the output redirection arguments separately. It's worth noting that, by default, shell commands write to file descriptor 1 (normally stdout) and file descriptor 2 (normally stderr).

   - exec 3>&1 will make file descriptor 3 available. When the script logs to file descriptor 3, it will output to the stdout logging of the Jenkins agent. Logs written to file descriptor 3 will show up in the Jenkins logs of the web UI.

   - 4>&2 will make file descriptor 4 available. When the script logs file descriptor 4, it will output to the stderr logging of the Jenkins agent. Logs written to file descriptor 4 will show up in the Jenkins logs of the web UI.

- 1> debug_output.log will change the behavior of file descriptor 1 (normally stdout). When a build log writes to stdout, it will be redirected to the debug_output.log file and overwrite the file if it exists.

- exec 2>&1 will change the behavior of file descriptor 2 (normally stderr). When a build log writes to stderr, it will be redirected to the debug_output.log file. Be aware that this is different from the first output redirect argument (not the same as 3>&1) because file descriptor 1 now points to a file.

5. Trap functions are a feature of bash that execute code under certain conditions. trap 'gzip debug_output.log' EXIT means that when the script exits, the last command run will compress and delete the debug_output.log file right as the script exits. This will occur even if the script exits on error.

6. agentout shows an example of using a bash function to write to the agent stdout. Writing to the agent stdout means it will show up in the Jenkins UI logs.

7. make build is an example of calling a build target in a Makefile.

8. agenterr shows an example of using a bash function to write to the agent stderr. Writing to the agent stderr means it will show up in the Jenkins UI logs. If make build exits with an error, then this will never be executed because of the shell options set at the beginning.

Finally, don't forget to collect the log as an artifact. Here's an example with the full script as a pipeline script:

```
// scripted Jenkinsfile pipeline source code
stage('Running a build') {
    node {
        try {
            sh('''\
                #!/bin/bash

                function agentout() { echo "$*" >&3; }
                function agenterr() { echo "$*" >&4; }

                set -exo pipefail

                exec 3>&1 4>&2 1> debug_output.log 2>&1

                trap 'gzip debug_output.log' EXIT

                agentout 'Logging to agent stdout: Building
software'

                make build

                agenterr 'Logging to agent stderr: Success'

                '''.stripIndent().trim())
        } finally {
            archiveArtifacts 'debug_output.log.gz'
            cleanWs()
        }
    }
}
```

The preceding pipeline script could be further optimized to only collect debug_output.log.gz when there's an error. Otherwise, there's not much use in a verbose success message.

Drawbacks: Writing logs to the agent disk

Occasionally, there are bugs reported within the Jenkins community issue tracker[17] where builds hang and there are random agent disconnects. If you've inherited an especially old version of Jenkins, then this feedback might make your build environment less stable. The workaround for the bugs is to put builds in a wrapper script that occasionally output some kind of text update to the agent log stream at least every minute. Please also keep in mind that secrets in the Jenkins environment are not filtered until the user attempts to view the console output in the controller

---

17 https://support.cloudbees.com/hc/en-us/articles/115001369667-Dedicated-SSH-agent-gets-disconnected

web UI. If a developer outputs a secret into the log file, then it will be a plain text secret within the log.

Summary

Jenkins deployment architecture and hosting is an important factor in how well it will perform when your organization starts to scale up usage for builds and deployment of software. Several bottleneck issues can be avoided simply by hosting it with a fast disk, proper systems configuration, and enough resources to perform most workloads. Planning out how users onboard jobs and organizing jobs into folders can also play a major role in Jenkins performance. Beyond those basics, any further optimizations don't have to be made right away. Some of them can be delayed until you start actually experiencing performance issues. This chapter tries to give you hints for where you can typically look for performance issues as well as how to resolve them. Furthermore, monitoring wasn't covered in this chapter, but extending Jenkins with application performance monitoring, such as using the jmxtrans agent jar with a time series database and a graphing solution such as Grafana, will give you specific insights into where your performance issues are occurring. Also, keep a lookout for performance-oriented talks every year at Jenkins' annual conference (now called DevOps World as of this writing), which I also discussed in the beginning and throughout the chapter. My hope is that I've given you enough information to think about planning a Jenkins architecture deployment, a reference for when you have Jenkins deployed, and quick tips to improve performance if you've had the unfortunate reality of inheriting a Jenkins instance.

In the next chapter, we'll be diving once again into Groovy code, as code reuse in pipelines is addressed by Jenkins Pipeline Shared Libraries.

10
Shared Libraries

As we start to use Groovy code to write the pipeline logic, it's inevitable that we'll end up repeating the same code across multiple pipelines. As programmers, we naturally look for ways to write functions, classes, and methods to organize repeated code. Shared libraries provide a way to encapsulate the repeated code and create helper functions, but that is only the beginning of what shared libraries can do.

The power of shared libraries is twofold. The first half of the chapter will show that there are two different ways of providing a shared library and four or five different ways of using a shared library7. Each of these options comes with slightly different capabilities, and this variety makes shared libraries useful in any Jenkins or pipeline configuration.

The second half of the chapter will show the different ways of using shared libraries. Exploring a commonly encountered scenario, we will write a global variable with helper functions to send a pre-formatted Slack message that always includes the Jenkins build URL. In a more advanced use case, we will see three different ways we can use shared libraries to create our own domain-specific languages (DSLs). The first DSL will feature an entirely new syntax using YAML, the second DSL will continue to use the Jenkins pipeline directives but in a limited scope, and the third DSL will override the built-in pipeline directives so that the existing pipelines start behaving differently.

Shared libraries are a vast topic with a lot to learn. Let's go through each step in detail so that we can fully understand the possibilities that shared libraries present.

In this chapter, we're going to cover the following main topics:

- Understanding the directory structure
- Creating a shared library
- Providing shared libraries
- Using shared libraries
- Use cases

Technical requirements

You need a Jenkins instance where you are an administrator. The one we made earlier in the book would work well, but any Jenkins instance is fine.

> **Not using the Jenkins that we made earlier in the book?**
>
> If you're using a different Jenkins, understand that admin user is an administrator and adder-admin is not an administrator. Also, there is a folder called adder where the adder-admin user has full permissions.

We will store a shared library in a GitHub repository, so you'll need a GitHub account. Other Git repositories such as Bitbucket or GitLab are also okay as long as you're comfortable with the Git commands.

We will create a shared library function to send a Slack message, so you'll need access to a Slack workspace where you can add an app and get a token. If your company doesn't use Slack or you don't have the permissions to add an app, you can create your own workspace for free at slack.com/get-started.

This chapter is largely about coding in Groovy for Jenkins, so a basic understanding of Java and Groovy is required. Since Groovy has many closing curly braces that take up the vertical space, they are compressed here onto single lines in order to fit the code into a page.

Correctly formatted files, along with the other files used in the chapter, are available on GitHub at https://github.com/PacktPublishing/Jenkins-Administrators-Guide/blob/main/ch10.

Understanding the directory structure

Let's get started by understanding what a shared library is and what it looks like.

A shared library is a collection of folders and files stored in a source code management system (SCM). These can be stored in any SCM, but we will use a Git repository in GitHub as we have been doing throughout the book.

A Git repository containing a shared library has three folders: vars, src, and resources:

- The vars directory is by far the most frequently used among the three directories. The common utility functions such as helper functions for Slack are stored in the vars directory. It may seem strange that the files containing function definitions are stored in a directory named vars, because vars is short for variables, not functions. This is because the files in the vars directory become available to Jenkinsfiles as variables. For example, if we have vars/slack.groovy that has a function named send(), we can call the function slack.send(), referencing the slack.groovy filename as a variable.

- As the library code gets more elaborate, the simple functions in the vars directory tend to become quite disorganized. src is the directory where the Groovy classes with the complex business logic are stored. The src directory uses the Java source structure. For example, if we are writing a Version Groovy class in the ca.lvin.books.jenkins package, the source file would be in src/ca/lvin/books/jenkins/Version.groovy.

- The resources directory contains the static files that can be referenced from the code in the shared library. For example, if there are several functions that need a reference to the list of hosts in our environment, the list can be saved in a file in this directory. We'll see how this works in the coming sections.

In summary, vars is the most frequently used directory, containing the functions that the Jenkinsfiles can call. src has the Groovy classes that contain more complex business logic. resources is where the static files are stored.

Take note that we will be using a simple class as an example of a Groovy class in the src directory to keep the focus on shared libraries. Take what you learn from this chapter and apply it to your own complex code.

Let's continue to create a shared library.

Creating a shared library

Our example library will have code that prints a random word from a file:

1. Go to GitHub and create a new repository named `jenkins-shared-library`, then clone the new empty repository:

$ git clone git@github.com:**<GitHub Username>**/jenkins-shared-library.git

2. Go into the cloned repository and create the three directories:

$ cd jenkins-shared-library

$ mkdir {vars,src,resources}

3. Add the following file in the resources directory. It has one word on each line:

resources/words.txt

```
Quick
brown
fox
jumps
over
the
lazy
dog
```

4. Add the following file to the src directory. The RandomWord class reads a list of words from `resources/words.txt`, then returns a random word when `get()` is called:

src/ca/lvin/books/jenkins/RandomWord.groovy

```
package ca.lvin.books.jenkins

class RandomWord {
    private def pipelineSteps
    private List<String> words
    private Random random

    public RandomWord(pipelineSteps) {

        this.pipelineSteps = pipelineSteps
        this.words = this.readResourcesAsList("words.txt")
        this.random = new Random()
    }

    public String get() {

        int wordIndex = this.random.nextInt(this.words.size())
        String capitalized = org.apache.commons.lang.WordUtils.
capitalizeFully(this.words[wordIndex])

        return capitalized
    }

    @NonCPS
    private List readResourcesAsList(String name) {

        String resource = this.pipelineSteps.libraryResource(name)
        return resource.split()
    }
}
```

There's a lot going on in this file:

- The files in the src directory follow Java's conventions, so the directory structure must match the package name, and the filename must match the class name. In this case, the package name ca.lvin.books.jenkins yielded the directory structure src/ca/lvin/books/jenkins/. Also, the class name RandomWord yielded the filename RandomWord.groovy.

- The file is a plain old Java object, commonly referred to as a POJO. Support for POJOs gives us an unprecedented amount of flexibility as we are no longer bound by the limits of a DSL. Also, because a POJO is a generic language tool that is not specific to Jenkins, we can develop it offline using an IDE.

- The constructor takes an argument named `pipelineSteps`. Since this file is a POJO, it doesn't have access to the Jenkins pipeline steps such as echo or sh. We need to call the `libraryResource` pipeline step in the `readResourcesAsList()` method to read the `words.txt` file from the resources directory, so `pipelineSteps` is passed to the constructor and saved as a member variable.

- The `get()` method uses a helper function from the Apache Commons library. Groovy on Jenkins comes with Apache Commons built in and gives us a powerful set of helper functions.

- Jenkins uses a continuation-passing style (CPS) transform to serialize the code, but not all methods can use CPS. In such cases, the methods must be annotated with `@NonCPS` as we did for the `readResourcesAsList()` method. CPS is covered more in detail in *Chapter 9, Reducing Bottlenecks*. Also, there is additional information about CPS in the Jenkins User Handbook (https://www.jenkins.io/doc/book/pipeline/cps-method-mismatches/).

- Add the following file to the vars directory. This creates a variable named randomWordFromVars that we can use in the Jenkinsfile. Calling randomWordFromVars.echo() picks a random word from resources/words.txt and prints it on the console output. We will see this in action soon:

vars/randomWordFromVars.groovy

```
import ca.lvin.books.jenkins.RandomWord

void echo() {

    RandomWord rw = new RandomWord(this)
    echo rw.get()
}
```

This file also has a few interesting points:

- The `import` statement doesn't need any special treatment – it's a simple, normal `import` statement. This means that the files inside the vars directory have native access to the classes in the src directory. Remember this, as we will see a counter-example soon.

- We are passing this to the constructor for the RandomWord class. In the current context, this points to the steps directive in the Jenkinsfile. As a result, RandomWord can use the `libraryResource` pipeline step as we saw in the RandomWord.groovy file.

- We are using the echo Jenkins pipeline step inside the function. Unlike the POJOs in the src directory, the variables in the vars directory have native access to the pipeline steps.

- The function is also named echo. Even though it's named the same as the echo pipeline step, it doesn't result in a collision because the echo function name is namespaced under the randomWordFromVars variable.

5. Finally, add a file of the same name but with the .txt extension. This contains the HTML code that shows up in the Pipeline Syntax documentation:

vars/randomWordFromVars.txt

```
<b>echo</b> method prints a random word from resources/words.txt
```

6. Make a commit with the three files and push the commit to the Git repository:

```
$ git add .
$ git commit -m "Initial four files"
$ git push origin main --set-upstream
```

The shared library is ready. Let's continue to see the various ways we can use it.

Providing shared libraries

As an administrator, there are two ways to make the shared libraries available to the pipelines in our Jenkins. Let's start with the folder-level shared library, which provides more limited access.

Folder-level shared libraries

A shared library can be made available to the pipelines *in a specific folder* by configuring it in the folder's configuration page. When a specific project's pipelines have repeated code, the helper functions that modularize the project's repeated code should be stored in a folder-level shared library, not a global shared library. We will see why soon.

Log in to Jenkins as the adder-admin user, click the adder folder, then click Configure on the left. Under Pipeline Libraries (gotta love the naming inconsistency), click the Add button to reveal a Library section. Configure the library as follows:

- Name: my-folder-shared-lib (This is the name that the Jenkinsfiles will use to point to this library.)

- Default version: main (This is the branch in our Git repository.)

- Retrieval method: Modern SCM

- Source Code Management: Git

  - Project Repository: `https://github.com/<GitHub Username>/jenkins-shared-library.git`

  If the Git repository is private, we can specify the keys in the Credentials field.

- Load implicitly **allows this library to be used by the pipelines without having to request it. While this may sound convenient, it can lead to variable name collisions if multiple libraries are loaded on a Jenkinsfile. Use this feature only if you want to override the built-in pipeline directives, as we will see in the** *Advanced – Custom DSL* **section.**

- Allow default version to be overridden **allows the Jenkinsfiles to specify a branch name rather than always loading the branch that we've set. This is useful for debugging a shared library code so in most cases it should be left checked. We can uncheck it if the shared library contains sensitive code that must not be overridden.**

Leave all other fields as their defaults and click Apply to save:

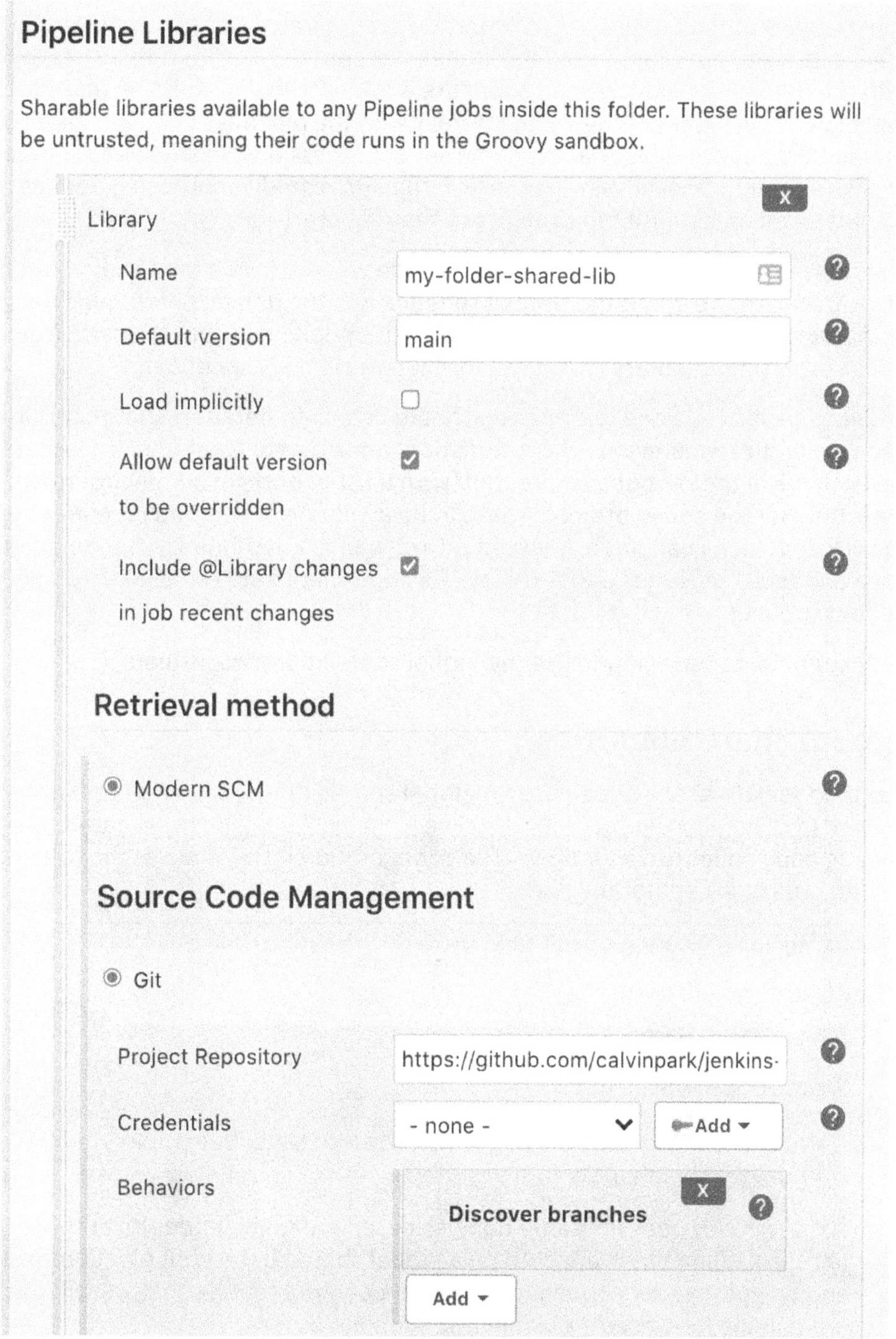

Pipeline Libraries

Sharable libraries available to any Pipeline jobs inside this folder. These libraries will be untrusted, meaning their code runs in the Groovy sandbox.

Library   X

    Name   `my-folder-shared-lib`

    Default version   `main`

    Load implicitly   ☐

    Allow default version   ☑
    to be overridden

    Include @Library changes   ☑
    in job recent changes

Retrieval method

◉ Modern SCM

Source Code Management

◉ Git

    Project Repository   `https://github.com/calvinpark/jenkins-`

    Credentials   - none -   Add ▼

    Behaviors   **Discover branches** X

    Add ▼

Figure 10.1 – Folder-level shared library configuration

The message at the top of the configuration section tells us that the shared libraries are available to any pipelines in this folder. In *Chapter 3, GitOps-Driven CI Pipeline with GitHub*, we saw how we can configure a folder with a permission matrix so that only a specific set of users can interact with the pipelines in it. The folder-level shared libraries are subject to the same set of permissions, so when adder-admin configures a folder-level shared library in the adder folder, only the pipelines inside the adder folder can use the code in the shared library.

The message continues to tell us that the code will run inside the Groovy sandbox. We will learn more about the Groovy sandbox and the dangers of running outside of the sandbox in *Chapter 11, Script Security*. It's good news that the code from a folder-level shared library is secured through the Groovy sandbox.

Project-specific code and information should be in a folder-level shared library because of the two previous characteristics. In the previous section, we discussed saving a file in the resources directory with a list of hosts in our environment. If the list is only for the adder project, it should be available only to the pipelines inside the adder folder. In addition, if adder-admin wants to modularize the repeated code from the adder pipelines into a library, the code should also be available only to the adder pipelines.

Let's continue to see when to use the global shared libraries instead.

Global shared libraries

Log in to Jenkins as the admin user. A global shared library is configured in the System Configuration under Global Pipeline Libraries. Click Add to reveal a Library section and configure it as follows. The configuration is the same as the folder-level shared library, except for the name:

- Name: `my-global-shared-lib`

- Default version: `main`

- Retrieval method: Modern SCM

- Source Code Management: Git

  - Project Repository: `https://github.com/<GitHub Username>/jenkins-shared-library.git`

- Load implicitly does the same thing as described in the folder-level shared libraries. When this is enabled on a global shared library, all pipelines in Jenkins end up getting new commands. This is a powerful feature that we will discuss more in the *Advanced – Custom DSL* section.

Let's continue to examine the warning message in the configuration section:

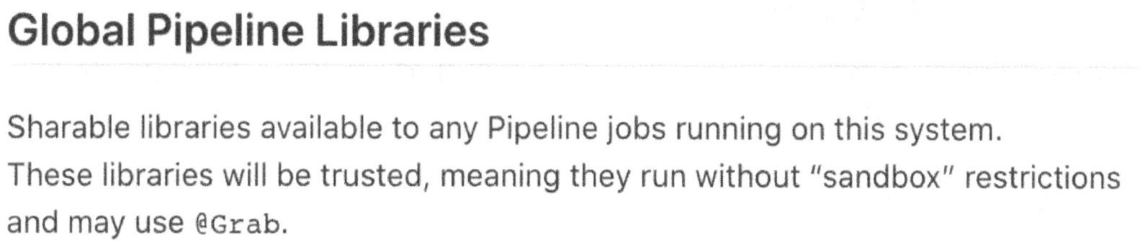

Global Pipeline Libraries

Sharable libraries available to any Pipeline jobs running on this system.
These libraries will be trusted, meaning they run without "sandbox" restrictions
and may use `@Grab`.

Figure 10.2 – Global shared library configuration warning

The message first tells us that the shared libraries are available to any pipeline on
this Jenkins. This is perhaps obvious, but it's important to remind ourselves that a
global shared library should not contain any methods or information specific to a
project. It goes without saying that the secrets also must not be stored in a global
shared library. As discussed, a file containing the list of hosts in our environment
should be in a folder-level shared library rather than a global shared library.

The second part of the message is important as it affects the security. *Everything
in the global shared libraries run outside of the sandbox*, which means that the
code inside a global shared library has *full access to Jenkins without any security
oversight*. This includes activities such as deleting a pipeline, decrypting secrets, or
shutting down Jenkins.

Since only the administrators can configure a global shared library, it's easy to
mistakenly think that these security risks don't pose a real threat. That is not the
case in reality, because we're setting the source of the global shared library as a Git
repository that can be unsecured. Anyone who can put code into the Git repository
can create a method in the global shared library. The methods in a global shared
library have full access to Jenkins, which includes decrypting secrets and deleting
pipelines. In other words, *anyone who can put code into the Git repository can steal
the secrets and destroy our Jenkins*. It is very important that the source Git repository
for a global shared library is secured at the same level as administrator access
to Jenkins.

The @Grab command mentioned in the message allows us to dynamically fetch a
third-party library. Since a global shared library has full access to Jenkins, using a
library from the internet is quite risky. This powerful feature is allowed only to the
global shared libraries, which is another reason why we must protect the source Git
repository carefully.

We have now configured the same Git repository as both a folder-level shared
library and a global shared library. Let's try using them to understand the difference
between the two.

Using shared libraries

There are two ways to use a shared library. Let's start with static loading, which provides more options.

Static loading

There are two syntax options to statically load a shared library in a Jenkinsfile. Let's start with the form for loading the classes in the src directory.

Annotating an import statement

Log in to Jenkins as adder-admin and create a new pipeline named global-shared-lib-src inside the adder folder. Enter the following code in the pipeline definition text box. Running the pipeline prints a random word from resources/words.txt twice, once in each stage:

Jenkinsfile.global.src

```
@Library('my-global-shared-lib')
import ca.lvin.books.jenkins.RandomWord

pipeline {
    agent any
    stages {
        stage('src test 1') {
            steps { script {
                RandomWord rw = new RandomWord(this)
                echo rw.get()
        } } }
        stage('vars test 1') {
            steps { script {

            randomWordFromVars.echo()
} } } } }
```

Let's examine the syntax. The first line is the code that loads the shared library. It may look like a declarative command to load the library, but it's actually an annotation like @NonCPS or @Deprecated. In our example, we are annotating the import statement in the next line to tell Jenkins that the ca.lvin.books.jenkins.RandomWord class should come from the my-global-shared-lib shared library. Once the class is successfully imported, it is used in the src test 1 stage to instantiate an object using the new keyword and, then we call the get() method from it.

A side effect of using the annotation is that all files in the vars directory become available to Jenkinsfile as variables. In the vars test 1 stage we use the randomWordFromVars variable and call the echo method from it. Even though the end result is the same as what we saw from src test 1, the mechanics are quite different. In src test 1, we imported a class from the src directory by annotating the import statement. This makes the RandomWord class available in this Jenkinsfile, and we're using the actual class in the src test 1 stage. In vars test 1 we don't use any classes. All the files in the vars directory became available to Jenkinsfile as variables, and we're just using one of the variables. Let's take a look at a different form to highlight the difference.

Annotating nothing but still using the side effect

Stay logged in to Jenkins as adder-admin and create a new pipeline named global-shared-lib-vars. Enter the following code in the pipeline definition text box:

Jenkinsfile.global.vars

```
@Library('my-global-shared-lib') _
pipeline {
    agent any
    stages {
        stage('src test 2') {
            steps { script {

                RandomWord rw = new RandomWord(this)
                echo rw.get()
        } } }
        stage('vars test 2') {
            steps { script {

                randomWordFromVars.echo()
} } } } }
```

The content of this file is the same as the previous pipeline except for the missing import statement and the new trailing underscore character. Recall that in the previous Jenkinsfile.global.src file, @Library was annotating the import statement to tell Jenkins where to look for the class. In the current file, @Library is annotating an underscore character, which does nothing. We're essentially not using the annotation feature, which tells Jenkins where to look for a class. We are, however, still benefiting from the side effect of using the annotation, which makes all files in the vars directory available to the Jenkinsfile as variables.

When we run this pipeline, the `src test 2` stage fails because Jenkins doesn't know where to find the RandomWord class, because we never imported it:

Dashboard ▸ **adder** ▸ **global-shared-lib-vars** ▸ **#1**

```
org.codehaus.groovy.control.MultipleCompilationErrorsException:
startup failed:
WorkflowScript: 7: unable to resolve class RandomWord
 @ line 7, column 28.
                 RandomWord rw = new RandomWord(this)
                     ^


WorkflowScript: 7: unable to resolve class RandomWord
 @ line 7, column 33.
                 RandomWord rw = new RandomWord(this)
                     ^
```

Figure 10.3 – The src test 2 stage fails to resolve class RandomWord because of the missing import

Once we delete the `src test 2` stage and run again, the `vars test 2` stage runs successfully because the annotation's side effect made all files in the vars directory available to Jenkins as variables. Again, this is done purely at a variable level, and no classes are involved. If we want to use the classes from the src directory in the Jenkinsfile, we must first import the classes by annotating them.

Here are a few more important points:

- Even though we can import the classes into our Jenkinsfile, oftentimes it leads to better organization if the classes are wrapped by the files in the vars directory for two reasons:

  - The vars directory already has access to the classes in the src directory. Recall that in vars/randomWordFromVars.groovy, the import statement doesn't have the @Library annotation, because the vars directory already has access to the classes in the src directory.

  - The vars directory has access to the pipeline steps. Recall that the POJO classes don't have access to the pipeline steps, so we had to provide it through the constructor. The vars directory on the other hand *does* have access to the pipeline steps, so we don't need to pass it again.

Since the vars directory has access to both the pipeline steps as well as the classes in the src directory, we can consider the variables as the bridge between the Jenkinsfile and the classes.

- When we click Replay, we are presented only with the modifiable text box for the Jenkinsfile, but not vars/randomWordFromVars.groovy nor src/ca/lvin/ books/jenkins/RandomWord.groovy. This is because the global shared library code runs with administrator permissions outside of the Groovy sandbox. A non-administrator user is not allowed to modify the library code, because that would allow them to run arbitrary code using administrator permissions outside of the sandbox:

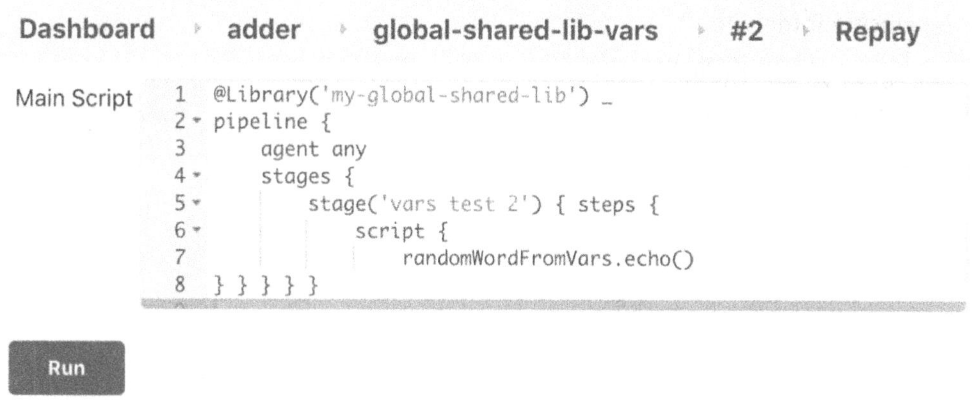

Dashboard   ›   adder   ›   global-shared-lib-vars   ›   #2   ›   Replay

```
Main Script   1   @Library('my-global-shared-lib') _
              2 ▾ pipeline {
              3       agent any
              4 ▾     stages {
              5 ▾         stage('vars test 2') { steps {
              6 ▾             script {
              7                   randomWordFromVars.echo()
              8   } } } } }
```

Run

Figure 10.4 – Replay doesn't allow us to modify a global shared library

- Because the global shared library code runs with administrator permissions, we have not been asked to get security approval for the method signatures used in the code for both the vars and src directories. We will see how to take advantage of this feature in *Chapter 11, Script Security*.

- We can see the documentation we've created in the *Creating a shared library* section by clicking Pipeline Syntax and Global Variable Reference. Scroll to the bottom to find the following message:

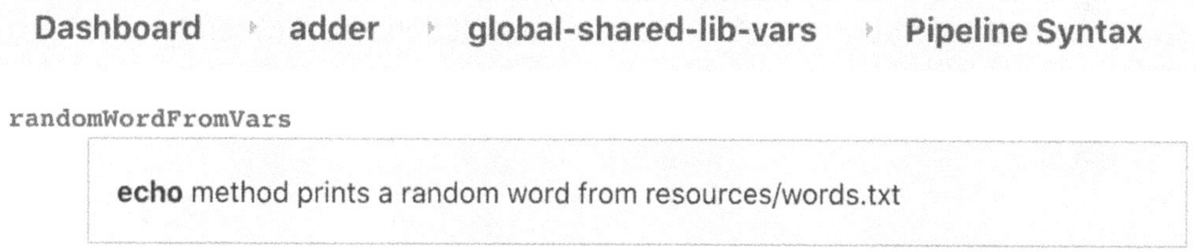

Dashboard   ›   adder   ›   global-shared-lib-vars   ›   Pipeline Syntax

randomWordFromVars

echo method prints a random word from resources/words.txt

Figure 10.5 – The Pipeline Syntax / Global Variable Reference page shows the content of <var>.txt files

Now, let's take a look at a folder-level shared library to see how it differs.

Folder-level shared library

Stay logged in to Jenkins as adder-admin and create a new pipeline named folder-level-shared-lib. It's important that we don't use the admin account because different permissions apply when an administrator saves a pipeline. We will see more on that in *Chapter 11, Script Security.*

Enter the following code in the pipeline definition text box. The code is the same as Jenkinsfile.global.src except for the library name. This time we're loading a folder-level shared library instead of a global shared library:

Jenkinsfile.folder.src

```
@Library('my-folder-shared-lib')
import ca.lvin.books.jenkins.RandomWord

pipeline {

    agent any

    stages {

        stage('src test 3') {
            steps { script {

                RandomWord rw = new RandomWord(this)
                echo rw.get()
        } } }
        stage('vars test 3') {
            steps { script {

                randomWordFromVars.echo()
} } } } }
```

Recall that both the global and the folder-level shared libraries use the same Git repository, github.com/<GitHub Username>/jenkins-shared-library. We are running *exactly* the same code with the only difference being that we load it as a folder-level shared library. Let's run it to see how it behaves differently:

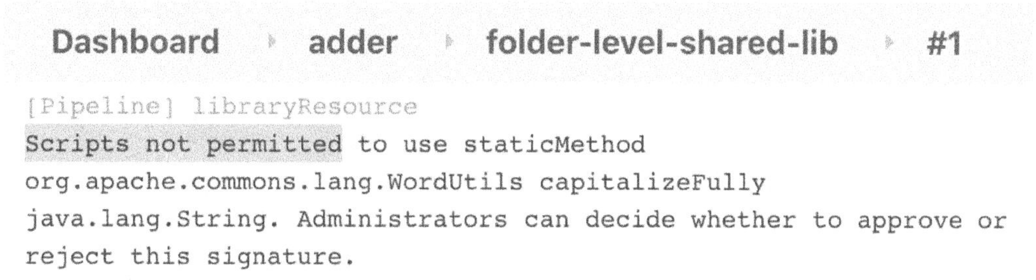

Dashboard ▸ **adder** ▸ **folder-level-shared-lib** ▸ **#1**

```
[Pipeline] libraryResource
Scripts not permitted to use staticMethod
org.apache.commons.lang.WordUtils capitalizeFully
java.lang.String. Administrators can decide whether to approve or
reject this signature.
```

Figure 10.6 – Method signatures require an administrator's approval when the shared library code is used as a folder-level shared library

Running the pipeline fails with an error message `Scripts not permitted`. As we discussed in the previous section, a folder-level shared library runs *inside the Groovy sandbox*, and the method signatures are subject to the administrators' review and approval. For now, let's approve them so we can focus on learning how the shared library works. We will learn more about method signature approval in *Chapter 11, Script Security*.

Open a new browser window in Incognito mode (or use a different browser) and log in to Jenkins as admin user. Click Manage Jenkins on the left, then click In-process Script Approval. We can see that the method signature of `staticMethod org.apache.commons.lang.WordUtils capitalizeFully java.lang.String` is waiting for review. Click Approve. Go back to the original browser window where we are logged in as adder-admin and run the pipeline again. The build finishes successfully.

Now, click the Replay button to see that we are presented with three editable text boxes that include the code from both the `src` directory and the `vars` directory. Herein lies the power of the folder-level shared libraries – they are readily available to be edited right on the Jenkins UI. This allows us to quickly modify the code and retry the build, rather than having to make changes to a file, make a Git commit, push the commit, then rebuild. Being able to update the code in the UI tightens the feedback loop and leads to a more pleasant development experience. Once we are done developing the library code on the UI, we can make a Git commit with the final code and push just once:

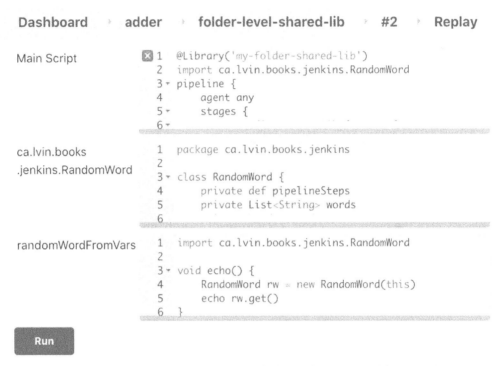

Figure 10.7 – Replay allows us to modify folder-level shared library code

As we discussed earlier, running outside of the Groovy sandbox as an administrator is a very dangerous thing to do. Therefore, a global shared library should be used only for these two purposes, and all others should use a folder-level shared library:

- Creating a helper method that *all* pipelines in Jenkins will use.

- Encapsulating a dangerous method call with proper checks so that the pipelines can use it safely. We'll see more on this in *Chapter 11, Script Security*.

Let's look at a couple additional syntax options for statically loading a shared library.

Loading a specific version and multiple libraries at once

`@Library('libname@version')` allows us to specify a Git branch, tag, or sha1 to load, as long as the shared library is configured with Allow default version to be overridden enabled. This is a good way to test a new feature.

`@Libary(['libname1', 'libname2'])` allows us to load multiple libraries in one statement, and it can be further augmented with @version on each libname.

These are the various ways of loading a global or folder-level shared library statically. Next, let's take a look at loading a shared library dynamically.

Dynamic loading

The static loading syntax works well when the libraries are already defined as a global or folder-level shared library. In some environments, administrators do not allow folder-level shared libraries, which means that we can't add any shared libraries to our projects. In that case, we can load a library dynamically.

Stay logged into Jenkins as adder-admin and create a new pipeline named dynamic-shared-lib. Enter the following code in the pipeline definition text box. Replace the GitHub clone URL with your own:

Jenkinsfile.dynamic

```
library(identifier: 'dynamic-lib@main',
        retriever: modernSCM([
            $class: 'GitSCMSource', credentialsId: '',
            remote: 'https://github.com/<GitHub Username>/
jenkins-shared-library.git']))

pipeline {

    agent any

    stages {
        stage('vars test 4') { steps {
            script {

                randomWordFromVars.echo()
} } } } }
```

Running the pipeline outputs a random word the same way as for the statically loaded libraries. However, notice that the library call doesn't begin with the @ symbol, and doesn't end with a trailing underscore character. Unlike the static loading, which uses a Java annotation, this library statement is a function call. As we can see in the code, we provide the details of the shared library as the function arguments, and Jenkins processes it to load the shared library content. Since this is a function call rather than an annotation, putting an import statement below the function call causes the build to fail because Jenkins cannot find the RandomWord class:

```
library(identifier: 'dynamic-lib@main',
        retriever: modernSCM([
            $class: 'GitSCMSource', credentialsId: '',
            remote: 'https://github.com/<GitHub Username>/
jenkins-shared-library.git']))
import ca.lvin.books.jenkins.RandomWord   // FAILS!
```

This means that a dynamically loaded shared library cannot be used to retrieve the classes in the src directory. There are ways to call the static methods inside the classes; however, it is not possible to take a class from the src directory and instantiate an object from it[1]. Thankfully, there is still a way to get access to the classes in the src directory from a dynamically loaded shared library.

Recall that the files inside the vars directory can use the classes in the src directory. In fact, vars test 4 in the dynamic-shared-lib pipeline did exactly that. We can't instantiate an object from the classes in the Jenkinsfile directly, but we can use a variable that can instantiate an object from the classes. In the previous section, we already discussed that the classes in the src directory should be used from the files in the vars directory, rather than directly from a Jenkinsfile. This makes the dynamically loaded libraries equally as capable as the statically loaded libraries.

The ability to load a shared library without needing it to be pre-configured provides great flexibility. The choice between loading the pre-configured libraries statically or loading one dynamically is a policy decision that can be made based on your organization's preference.

That covers the different ways of providing and using a shared library. We've learned a lot of different ways, but in reality, the shared libraries will be used like this in most cases:

- The vars directory from the global shared libraries will provide the globally useful helper functions.

- The vars directory from the folder-level shared libraries will provide the modularization of the repeated code in a given project.

- The shared libraries will be loaded with this syntax:

  `@Library('lib-name') _`

  That is, unless the admins have blocked the folder-level shared libraries and we're forced to use dynamic loading.

Let's put aside the academic material that we've seen so far, and see some real-life examples of a shared library.

Use cases

We have divided the *Use cases* section into two parts. In the first part, we will build a library for sending pre-formatted Slack messages. This is the most commonly used shared library code pattern for providing a globally useful helper function. In the second part, we will develop our own DSL using the shared libraries in three different ways.

---

1   https://www.jenkins.io/doc/book/pipeline/shared-libraries/#loading-libraries-dynamically

This is a more advanced use case that will be useful in scaling your Jenkins instance. Let's get started.

Code reuse via global variables – Pre-formatted Slack messages

> **Required plugins**
> Slack Notification

Let's create a helper function that sends a Slack message in a standard format.

Prepare the Jenkins integration for Slack

Before starting, we need to configure Slack with Jenkins CI app so that it can receive messages from Jenkins:

1. On Slack, click Apps, search for Jenkins CI, then click Add:

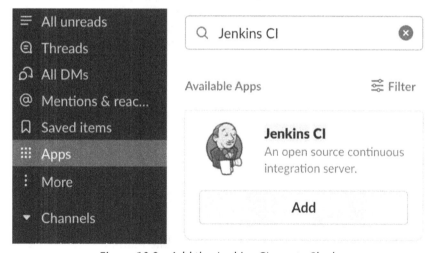

Figure 10.8 – Add the Jenkins CI app to Slack

2. When the web page for the Slack Jenkins CI app opens, click Add to Slack.

> **Slack app**
> There is a newer way of connecting to Slack by creating a custom Slack app. We will use the traditional method to keep the process simple so that we can continue to focus on shared libraries.

3. On the next page, choose #general for Post to Channel, **then click** Add Jenkins CI integration. Even though we're choosing #general, we can use this to send messages to all channels.

4. The next page gives us a step-by-step guide on how to configure the app. The most important part for us is the Token field in the Integration Settings toward the bottom of the page. Your token will be different from the example:

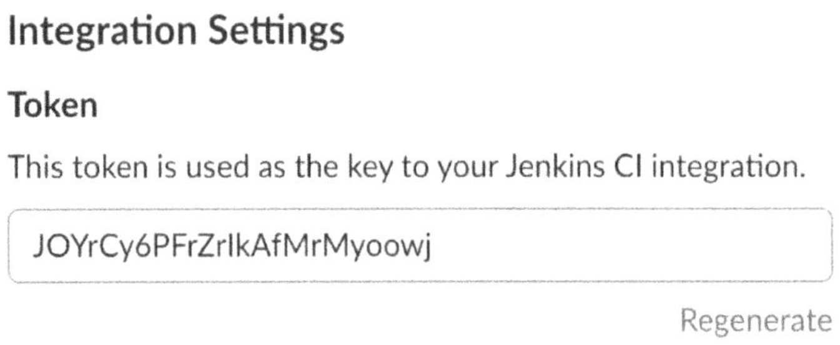

Integration Settings

Token

This token is used as the key to your Jenkins CI integration.

```
JOYrCy6PFrZrlkAfMrMyoowj
```

Regenerate

Figure 10.9 – The Slack token from the Jenkins CI app configuration

Slack is ready to receive messages from Jenkins. Let's continue to save the token in Jenkins and use it to send a Slack message.

5. Log in to Jenkins as the admin user. Go to the Global Credentials **page and make a new Jenkins credential as follows:**

- Kind: Secret text
- Scope: Global (Jenkins, nodes, items, all child items, etc.)
- Secret: <the token from above>
- ID: slack-token
- Description: slack-token

6. Let's test the connection. Create a new pipeline named slack and enter the following code in the pipeline definition text box. Here, we're using the slackSend() function that's provided by the Slack Notification plugin to send a message. The name of the function, along with all other functions and variables, can be found in https://<JENKINS URL>/pipeline-syntax/html. Be sure to replace <your Slack workspace> with your own workspace name:

Jenkinsfile.slack.1

```
pipeline {
    agent any
    stages {
        stage('Hello Slack!') {
            steps {
                slackSend(
                    baseUrl: "https://<your Slack workspace>.
slack.com/services/hooks/jenkins-ci/",
                    tokenCredentialId: "slack-token",
                    channel: "#general",
                    message: "Hello Slack!")
} } } }
```

7. Run the pipeline and we can see the message on our Slack's #general channel. Sending messages to other channels should work as well:

Figure 10.10 – Slack message sent from Jenkins

Great! Now we have a simple pipeline that we can use to send Slack messages. Let's continue to see how we can improve this.

Shared library for sending pre-formatted Slack messages

Now, let's use the global shared library to improve the pipeline code in three ways. First, we'll hide baseUrl and tokenCredentialId because they will be the same for all pipelines in this Jenkins. Second, we'll have a template for the messages so that the build URL is always included in the message. Third, we'll provide an optional parameter for specifying the color of the message status bar. Let's get started:

1. Create a new file called vars/slack.groovy in our jenkins-shared-library Git repository. Let's start by creating a low-level function. The send() function contains a map that has the default parameters that are common to all Slack messages. Later, the map from the function parameter is merged into the default parameters, then slackSend() is called with the merged map:

vars/slack.groovy

```
void send(Map params=[:]) {
    Map defaultParams = [
        baseUrl: "https://<your Slack workspace>.slack.com/
services/hooks/jenkins-ci/",
        tokenCredentialId: "slack-token",
    ]
    slackSend(defaultParams << params)
}
```

Add the file to the Git repository, make a commit, and push.

2. Let's go back to our slack pipeline and test the new library variable. Enter the following code in the pipeline definition text box and run the pipeline. Here, we are calling the send function from the previous file while providing a map that specifies the channel and the message content:

Jenkinsfile.slack.2

```
@Library('my-global-shared-lib') _
pipeline {
    agent any
    stages {
        stage('Hello Slack!') {
            steps { script {
                slack.send(
                    channel: "#general",
                    message: "Hello slack.send()!")
} } } } }
```

We can see that the message was successfully sent to our Slack's #general channel.

3. Now let's make a more user-friendly wrapper function so that the Jenkinsfile doesn't need to pass a map. Add the following function to vars/slack.groovy:

vars/slack.groovy

```groovy
void send(String channel, String message) {
    send(channel: channel, message: message)
}
void send(Map params=[:]) {
    Map defaultParams = [
        baseUrl: "https://<your Slack workspace>.slack.com/
services/hooks/jenkins-ci/",
        tokenCredentialId: "slack-token",
    ]

    slackSend(defaultParams << params)
}
```

Take note that we're not replacing the existing function created in *step 1*. We are overloading the function with a different signature to accept two String parameters, then calling the original function with a map created from the two String parameters.

Add the file, make a commit, and push.

4. Go back to the Slack pipeline and run the following code:

Jenkinsfile.slack.3

```groovy
@Library('my-global-shared-lib') _
pipeline {
    agent any

    stages {
        stage('Hello Slack!') {
            steps { script {

                slack.send(
                    "#general",
                    "Hello slack.send(String, String)!")
} } } } }
```

The message was sent successfully, and we didn't have to pass a map in the Jenkinsfile. The Jenkinsfile is looking much cleaner.

5. Let's add the build URL to every message, so that we can always identify which build from which pipeline sent the message. Modify the function from *step 3* as follows. Regardless of what the message is, we are now prepending the build URL in the first line:

vars/slack.groovy

```
void send(String channel, String message) {
    send(channel: channel,
        message: "${BUILD_URL}\n${message}")
}
```

Add the file, make a commit, and push.

6. Go back to the `slack` pipeline and run it again without making any changes. This time, the library function will add the build URL in the message:

 jenkins APP 8:54 PM

https://jenkins-firewalled.lvin.ca/job/slack/5/

Hello slack.send(String, String)!

Figure 10.11 – The message was reformatted to include the build URL before sending

Perhaps this is obvious, but realize that we ran the same pipeline and received a different result, because the library function it uses has changed. Creating a useful set of library functions and having the users adopt them is a good way to raise the overall quality of engineering.

7. Finally, let's add an option to specify the color of the message status bar. Add the following function to vars/slack.groovy. Be sure to add a new function instead of updating the existing ones:

vars/slack.groovy

```
void send(String channel, String message, String color) {
    send(channel: channel,
        message: "${BUILD_URL}\n${message}",
        color: color)
}
```

As with *step 3*, we are overloading the function with a different signature to accept three `String` parameters.

Add the file, make a commit, and push.

8. Go back to the `slack` pipeline and run the following code. We're calling `slack. send()` twice, first with two parameters and then again with three parameters:

Jenkinsfile.slack.4

```
@Library('my-global-shared-lib') _
pipeline {
    agent any
    stages {
        stage('Hello Slack!') {
            steps { script {
                slack.send("#general", "Hello default!")
                slack.send("#general", "Hello color!",
                        "#009900")
} } } } }
```

Two messages were sent this time, one with the default gray status bar and another with a custom green status bar. Since we have three functions of the same name, Groovy checks how many String parameters we've provided and calls the correct one:

 jenkins APP 8:55 PM

https://jenkins-firewalled.lvin.ca/job/slack/8/
Hello default!

https://jenkins-firewalled.lvin.ca/job/slack/8/
Hello color!

Figure 10.12 – Two messages were sent using two different functions based on the parameters

9. Since we've created several overloading functions, which could get confusing, it would be a good practice to document the function signatures in vars/slack. txt to guide the users.

10. Before we wrap up, let's learn one additional feature. We have been calling slack.functionName(). What if we want to just call slack() instead? For that, create a special function named call().

Add the following function to vars/slack.groovy. The call() function is a simple wrapper for the send() function:

vars/slack.groovy

```
void call(String channel, String message) {
    send(channel, message)
}
```

Add the file, make a commit, and push.

11. Go back to the slack pipeline and run the following code, which calls slack() rather than slack.send():

Jenkinsfile.slack.5

```
@Library('my-global-shared-lib') _

pipeline {
    agent any
    stages {
        stage('Hello Slack!') {
            steps { script {
                slack("#general", "Hello slack()!")
} } } } }
```

This time we were able to send a message while using the filename as the function name, without having to call a function inside the file.

This is the basics of a shared library. It's not difficult to imagine that the same technique can be used to provide a uniform Git tagging library, Jira ticket updating library, email notification library, Artifactory operations library, or even a general utility library for text and file manipulations. For most people, this use pattern will be sufficient for everything. For the curious readers, let's continue to see what else we can do with shared libraries.

Advanced – Custom DSL

> **Required plugins**
> Pipeline Utility Steps

In the previous example, we saw a way to modularize a commonly used feature into a function. What if we want to modularize the entire pipeline? If many of our pipelines are similar, can we turn the pipeline itself into a function? Yes, we can! Let's see how to do this.

Since we're past the beginner section, I'll stop telling you every keystroke to type and instead focus on the main ideas. Log in to Jenkins as adder-admin user and make a new pipeline named my-dsl inside the adder folder. This example is specific to the adder project, so we will be using a folder-level shared library. As we learned earlier, the code in a folder-level library runs inside the Groovy sandbox and is subject to an administrator's approval. Open a different web browser and log in to Jenkins as the admin user so that you can approve the method signatures easily. We will remove the approvals at the end.

Custom DSL

Add the following code as the pipeline definition. We are creating a string in YAML format with information about two stages. Each stage specifies its name and its build command. Afterward, we pass it to myDsl() from the my-folder-shared-lib library. Notice that the pipeline directive is missing entirely. If we had configured the shared library to be loaded automatically, even the first line wouldn't be necessary:

Jenkinsfile.myDsl

```
@Library('my-folder-shared-lib') _
String spec = '''
- stage_name: First stage
  build_cmd: echo "Hello myDsl!"

- stage_name: Second stage
  build_cmd: echo "Hello dynamic stages!"
'''

myDsl(spec)
```

Let's see what the shared library variable looks like. As we saw at the end of the last section, a function named call() allows the filename to be used as a function name, which is myDsl in this case. As a parameter we are receiving a String that contains the YAML. We parse the YAML content into a list of maps, then loop through the list to feed its content into a makeStage() helper function. I have skipped all validations to keep the example simple:

vars/myDsl.groovy

```
void call(String yamlSpec) {
    List<Map> spec = readYaml(text: yamlSpec)
    pipeline {
        agent any
        stages {
            stage("Init") {
                steps { script {
                    for (Map stg in spec) {
                        makeStage(stg)
} } } } } } }

void makeStage(Map stg){
    stage(stg.stage_name) {
        sh "${stg.build_cmd}"
} }
```

Save the shared library and run it a few times to approve the method signatures. When the pipeline finally executes all the steps, the build log shows that it successfully created two stages and ran the build command for each stage:

Dashboard › **adder** › **my-dsl** › **#21**

```
[Pipeline] stage
[Pipeline] { (First stage)
[Pipeline] sh
+ echo Hello myDsl!
Hello myDsl!
[Pipeline] }
[Pipeline] // stage
[Pipeline] stage
[Pipeline] { (Second stage)
[Pipeline] sh
+ echo Hello dynamic stages!
Hello dynamic stages!
```

Figure 10.13 – A custom DSL in the pipeline was successfully processed by the function in the folder-level shared library

This is of course a skeleton example, but we can see that we're able to develop our own DSL using the shared libraries. Developers can provide the values for the keys that we've defined, then we can process the input based on our own rules. We can even imagine a scenario where a product's Git repository doesn't even have a Jenkinsfile but instead has an actual YAML file with the key/value pairs for our own DSL. The pipelines can use a generic Jenkinsfile that loads the YAML file from a product's Git repository. This would be a perfect solution for a developer who needs to use Jenkins but just hates writing a Jenkinsfile (there are always a few).

> **Ode to Jenkins**
>
> On a side note, go back to vars/myDsl.groovy and see how naturally the Groovy code weaves in and out of the Jenkinsfile pipeline code. There are many CI/CD solutions in the market, but I can't think of anything that allows such a smooth transition between its DSL and non-DSL code. This is the true power of Jenkins that no others come even close.

One common pitfall to watch out for is that the pipeline directive can be defined only once. If myDsl() were called with the following Jenkinsfile, it would fail to run since the pipeline directive is defined twice, once in the Jenkinsfile and another in the shared library function:

```
@Library('my-folder-shared-lib') _
pipeline {
    agent any
    stages {
        stage('My DSL') {
            steps { script {
                myDsl(...)
} } } } }
```

Next, we'll take a look at a different kind of custom DSL that still uses the Jenkins pipeline syntax, but in a more limited scope.

Custom DSL with native Jenkins steps using Closure

We can also have a DSL for someone who's more familiar with the Jenkinsfile syntax. Remember how cumbersome the process was for setting up Docker-outside-of-Docker (DooD) in *Chapter 4, GitOps-Driven CD Pipeline with Docker Hub and More Jenkinsfile Features*? We can encapsulate the whole process and provide it as a service, while still allowing the developers to use the native Jenkinsfile steps using a `Closure`. The code for the two stages for DooD is from *Chapter 4, GitOps-Driven CD Pipeline with Docker Hub and More Jenkinsfile Features*:

vars/doodaaS.groovy

```
def call(Closure steps_closure) {
  pipeline {
    agent none
    stages {
      stage('Get Docker group') {
        agent { label 'firewalled-firewalled-agent' }
        steps { script {
          docker_group = sh (
            script: "stat -c '%g' /var/run/docker.sock",
            returnStdout: true).trim()
      } } }
      stage('Build') {
        agent {
          docker {
            label 'firewalled-firewalled-agent'
            image 'docker:dind'
            args "-v /var/run/docker.sock:/var/run/docker.sock:rw
--group-add ${docker_group}"
          } }
          steps { script {
            steps_closure.call()
} } } } } }
```

Developers can provide a Jenkinsfile with just the build steps:

Jenkinsfile.doodaaS

```
@Library('my-folder-shared-lib') _
Closure steps_closure = {
    echo "Hello closure & DooD!"
    sh "docker ps"
    slack('#general', 'This works too!')
}
doodaaS(steps_closure)
```

The closure contains the actual Jenkinsfile directives, which are evaluated as-is. This way, developers can still use all the Jenkinsfile directives including the shared library functions but do not need to worry about setting up DooD for the agent:

Dashboard ▸ **adder** ▸ **my-dsl-2** ▸ **#12**

```
[Pipeline] echo
Hello closure & DooD!
[Pipeline] sh
+ docker ps
CONTAINER ID    IMAGE          COMMAND              CREATED
STATUS                  PORTS           NAMES
39884bf1a0c8   docker:dind    "dockerd-entrypoint.…"   1 second
ago    Up Less than a second    2375-2376/tcp   youthful_ramanujan
[Pipeline] slackSend
Slack Send Pipeline step running, values are - baseUrl:
https://jenkins-book.slack.com/services/hooks/jenkins-ci/,
teamDomain: <empty>, channel: #general, color: <empty>, botUser:
false, tokenCredentialId: slack-token, notifyCommitters: false,
iconEmoji: <empty>, username: <empty>, timestamp: <empty>
```

Figure 10.14 – DooD is set up and run through the shared library function

This technique can be further expanded to handle the resource locks. For example, in a testing environment where there are physical USB devices attached, it's important that the test pipelines use the Lockable Resources plugin properly so that the builds do not step on each other. Since the pipeline owners have varying levels of expertise in writing Jenkinsfiles, it's difficult to ensure that the pipeline owners write a Jenkinsfile that properly requests for a lock and always releases the lock at the end. The problem can get worse if there is a cleanup process that must run when a lock is released, since some Jenkinsfiles may not handle the failure cases gracefully and end the build without calling the cleanup process.

In this case, we can provide a wrapper that manages the USB devices before and after the build. Similar to how we handled the DooD configuration, we can request for a lock and release it at the end on behalf of the pipelines. This way, the builds-to-device-mapping management is taken out of the hands of the testers entirely. In a large org, managing the test devices can be delegated to a dedicated team of engineers who specialize in handling the test equipment.

Let's look at one more type of custom DSL.

Custom DSL overriding the built-in pipeline directives

So far we've seen two different types of custom DSLs: one that uses an entirely new syntax based on YAML and another that uses the Jenkins pipeline directives but only asks for the steps closure rather than the entire pipeline. What if we approach this from the other direction? Can we override the built-in pipeline directives so that the existing pipelines start acting differently? With the flexibility and the versatility of Jenkins, yes, we can!

Let's consider the archiveArtifacts step. This step takes a file from the workspace and saves it in the controller along with the build history. For example, a postmerge pipeline can build a Debian package, and at the end of the build, the .deb file can be saved in the controller using the archiveArtifacts step. While this is a convenient way to save and associate the artifacts with a specific build, it quickly fills up the controller's storage space. In an enterprise setting, an external artifact management solution such as JFrog Artifactory or Sonatype Nexus is usually available, and the developers are expected to save the artifacts in the external artifact store rather than on Jenkins itself. This means that we don't want the developers to use the archiveArtifacts directive at all.

We can effectively disable the directive by creating a variable of the same name in a global shared library like the following. We also want this to apply to all pipelines regardless of whether they want it or not, so the global shared library containing this code would be configured with Load implicitly enabled. Once the file is added to a global shared library and the library is configured to load implicitly, the archiveArtifacts calls in the existing pipelines will be redirected to call our shared library variable instead, then fail with the notice that the directive is disabled:

vars/archiveArtifacts.groovy

```
void call(Map params = [:]) {
    error("archiveArtifacts step is disabled. " +
          "Use rtUpload instead to upload to Artifactory.")
}
```

Calling the archiveArtifacts directive fails with the error message as expected:

Dashboard   ›   **archiveArtifacts-disabled**   ›   **#2**

```
ERROR: archiveArtifacts step is disabled. Use rtUpload
instead to upload to Artifactory.
Finished: FAILURE
```

Figure 10.15 – The archiveArtifacts call is redirected to the variable in the global shared library

Instead of simply failing the build, a nicer option would be to handle the uploads to Artifactory for them. The following example doesn't use the real syntax for the rtUpload step, but it's not difficult to see that we can use the data from the input map params to upload the requested files to Artifactory:

vars/archiveArtifacts.groovy

```groovy
void call(Map params = [:]) {

  unstable("archiveArtifacts step is disabled. " +
          "Uploading to the Artifactory instead.")

  rtUpload(
    serverId: 'artifactory',
    spec: '''{
      "files": [ {
        "pattern": params.artifacts,
        "target": "jenkins-artifacts/${env.BUILD_TAG}"
      } ] }'''
  ) }
```

In the previous examples, we intercepted the archiveArtifacts calls and did something different instead. The next example shows that we can also call *the directive that we're overriding* by using a magic variable named steps.

Here, the override variable intercepts the input calls, wraps them with a timeout of 1 hour, then calls the actual input pipeline step. If our override variable called input() instead of steps.input(), our own override variable would be called instead of the pipeline step, resulting in recursion and eventually a stack overflow. The steps magic variable allows us to specify the actual pipeline step so that we can embellish the pipeline steps with additional features:

vars/input.groovy

```groovy
def call(Map params = [:]) {
    timeout(time: 1, unit: 'HOURS') {
        return steps.input(params)
    }  }
```

Using this technique, we can create a *custom DSL* that uses the Jenkins pipeline syntax by changing the meaning of each directive.

With the power to create custom DSLs with varying degrees of freedom as shown in the preceding three examples, the only limit is our imagination.

Summary

We started this chapter by learning what a shared library is. We went through the three directories in a shared library and discussed how the vars directory holds the helper functions, the src directory holds the Groovy class files, and the resources directory holds the static files.

We created a shared library in a GitHub repository by adding a sample code that reads a file from the resources directory, then prints a random line from the file.

Then we learned the two different ways of providing a shared library. We learned that a global shared library should be used to provide globally useful helper functions, and that a folder-level shared library should be used for anything else.

Then we went through the five different ways of loading a shared library. We saw how we can load a library statically, dynamically, and with or without importing a class in the src directory. We saw how each one differs and discussed when we should use one over another.

Once we finished learning what a shared library is, how to provide it, and how to load it, we moved on to create a wrapper library for sending pre-formatted Slack messages. We started by configuring Slack with the Jenkins CI app so that it can receive messages from Jenkins. We saved a token from the app as a Jenkins credential, then created a base function that sent a Slack message from Jenkins. We then overloaded the function by using a different signature, and even added a variant that allows us to change the color of the message status bar. Once we tested that each variant works correctly, we learned how to use the call() method to use the Groovy filename as a function name.

Finally, we learned three different ways of creating our own DSL. The first example used a brand new syntax using YAML. The second example still used the Jenkins pipeline syntax, but it only asked for the steps closure rather than the entire pipeline. Finally, the last example showed a way to override the built-in pipeline directives so that the existing pipelines behave differently.

As we discussed in the introduction of the chapter, shared libraries are amazingly flexible and provide an unprecedented amount of customizability. Using shared libraries effectively will be the key to scaling our Jenkins instance as a managed service rather than a collection of pipelines of varying quality. This chapter provides a good starting point for developing your own shared libraries.

In the next chapter, we will learn about the Jenkins security model by going through the features of the Script Security plugin in detail. It builds on the learnings from this chapter by using a global shared library to wrap dangerous method calls. See you there!

11
Script Security

The power and the flexibility of Jenkins come from the pipelines that can run Groovy code – a fully featured language independent of Jenkins. Since Jenkins pipelines allow Groovy code to be executed, a pipeline can do nearly anything that Groovy can do. The *Difference from plain Groovy* section in the Jenkins User Handbook lists just one difference, that some Groovy idioms are not fully supported because the pipeline code must be serialized[1]. Allowing us to use a real programming language, as opposed to twisting an inflexible DSL that is not designed to be a general-purpose language, provides amazing power and flexibility. With great power, of course, comes great responsibility. Take this pipeline, for example:

```
pipeline {

    agent any

    stages { stage ('Destroy') {
        steps { script {

            Jenkins.instance.items.each {
                if (it.name != env.JOB_NAME)
                    it.delete()

} } } } }
```

This allows us to delete all pipelines and folders at the convenience of a single click. Thankfully, Jenkins comes with a built-in security mechanism to prevent these disasters through the use of the Script Security plugin. Running the pipeline fails with an error, Scripts not permitted to use ... .

---

1   https://www.jenkins.io/doc/book/pipeline/syntax/#differences-from-plain-groovy

Let's learn how to use the Script Security plugin effectively to manage these risks.

In this chapter, we're going to cover the following main topics:

- Administrator versus non-administrator
- Outside the Groovy sandbox
- Inside the Groovy sandbox
- Approve assuming permission check
- Thoughts on disabling Script Security

Technical requirements

We will need the Jenkins that we built in *Chapter 3, GitOps-Driven CI Pipeline with GitHub*. In particular, we'll examine the users of the two projects, adder and subtractor, for their security restrictions, so you'll need to follow the directions in *Chapter 3, GitOps-Driven CI Pipeline with GitHub*, to set up adder and subtractor projects and users.

The files for this chapter are available in the GitHub repository at https://github.com/PacktPublishing/Jenkins-Administrators-Guide/blob/main/ch11.

Administrator versus non-administrator

There are several key concepts in understanding the Jenkins security model.

The first concept is *administrator* versus *non-administrator*. Jenkins security is structured around approving or denying pipeline codes and method signatures, and the administrators are the only users who can approve or deny them.

A user is an administrator if the Overall / Administrator box is checked on the Global Security page:

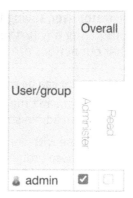

Figure 11.1 – admin user is an administrator because it has Overall/Administrator permission

In *Chapter 3, GitOps-Driven CI Pipeline with GitHub*, we have created project admin users – adder-admin and subtractor-admin. They are named *-admin and have all the permissions for their respective folders, but they do not have the Overall / Administrator permission, so they're not administrators. In our setup, admin is the only administrator user, and therefore only the admin user can approve or deny codes and method signatures.

The second concept is Groovy sandbox. Let's continue to examine it further.

Outside the Groovy sandbox

A pipeline runs either inside or outside of the sandbox.

Jenkins does not offer protection against dangerous code outside of the sandbox, hence, *it is very risky to be outside of the sandbox*. Areas outside of the sandbox are meant to be used by administrators for Jenkins administrative tasks, not for running software build/test/release pipelines. Since it has a very limited scope of use, there are only two ways to run outside of the sandbox.

Direct pipeline

A pipeline with its content directly written into the UI, as opposed to a pipeline that loads a Jenkinsfile from an SCM as we saw in *Chapter 3, GitOps-Driven CI Pipeline with GitHub*, and *Chapter 4, GitOps-Driven CD Pipeline with Docker Hub and More Jenkinsfile Features*, can run outside of the Groovy sandbox. We can create a new pipeline, then in the Pipeline / Definition drop-down menu, choose Pipeline script to get a textbox titled Script. Afterward, uncheck the box for Use Groovy Sandbox to have the pipeline run outside of the sandbox.

If we are logged in as an administrator user, saving the pipeline will allow it to run outside of the sandbox without requiring any further approvals. We are already an administrator, so it's assumed that our actions are already approved.

A pipeline saved as an administrator can be run by non-administrator users

Typically, a non-administrator user can't run a pipeline outside of the sandbox without approval from an administrator.

Suppose that we created a pipeline with dangerous calls and saved it as an administrator so that it doesn't require approval. Once saved, this pipeline can be run by a non-administrator user, allowing the non-administrator to execute code outside of the sandbox.

Be sure to protect the pipelines saved as an administrator through the permission restrictions that we saw in *Chapter 3, GitOps-Driven CI Pipeline with GitHub*.

If we are logged in as a non-administrator user, Jenkins warns us that A Jenkins 20-9 administrator will need to approve this script before it can be used, **as seen in** *Figure 11.2*:

Pipeline script   ⌄

Script
```
 1 ▾ pipeline {
 2        agent any
 3 ▾      stages {
 4 ▾          stage ('Destroy') {
 5 ▾              steps {
 6 ▾                  script {
 7 ▾                      Jenkins.instance.items.each {
 8                            if (it.name != env.JOB_NAME)
 9                                it.delete()
10                        }
11                    }
12                }
13            }
14        }
15 }
```

A Jenkins administrator will need to approve this script before it can be used.

☐ Use Groovy Sandbox

Figure 11.2 – A pipeline running outside of the Groovy sandbox that deletes all pipelines and folders

Saving this pipeline code as a non-administrator user and running it will fail with the message script not yet approved for use:

```
20:38:25  Started by user adder-admin

20:38:25  org.jenkinsci.plugins.scriptsecurity.scripts.
UnapprovedUsageException: script not yet approved for use

[...]

20:38:25  Finished: FAILURE
```

An administrator user can review this code in Manage Jenkins, In process Script Approval. Once approved, the pipeline can run the approved code to delete all pipelines and folders:

```
 Approve / Deny  Groovy script from adder-admin in adder » destroyer:
┌─────────────────────────────────────────────────────────────────┐
│ pipeline {                                                        │
│     agent any                                                     │
│     stages {                                                      │
│         stage ('Destroy') {                                       │
│             steps {                                               │
│                 script {                                          │
│                     Jenkins.instance.items.each {                │
│                         if (it.name != env.JOB_NAME)             │
│                             it.delete()                          │
│                     }                                             │
│                 }                                                 │
│             }                                                     │
│         }                                                         │
│     }                                                             │
│ }                                                                 │
└─────────────────────────────────────────────────────────────────┘
```

Figure 11.3 – Running a pipeline outside of the sandbox prompts administrators to either approve or deny the entire script as a whole

An additional protection that the Script Security plugin provides is requiring a re-approval when the script changes. If an approved script changes even by a single character, it is now considered a new script that requires a new approval. This mechanism makes it very difficult for a non-administrator user to develop a pipeline code that runs outside of the sandbox because the non-administrator user can't run the pipeline to test the content without asking for a re-approval each time.

Now, let's take a step back and think about why the Jenkins developers made it so difficult. The simple answer is that non-administrator users simply have no business making pipelines that run outside of the sandbox. Areas outside of the sandbox are meant to be used by administrators for Jenkins administrative tasks, not for running software build/test/release pipelines. If a non-administrator user is asking for an approval to run a pipeline outside of the sandbox, the request should be denied. There are *no reasons* for a non-administrator user to run a pipeline outside of the sandbox. None. Zero. Never approve these. Always deny all of them. If you find yourself needing to approve these, re-examine your workflow to make sure that the right set of people are administrators.

In case you're not convinced, here are several reasons why approving a script is dangerous:

- An approved script can be run by *anyone with a run permission to the pipeline.* The destroyer pipeline in Figure 11.3 is inside the adder folder, and adder-user has permissions to run pipelines in the adder folder, therefore adder-user can run the destroyer pipeline. If this script is approved, adder-user who is not an administrator, not even a project admin, can execute this script to delete all pipelines and folders.

- The adder/destroyer pipeline runs the approved script *with administrator privileges.* This allows adder-admin and adder-user, who don't even have the read permissions for the subtractor folder, to delete the subtractor folder along with its contents.

- A script is approved *globally across all projects and all users.* This allows subtractor-admin to create its own destroyer pipeline in the subtractor folder with the same content, allowing subtractor-user to run it to delete all pipelines and folders, including the adder folder. When the subtractor/destroyer pipeline runs, the Script Approval page doesn't prompt administrators to approve the subtractor/destroyer pipeline's script because it's already approved *globally across all projects and all users.*

- Recall that changing even a single character of an approved script requires a new approval. The gotcha is that approving a new version of a script *doesn't remove the approval of the old versions* of the script. There are no old or new versions of a script – they are all independent scripts. Administrative control for disapproving a specific script doesn't exist. In order to disapprove an old script, *all* previous script approvals must be removed.

Let's gather these issues and see how they could cause a problem in the real world. Consider the following disaster scenario.

The adder-admin user requests for an approval of a script, which is necessary for managing the adder project, but is also potentially destructive. An administrator approves the script on the condition that adder-admin makes the pipeline using this script *inaccessible to anyone except for* adder-admin himself. adder-admin restricts pipeline access as requested by the administrator, uses the pipeline with the approved script to manage the adder project, and saves the content of the script in a Git repository for the record. Later, adder-admin finds a bug in the code and continues to improve the script while getting the new versions approved.

subtractor-admin sees the useful script in the Git repository and creates a new pipeline under the subtractor directory with the same script content. Unfortunately, the new pipeline uses an older buggy version of the script that is destructive beyond its intended scope. Furthermore, the subtractor-admin user is *not asked to restrict the permission* of the new pipeline because the *administrators are not prompted to approve the script* – it was already approved *globally across all projects and all users*. A curious subtractor-user sees the interesting new pipeline and runs it to break something outside of the subtractor folder.

The Jenkins administrators who thought they approved the script only for the adder-admin user's specific pipeline look for clues on how this script was executed, but they are stumped when there is no recent build history on the adder-admin's pipeline. After digging deeper and finding that a pipeline in the subtractor folder is using an old buggy version of a script, the administrators try to remove the approval of that script. However, *the only option is to remove all previous script approvals*. Now, administrators are left with the choice of either leaving the dangerous script approved or removing all previous approvals, thereby causing disruption across all projects and requiring reapproval for all pipelines running outside the sandbox.

Do not approve any pipelines to run outside of the sandbox. Instead, we can use the global shared library to wrap the dangerous method calls.

Global shared library

This section requires a good understanding of Jenkins Shared Libraries. You can learn more about these in *Chapter 10, Shared Libraries*.

Global shared library methods always run outside of the sandbox because only administrators can configure global shared libraries. As with the script approval process, Jenkins doesn't require us to get approval from an administrator because we are already an administrator. This allows an administrator to encapsulate an unsafe method with permission checks in a library function.

For example, the following library method validates the fact that a shared library method was called with the following conditions met:

- It is called from a pipeline from a list of approved pipelines.
- It is called by a user from a list of approved users.
- It is called against a target from a list of approved targets:

vars/validatePipelineUserTarget.groovy

```
def call (String libMethodName,
        List<String> approvedPipelines,
        List<String> approvedUsers,
        List<String> approvedTargets,
        String target) {

    if (!approvedPipelines.contains(env.JOB_NAME))
        error("${env.JOB_NAME} is not approved to call\
            ${libMethodName} global shared library method")

    String userId = currentBuild.rawBuild.getCause(
                        hudson.model.Cause$UserIdCause).userId

    if (!approvedUsers.contains(userId))
        error("${userId} is not approved to call\
            ${libMethodName} global shared library method")

    if (!approvedTargets.contains(target))
        error("${libMethodName} global shared library method\
            isn't approved to be called with ${target}")

}
```

Using validatePipelineUserTarget as a validation step, deleteAdderPipeline can be a global shared library wrapper for deleting adder pipelines, with the following requirements:

- The deleteAdderPipeline method runs only when it's called from a pipeline named delete-adder-pipeline.

- The build is started by the adder-admin user.

- The build is requesting to delete one of these three pipelines: adder/premerge, adder/postmerge, or adder/delete-me.

This is what the wrapper looks like:

vars/deleteAdderPipeline.groovy

```
def call (String pipelineToDelete) {
    // Results in 'deleteAdderPipeline'
    String currentMethodName =
        this.class.getSimpleName() - ".groovy"
    validatePipelineUserTarget(
        currentMethodName,
        ["delete-adder-pipeline"],
        ["adder-admin"],
        ["adder/premerge",
         "adder/postmerge",
         "adder/delete-me"],
        pipelineToDelete)

    def pipeline =
        Jenkins.instance.getItemByFullName(pipelineToDelete)

    pipeline.delete()
}
```

With the shared library methods in place, adder-admin can create a pipeline named delete-adder-pipeline and call the deleteAdderPipeline method, as seen here. This allows an administrator to encapsulate an unsafe method with permission checks in a library function. More specifically, we've wrapped a dangerous method call, delete(), with a validation logic for restricting which pipeline can make this call, who can trigger a build for the pipeline, and which target pipelines it can delete:

Jenkinsfile.delete-adder-pipeline

```
@Library('shared-lib') _
pipeline {
    agent any
    parameters {
        string(name: 'pipelineName',
               description: 'Pipeline name to delete')
    }
    stages {
        stage('Delete pipeline') {
            steps {
                deleteAdderPipeline(params.pipelineName)
} } } }
```

When adder-admin runs the pipeline to delete adder/delete-me, all validations in the wrapper pass and adder/delete-me is successfully deleted. However, when adder-user runs the pipeline, the user validation in the wrapper fails because only adder-admin can run this pipeline:

Dashboard   ›   adder   ›   delete-adder-pipeline   ›   #28

⬤ Console Output

```
Started by user adder-user
[Pipeline] error
[Pipeline] }
[Pipeline] // stage
[Pipeline] }
[Pipeline] // node
[Pipeline] End of Pipeline
ERROR: adder-user is not approved to call deleteAdderPipeline
global shared library method
Finished: FAILURE
```

Figure 11.4 – The delete-adder-pipeline build fails when not all the safety requirements are met. The screenshot was modified to highlight the relevant information

Another benefit of using global shared libraries to wrap dangerous method calls is that the unsafe methods used in the libraries are not added to the list of approved scripts or methods. Run delete-adder-pipeline and notice that the build ran without asking for a script approval or a method signature approval. This helps to minimize the number of approved scripts and method signatures, which results in a more secure environment. For instance, the disaster scenario from the previous section cannot occur for the following reasons:

- When the subtractor-admin user copies the library call into their own pipeline, the pipeline build fails when the library checks for the originating pipeline name and the build user.

- If subtractor-admin looks up the content of the library method (specifically, Jenkins.instance.getItemByFullName(pipelineToDelete).delete()) and writes the same code directly into their own pipeline, it still fails to run because the method calls in the global shared libraries are not added to the approved scripts or methods list.

In an environment where multiple teams share one Jenkins, this is the best way to manage the administrative pipelines.

With the dangers of running pipelines outside of the Groovy sandbox understood, let's see how things work inside the sandbox.

Inside the Groovy sandbox

All other pipelines run inside the Groovy sandbox:

- A direct pipeline with Use Groovy Sandbox checked
- A pipeline from SCM, multibranch pipeline, or any other pipeline type that loads a Jenkinsfile
- A folder-level shared library

An approval process for inside the sandbox is for *each method signature*, as opposed to the entire script as a whole. Let's go back to the destroyer script that we saw in the introduction and run it inside the sandbox. The build fails this time with a different error message:

```
15:50:28  Started by user subtractor-admin
[...]
15:50:30  Scripts not permitted to use staticMethod jenkins.
model.Jenkins getInstance. Administrators can decide whether to
approve or reject this signature.
[...]
15:50:30  Finished: FAILURE
```

When an administrator checks the Script Approval page, only the specific signature of the method is pending for a review:

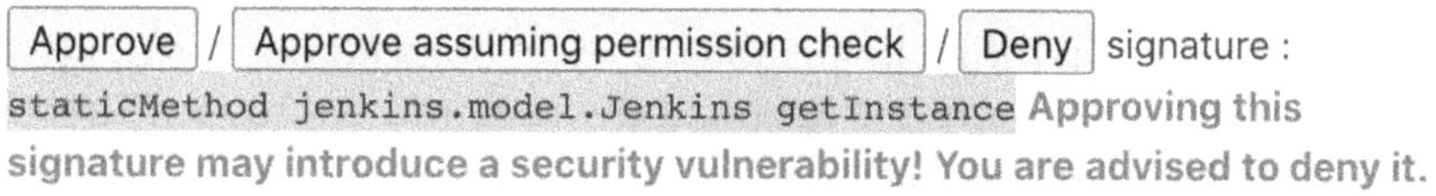

Approve / Approve assuming permission check / Deny signature :
staticMethod jenkins.model.Jenkins getInstance Approving this
signature may introduce a security vulnerability! You are advised to deny it.

Figure 11.5 – Running a method inside the sandbox prompts administrators to either approve or deny just that method signature

The red warning message is based on a deny-list file (https://github.com/jenkinsci/script-security-plugin/blob/master/src/main/resources/org/jenkinsci/plugins/scriptsecurity/sandbox/whitelists/blacklist), which lists the particularly dangerous method calls. Let's approve it to learn how it works – we can clear the approvals later. Running the build again fails, this time with the error message Scripts not permitted to use method hudson.model.ItemGroup getItems. Approve the method signature as an administrator and run it again. The build fails for method hudson.model.Item getName. Approve and run it again, and this time it fails for method hudson.model.Item delete. *Do not approve this method!* Before continuing, think about what you were going to do just now. We have approved three methods in a row, and you might have felt that we were going to continue approving the fourth without even looking at the signature. The signature is delete(), which is obviously destructive and shouldn't be approved. Did you notice the signature?

Herein lies one of the biggest challenges in using Script Security effectively. It's easy for an administrator to mistakenly approve a dangerous method. At least for the getInstance() method, Script Security gave us a warning in big red letters as we saw in *Figure 11.5*. In the case of the delete() method, there isn't even a warning, which means administrators really need to review each signature carefully before approving:

Figure 11.6 – Script Security doesn't warn about the dangers of approving the delete() method

Another problem is that we had to run the build three times to review the three method signatures, rather than being able to run it once (or not run at all and have Jenkins auto-detect) to generate a list of all the required approvals. There's also no easy way to pre-approve a signature without having it be called from a pipeline. This is an unfortunate whack-a-mole that is required for using Groovy sandbox, and this pattern of approving signatures one after another gets us in a habit of approving them without examining them carefully. This is usually a problem only during the initial few days of Jenkins and pipeline creations, but it's understandably frustrating for both administrators and users.

Bad UX and frustrations aside, a more important problem is that an approved method signature is approved *globally across all projects and all users*, just like the script approval. We want to give adder-admin the permissions to delete the pipelines inside the adder folder, but not give the permissions to delete the pipelines outside of the adder folder. If you click Approve, adder-admin and adder-user are granted permissions to delete *all* pipelines. subtractor-admin can also create a new pipeline inside the subtractor folder with the same code, and subtractor-admin and subtractor-user will also be able to delete all pipelines.

Delete is obviously a destructive method, so an administrator should deny this signature. The project admin and administrators can discuss how to wrap this safely in a global shared library call.

Before we conclude this section, here is a word of caution: if you find yourself having to approve method signatures frequently, it's likely that the pipeline that's using the methods is too complex. No approvals are required if the build logic is moved into a script and the script is run using the sh step. While it's a blessing that Jenkins provides a way to use a full-blown programming language, the official documentation continuously stresses the need to use Jenkinsfile as a way to call your build scripts, not the build script itself (https://www.jenkins.io/doc/book/pipeline/pipeline-best-practices/#making-sure-to-use-groovy-code-in-pipelines-as-glue). If you get frequent requests to approve method signatures, educate users to move the logic into a separate script instead.

Now, let's move on to examine an advanced permissions management feature that Jenkins provides.

Approve assuming permission check

A method call for deleting a pipeline is dangerous yet necessary for managing Jenkins, so that the project admins can programmatically delete unused pipelines. What if there were a way to allow deleting only the pipelines that a user has the permissions to delete? That is where the Approve assuming permission check button comes in.

> **Feel free to skip this section**
>
> The button fills an important gap in Jenkins security; however, there are a few challenges in making the button useful. Getting this to work and understanding why it works requires an extensive understanding of the Jenkins security model. Even if you do get it to work, there are so few methods that implement this feature that it's not really useful. Given the steep learning curve and low utility, my practical recommendation is to use the global shared library wrappers from the previous section instead. Go ahead and follow along for fun (oh the things we do for fun), but don't expect a solution that you can use. I've considered cutting this section short since it takes up half of the chapter, but I couldn't find a coherent article anywhere that explains this button end to end, so I am including this section for completeness. I am yet to see anyone use this feature effectively, but if you are using it effectively, please contact me for a free copy of this book. I would love to chat and learn how you're using it. Special thanks to Madhusudan N for helping me with this section.

By the name of it, the Approve assuming permission check button seems like Jenkins will check the user permissions before executing a method. For example, Jenkins would check the permissions of adder-admin and allow them to delete adder pipelines but not subtractor pipelines. It seems very straightforward, but in reality, it needs *a lot* more configurations to work. Let's learn about the security model in Jenkins to see why.

Identity crisis – everyone is a SYSTEM user

This is going to be confusing even to the people who are already familiar with Jenkins. I'll do my best to simplify the concept by making up the terms, before-build-user and during-build-user. Though these are not official terms, they are the best words to describe the situation.

The normal users that we're all familiar with – admin, adder-admin, adder-user, and so on, are all before-build-users. The permissions that we configured in *Chapter 3, GitOps-Driven CI Pipeline with GitHub*, as set out in the following table, are for the before-build-users, so they are *only relevant before a build starts*:

	adder folder		subtractor folder	
	Read, Build	Create, Delete	Read, Build	Create, Delete
admin	Y	Y	Y	Y
adder-admin	Y	Y		
adder-user	Y			
subtractor-admin			Y	Y
subtractor-user			Y	

Once a build starts, our identity changes (unless we configure otherwise, which we will see in a bit). We are no longer the before-build-user that we've logged in as, but instead, we all become one special user called a SYSTEM user. This is the *during-build-user*, which everyone becomes during a build.

This is quite different from the familiar Linux permissions model. In Linux, the user who initiates a command is the same as the user who executes the command. In the following example, robot_acct is the user who initiates the id command, and robot_acct is also the user who executes the id command, resulting in an output that shows the ID information of robot_acct:

robot_acct@firewalled-controller:~$ id
 uid=123(**robot_acct**) gid=30(dip) groups=30(dip),998(docker)

Jenkins, by default, does not use this model. This is what happens instead:

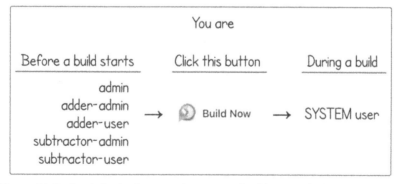

Figure 11.7 – By default, the user that runs a build is not the logged-in user

Why does this switch happen? Why would a build run as a different user? Most importantly, why does everyone become the *same* SYSTEM user?

One advantage of switching everyone to become the same user during a build is that it makes pipelines more predictable. Recall that in *Chapter 3, GitOps-Driven CI Pipeline with GitHub*, we created a premerge pipeline that can be run with any arbitrary branch. This pipeline build can be triggered by admin, adder-admin, adder-user, and also the GHPRB plugin. The four users (or three users and a plugin) have different permissions: adder-user and the GHPRB plugin can't delete the pipeline, while admin and adder-admin can. Despite the different levels of permissions, we can reliably expect the entire pipeline code to be executed the same way regardless of who triggers a build. With a guarantee that a pipeline behaves the same way regardless of who initiates the build, we end up with a simpler set of predictable pipelines. In other words, switching everyone to become the same during-build-user SYSTEM user fixes the *it works when I do it, but not when you do it* problem.

The disadvantage, of course, is that everyone's now the same during-build-user SYSTEM user. How can we have any security restrictions if everyone's the same user? Let's continue to understand the permission model for the SYSTEM user.

Where the SYSTEM user can do things

Permission is usually about *what* a user can do, and *where* the user can do it.

Recall that the SYSTEM user can run both adder pipelines and subtractor pipelines. Jenkins uses the SYSTEM user for all builds, which means that the SYSTEM user must have access to all pipelines. In reality, it's even farther reaching than that – the SYSTEM user has access to all folders, pipelines, system configurations, credential management, agents, and literally everything in Jenkins.

On a more technical level, the SYSTEM user isn't a *user* in Jenkins, but it is instead the actual Jenkins process – some documentations call it an *internal user* and others call it a *virtual user*. The Jenkins process needs to manage the entire Jenkins, so it makes sense that the SYSTEM user has access to everything in Jenkins.

In summary, when it comes to the SYSTEM user, *where* it can do things is *everywhere*, and this can't be changed. Thankfully, *what* it can do is still limited.

What the SYSTEM user can do everywhere

The SYSTEM user can run two categories of commands: Pipeline Steps and Groovy scripts. Let's see what these are and how we can control them.

Pipeline steps

The SYSTEM user, by default, can do anything listed in the Pipeline Steps Reference (https://www.jenkins.io/doc/pipeline/steps/). This is a big list of hundreds of actions which includes many fundamental actions such as dir or sh. One surprising action that's included in the list is build: this allows a pipeline to start a build of a different pipeline. Coupled with the fact that the SYSTEM user has permissions *everywhere*, we can create the following pipeline in the adder folder that builds subtractor/premerge while bypassing the permission restrictions set to before-build-users:

adder/Jenkinsfile.run-subtractor-premerge

```
pipeline {
  agent any
  stages {
    stage('Build subtractor') {
      steps {
        build job: '../subtractor/premerge',
            parameters: [string(name: 'REF', value: "main")]
} } } }
```

When adder-user runs adder/run-subtractor-premerge, they become the SYSTEM user during the build, and as the SYSTEM user, they build the subtractor/premerge pipeline:

Dashboard › adder › run-subtractor-premerge › #7

```
Started by user adder-user
Running in Durability level: PERFORMANCE_OPTIMIZED
[Pipeline] Start of Pipeline
[Pipeline] node
Running on firewalled-firewalled-agent in
/home/robot_acct/firewalled-firewalled-
agent/workspace/adder/run-subtractor-premerge
[Pipeline] {
[Pipeline] stage
[Pipeline] { (Build subtractor)
[Pipeline] build (Building subtractor » premerge)
Scheduling project: subtractor » premerge
Starting building: subtractor » premerge #10
[Pipeline] }
[Pipeline] // stage
[Pipeline] }
[Pipeline] // node
[Pipeline] End of Pipeline
Finished: SUCCESS
```

Figure 11.8 – adder-user becomes the SYSTEM user during a build
and runs the subtractor/premerge pipeline

Here, we can see that the subtractor/premerge pipeline was started by adder-user, again because during-build-user is the SYSTEM user:

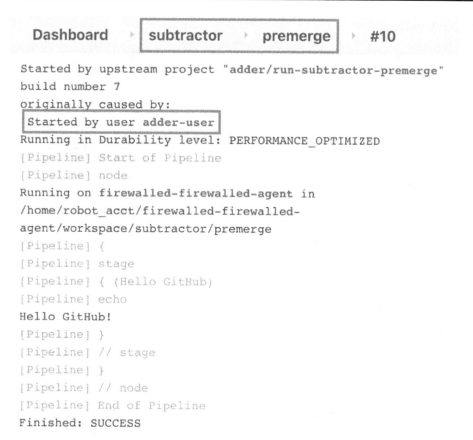

```
Dashboard  ▸  subtractor  ▸  premerge  ▸  #10

Started by upstream project "adder/run-subtractor-premerge"
build number 7
originally caused by:
Started by user adder-user
Running in Durability level: PERFORMANCE_OPTIMIZED
[Pipeline] Start of Pipeline
[Pipeline] node
Running on firewalled-firewalled-agent in
/home/robot_acct/firewalled-firewalled-
agent/workspace/subtractor/premerge
[Pipeline] {
[Pipeline] stage
[Pipeline] { (Hello GitHub)
[Pipeline] echo
Hello GitHub!
[Pipeline] }
[Pipeline] // stage
[Pipeline] }
[Pipeline] // node
[Pipeline] End of Pipeline
Finished: SUCCESS
```

Figure 11.9 – The SYSTEM user can build any pipeline, and everyone is a SYSTEM user!

This demonstrates the fact that the SYSTEM user has access everywhere and can run any pipeline steps. It's scary to learn that the permission configurations are not enforced during a build and the different projects are not fully isolated from one another.

Most pipeline steps in the Pipeline Steps Reference are enabled by installing plugins. With the SYSTEM user's permissions understood, we need to be mindful of which plugins we install because some plugins can enable a step that crosses the permission boundaries.

What else can a SYSTEM user do everywhere?

Groovy scripts

Another set of actions that a SYSTEM user can do is to execute Groovy methods *that are approved* by an administrator. This is the *only* aspect of a SYSTEM user's permissions that we can limit.

There are over a thousand pre-approved method signatures listed in the approve list (https://github.com/jenkinsci/script-security-plugin/tree/master/src/main/resources/ org/jenkinsci/plugins/scriptsecurity/sandbox/whitelists), and they can be used in a pipeline without any additional approvals. For example, toInteger("4") would run without requiring approval because that method signature is already in the approve list. However, a method that is not yet approved fails to run until an administrator approves it. We already saw how this works in the *Inside the Groovy sandbox* section.

In the past few sections, we've explored who the SYSTEM user is and what they can do. Let's recap before going back to the original topic:

- Before-build-user is our login user.
- During-build-user is the SYSTEM user for everyone by default.
- The SYSTEM user can run pipeline steps and approved Groovy methods everywhere.

Now, let's come back and think about what Approve assuming permission check means. *Permission* here means the permission of a before-build-user. When we approve the delete() method signature with the Approve assuming permission check button, we're asking Script Security to check whether a before-build-user has the permission to delete a pipeline, and then execute if it does. This way, adder-admin can delete only adder pipelines, subtractor-admin can delete only subtractor pipelines, and adder-user and subtractor-user can't delete any pipelines, as per their permissions.

But wait a minute. When Script Security encounters a delete() method call, we're already in the middle of a build. In the middle of a build, we're running as the during-build-user, not the before-build-user. Script Security doesn't know about our before-build-user, so is unable to check the permissions of our before-build-user.

In other words, we want Script Security to check the permissions of the before-build-user, but Script Security can only check the permissions of the during-build-user.

How do we bridge this gap? We use a plugin to assign the before-build-user (as opposed to the default SYSTEM user) to the during-build-user. Let's see how we can use the Authorize Project plugin for this.

Understanding why the Authorize Project plugin is needed

Required plugins
Authorize Project

You have probably seen a warning message in the notification area, as in *Figure 11.10*. The message is a bit difficult to understand at first, but we can start to see what the message is telling us now that we know that the builds run as a SYSTEM user:

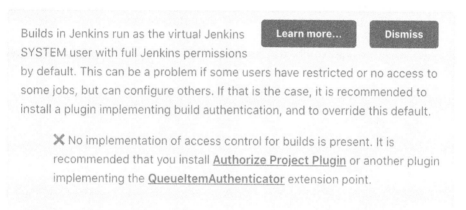

Figure 11.10 – *A warning message regarding the dangers of a SYSTEM user*

The first sentence confirms what we had discussed in the previous section – that the builds, by default, run as a SYSTEM user with too much permission.

The second sentence is a bit confusing to understand, but it precisely describes our project configuration. In our case, adder-admin and `subtractor-admin` *have restricted or no access to* subtractor and adder jobs, respectively, *but can configure* adder and subtractor jobs. As the sentence says, *this can be a problem* as we saw with the adder/run-subtractor-premerge pipeline, which allowed the users to bypass the permission restrictions.

The paragraph with a red X tells us that installing the Authorize Project plugin will implement *access control for builds*. Remember this information as we'll need it very soon.

Continue by clicking Learn more... to arrive at the Access Control for Builds page (https://www.jenkins.io/doc/book/system-administration/security/build-authorization/), which has a few more messages:

The permission Agent/Build requires access control for builds to be set up, as the build's authentication is checked, and not the user starting the build.

Figure 11.11 – An explanation of why Agent/Build permission requires the Authorize Project plugin

Let's examine further to understand what this means. Take a look at Agent/Build permission in Global Security. Putting a mouse over Agent/Build permission tells us that this permission allows users to *run jobs as them on agents*. Recall that a SYSTEM user has access everywhere, which includes being able to run builds on any agents. Our before-build-users, on the other hand, do not have the permissions to run builds on any agents. If we assign before-build-users to during-build-users, the builds cannot run on any agents until this permission is granted. Okay, so we want to enable this permission:

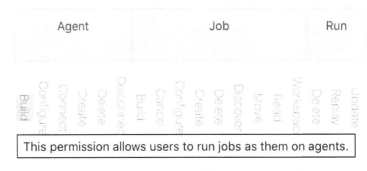

Figure 11.12 – Agent/Build permission tooltip

The page goes on to say that the solution is to install the Authorize Project plugin. Install the plugin if you haven't already done so and then continue to see how we can use it.

Configuring the Authorize Project plugin

Messages on the plugin home page (https://plugins.jenkins.io/authorize-project/) tell us that the Authorize Project plugin allows us to *Run as the user who triggered the build*, which is what we want. Under it, a bullet point says that it *Does not work for scheduled, or polled builds*. This means that the premerge and the postmerge pipelines can't use this feature, which is a bummer.

The warning message on the notification area has changed to reflect the plugin install:

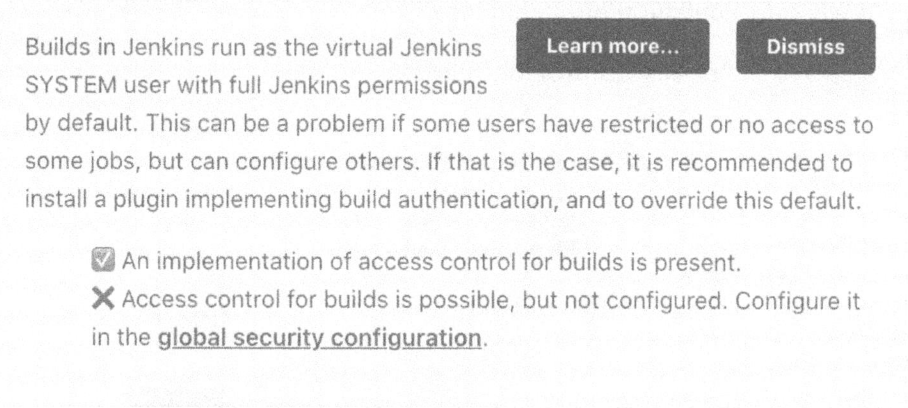

Builds in Jenkins run as the virtual Jenkins [Learn more...] [Dismiss]
SYSTEM user with full Jenkins permissions
by default. This can be a problem if some users have restricted or no access to
some jobs, but can configure others. If that is the case, it is recommended to
install a plugin implementing build authentication, and to override this default.

☑ An implementation of access control for builds is present.
✗ Access control for builds is possible, but not configured. Configure it
in the **global security configuration**.

Figure 11.13 – The plugin is installed, but not configured

Click the link in the message to go to Global Security. **Under** Access Control for Builds,
clicking the Add **drop-down menu shows you two configurable items:**

- Per-project configurable Build Authorization **allows each project to change the
  user who will be running the build. Check** Run as User who Triggered Build.

- Project default Build Authorization **should remain as SYSTEM. The other three
  options don't really work for us:**

  - We saw a message on the plugin home page that running as user who
    triggered build **breaks premerge and postmerge pipelines, so we can't
    use that option.**

  - We have not given any permissions to anonymous users, so running as
    anonymous **is not applicable in our Jenkins.**

  - Finally, running as specific user **doesn't really make sense as a default
    since our goal is to use our own users, not share a specific user.**

> **Order matters!**
> Be sure to put Per-project configurable Build Authorization
> above Project default Build Authorization so that the per-project
> configurations are processed first.

Here is what the configuration should look like:

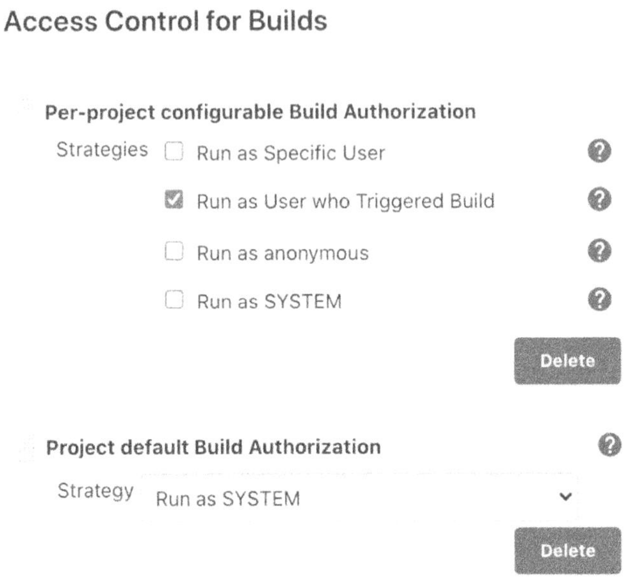

Access Control for Builds

Per-project configurable Build Authorization

Strategies ☐ Run as Specific User   ❓

            ☑ Run as User who Triggered Build   ❓

            ☐ Run as anonymous   ❓

            ☐ Run as SYSTEM   ❓

                                                        [Delete]

Project default Build Authorization   ❓

Strategy   Run as SYSTEM   ⌄

                                                        [Delete]

Figure 11.14 – Access Control for Builds configuration

Also, while we're here, go to the Authorization **section and give** Agent/Build **permission to** Authenticated Users:

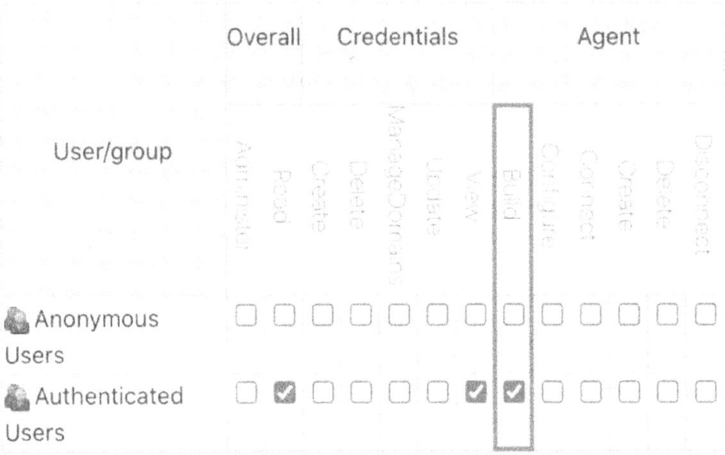

Figure 11.15 – Give Agent/Build permission to Authenticated Users

Click Save to exit.

We have gone through the journey of trying to understand how Jenkins manages identities. With the Authorize Project plugin installed and configured, we're now able to run a build as the logged-in user. Let's resume examining the Approve assuming permission check **feature button.**

Creating pipeline-lister

We began this section by discussing the delete() method, but let's instead create a less destructive example so that we can run it multiple times to test.

Follow these steps to create a pipeline that lists the names of all pipelines that the logged-in user has the permissions to see:

1. Log in to Jenkins as admin user.

2. On the Jenkins home page, click New Item to make a new pipeline on the top level, outside the adder or subtractor folders.

3. Name it pipeline-lister, choose Pipeline, and then click OK.

4. Check the box for Enable project-based security, and then give all permissions to Authenticated Users. This allows everyone to see, configure, and run the pipeline.

5. Without modifying anything else, click Save to exit.

Step 4 is important because it allows all logged-in users to see and run the pipeline. In the past, we have configured the default permissions such that a user can't see any folders or pipelines unless they were explicitly granted permissions. We want this pipeline to be accessible to all users so that everyone can run it.

In addition, it's important that a non-administrator user configures the pipeline. Recall from the previous *Direct pipeline* section that if we save a pipeline as an administrator user, Jenkins doesn't require us to get approval for the pipeline functions. We *want* our pipeline functions to require approval so that we can approve them using permission checks. By granting all permissions to all logged-in users, the non-administrator users can configure the pipeline.

Let's continue:

1. Open a different web browser than the usual one and log in to Jenkins as the adder-admin user. A different web browser is needed to have live sessions for both adder-admin and admin users at the same time. On the left browser as adder-admin, we should see the adder folder and the pipeline-lister pipeline. On the right browser as admin, we should see both adder and subtractor folders as well as the pipeline-lister pipeline:

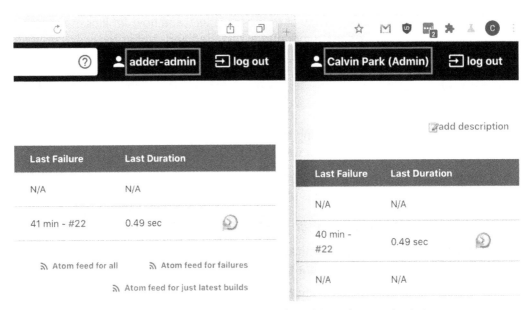

Figure 11.16 – Two web browsers open for adder-admin and admin users

2. On the left browser as adder–admin, click pipeline-lister and then Configure.

3. In Pipeline/Definition/Script, enter the following pipeline script and click Save. As we discussed, the code gathers a list of all pipelines and prints their names:

```
pipeline {
  agent any
  stages { stage ('List all pipelines') {
    steps { script {
      Jenkins.instance.getAllItems(Job.class).each {
        println it.fullName
} } } } } }
```

The point to watch out for is that this pipeline will print different names depending on who initiates the build. We're almost there – let's continue.

4. Click Build now to run the pipeline. The build will fail but that's okay – we only need the build log, which we will use to compare shortly.

5. On the left side, click Authorization, and check the Configure Build Authorization checkbox. Authorize Strategy should already be Run as User who Triggered Build. Click Save to exit.

6. Click Build Now. This will fail with a message Scripts not permitted to use

staticMethod jenkins.model.Jenkins getInstance. Observe that the second line of the console output says Running as adder-admin. This line didn't exist in the build from *step 4* before we configured Authorize Projects Plugin:

● **Console Output**

```
Started by user adder-admin
Running in Durability level: PERFORMANCE_O
[Pipeline] Start of Pipeline
```

● **Console Output**

```
Started by user adder-admin
Running as adder-admin
Running in Durability level: PERFORMANCE_OPTIMIZED
```

Figure 11.17 – The image on the left shows a build that ran as a SYSTEM user. The image on the right shows a build that ran as the user who initiated the build

7. On the right browser as the admin user, click Manage Jenkins, followed by In-process Script Approval.

8. There should be three boxes to Approve, Approve assuming permission check, or Deny the signature staticMethod jenkins.model.Jenkins getInstance. Click Approve assuming permission check. The signature will be added to a box titled Signatures already approved assuming permission check:

Approve / Approve assuming permission check / Deny signature : staticMethod jenkins.model.Jenkins getInstance Approving this signature may introduce a security vulnerability! You are advised to deny it.

Figure 11.18 – Click Approve assuming permission check

9. On the left browser as the adder-admin user, click Build Now. This time it will fail on method hudson.model.ItemGroup getAllItems java.lang.Class. On the right browser as admin, click Approve assuming permission check again. Do this a few more times until the build finally succeeds.

10. Upon a successful build, look at the console output to see that three pipeline names are listed: adder/postmerge, adder/premerge, and pipeline-lister. Notice that subtractor pipelines are missing, because the adder-admin user does not have the permissions to see them.

11. Log out, then log in again as subtractor-user, and then run pipeline-lister. The console output lists three pipelines: pipeline-lister, subtractor/postmerge, and subtractor/premerge. Again, notice that that adder pipelines are missing for the same reason.

12. On the right browser as the admin user, run pipeline-lister. The console

output lists all pipelines in Jenkins including both adder and subtractor pipelines, because the admin user has the permissions to see all pipelines.

Steps 10, 11, and *12* demonstrate that the same pipeline produces different results depending on who started the build because the Groovy method signatures were approved assuming permission check. This can be extended further to allow a project admin to delete only those pipelines that they have permission to delete.

This concludes learning about how Approve assuming permission check can be used to limit the scope of Groovy method calls based on the logged-in user's permissions.

Now, let's step back and think about whether any of this is actually necessary.

Thoughts on disabling Script Security

The top 10 search results for jenkins script security almost always contain a way to disable Script Security. While CloudBees, the maintainer of Jenkins, strongly discourages tampering with, disabling, or destroying Script Security, I think it may make sense in some specific settings.

Some of the most common attack vectors to Jenkins are as follows:

- Attackers saturating the server capacity (for example, DDoS'ing) to cause service disruption
- Attackers with no access hacking to gain access
- Attackers with limited access hacking to gain elevated privileges

The first issue concerns the infrastructure level, which is beyond Jenkins software's domain of control. The second issue is handled well through the global security model, leveraging AD, OAuth, or other established authentication systems. The third issue is what Script Security addresses.

In a non-trusted environment, such as running on AWS to build an open source software on GitHub, the third issue is very real. Anyone who's allowed to trigger a build through a PR can execute code on Jenkins, and Script Security prevents them from incorrectly escalating the privilege from limited access.

This concern is much less significant in a trusted environment. In a typical corporate setting, the corporate firewall ensures that the Jenkins instance is accessible only by trusted users, in other words, employees. In some cases, there are concerns about protecting Jenkins from contractors, which would demand Script Security's protection. In other cases where only employees have access, the concerns of users sabotaging the Jenkins instance are low. Sometimes, a Jenkins instance is shared by multiple non-trusting groups, as ours did for adder and subtractor, and a shared

Jenkins like this may benefit from Script Security. In such a case, however, a better alternative is to simply create another Jenkins instance dedicated for each trusting group of people. For example, give the adder project its own Jenkins instance, and give the subtractor project its own Jenkins instance. As we saw in *Chapter 2, Jenkins with Docker on HTTPS on AWS and inside a Corporate Firewall*, the agents and the Docker cloud hosts can be shared across multiple Jenkins instances, which minimizes the need for additional infrastructure.

In fact, a lot of problems are eliminated if a Jenkins instance is used by just one team, allowing for a greatly simplified configuration. In a typical two-pizza team[2] of six people, there is usually one Build and Release Engineer who manages Jenkins. That person can be a global administrator and the rest of the team can be normal users. By removing the project admin role, we no longer have to worry about a half-way admin who can create or delete pipelines but can't approve Groovy method signatures or create a global shared library. The admin is an actual administrator, true to its name. The direct pipelines created by the admin are approved automatically because the admin is an actual administrator. Creating a global shared library encapsulation is no longer necessary because the non-administrator users don't need to perform administrative tasks. It was necessary only because there was a half-way admin/user who needs to perform administrative tasks but does not have permission to do so. The admin also doesn't need to worry about mistakenly destroying other projects because there are no other projects.

Therefore, in a setting where all of the following is true, Script Security provides few benefits:

- Jenkins is behind a corporate firewall.
- Jenkins is backed up regularly to protect against mistakes or hardware failures.
- Jenkins is used by only one team.
- All team members are trusted.

Combined with the required whack-a-mole approvals for method signatures, it's understandable that people look for ways to disable Script Security entirely.

---

2 https://docs.aws.amazon.com/whitepapers/latest/introduction-devops-aws/two-pizza-teams.html

Summary

In this chapter, we learned the ins and outs of the Script Security plugin and Jenkins' security model as a whole.

We started off by learning about the important distinction between an administrator and a non-administrator.

Then, we learned about the Groovy sandbox. First, we saw how to run a pipeline outside of the sandbox using a direct pipeline, and then went through a disaster scenario where approving direct pipeline scripts can lead to problems. Again, never approve them. Then we learned how to use a Global Shared Library to wrap a dangerous method call with the necessary checks and restrictions for safe use. We also learned how the method signature approval works while running a pipeline inside the sandbox.

Then we went through the epic journey of understanding how the Approve assuming permission check button works. We first learned about who the SYSTEM user is and what they can do with the horror of realizing that the permission restrictions that we had configured in *Chapter 3, GitOps-Driven CI Pipeline with GitHub*, aren't enforced during a build. Then we examined the warning message about the dangers of a SYSTEM user's unchecked access to everything. We installed and configured the Authorize Project plugin and then created a `pipeline-lister` pipeline which uses the plugin features. After juggling two browsers to fully configure the pipeline code and approve the method signatures assuming permission check, we saw that it prints the names of only the pipelines that the user who initiated the build has permission to see.

And finally we discussed when none of this may be necessary.

This concludes *Part 3, Advanced Topics*. I hope that you enjoyed the depth of our discussions here.

Index

A

Packt.com

Subscribe to our online digital library for full access to over 7,000 books and videos, as well as industry leading tools to help you plan your personal development and advance your career. For more information, please visit our website.

Why subscribe?

- Spend less time learning and more time coding with practical eBooks and Videos from over 4,000 industry professionals

- Improve your learning with Skill Plans built especially for you

- Get a free eBook or video every month

- Fully searchable for easy access to vital information

- Copy and paste, print, and bookmark content

Did you know that Packt offers eBook versions of every book published, with PDF and ePub files available? You can upgrade to the eBook version at packt.com and as a print book customer, you are entitled to a discount on the eBook copy. Get in touch with us at customercare@packtpub.com for more details.

At www.packt.com, you can also read a collection of free technical articles, sign up for a range of free newsletters, and receive exclusive discounts and offers on Packt books and eBooks.

Other Books You May Enjoy

If you enjoyed this book, you may be interested in these other books by Packt:

Repeatability, Reliability, and Scalability through GitOps

https://packt.link/9781801077798

Bryan Feuling

ISBN: 9781801077798

- Explore a variety of common industry tools for GitOps
- Understand continuous deployment, continuous delivery, and why they are important
- Gain a practical understanding of using GitOps as an engineering organization
- Become well-versed with using GitOps and Kubernetes together
- Leverage Git events for automated deployments
- Implement GitOps best practices and find out how to avoid GitOps pitfalls

Modern DevOps Practices

https://packt.link/9781800562387

Gaurav Agarwal

ISBN: 9781800562387

- Become well-versed with AWS ECS, Google Cloud Run, and Knative
- Discover how to build and manage secure Docker images efficiently
- Understand continuous integration with Jenkins on Kubernetes and GitHub actions Get to grips with using Spinnaker for continuous deployment/delivery
- Manage immutable infrastructure on the cloud with Packer, Terraform, and Ansible
- Explore the world of GitOps with GitHub actions, Terraform, and Flux CD

Packt is searching for authors like you

If you're interested in becoming an author for Packt, please visit authors.packtpub.com and apply today. We have worked with thousands of developers and tech professionals, just like you, to help them share their insight with the global tech community. You can make a general application, apply for a specific hot topic that we are recruiting an author for, or submit your own idea.

Share Your Thoughts

Now you've finished, we'd love to hear your thoughts! Scan the QR code below to go straight to the Amazon review page for this book and share your feedback or leave a review on the site that you purchased it from.

https://packt.link/r/1838824324

Your review is important to us and the tech community and will help us make sure we're delivering excellent quality content.

www.ingramcontent.com/pod-product-compliance
Lightning Source LLC
Chambersburg PA
CBHW081500050326
40690CB00015B/2868